27th Micromechanics and Microsystems Europe Workshop 2016

Journal of Physics: Conference Series Volume 757

Cork, Ireland
28 – 30 August 2016

Editors:

Shashank N. Kane
Ashutosh Mishra
Anup K. Dutta

ISBN: 978-1-5108-3225-1
ISSN: 1742-6588

Printed from e-media with permission by:

Curran Associates, Inc.
57 Morehouse Lane
Red Hook, NY 12571

Some format issues inherent in the e-media version may also appear in this print version.

Printed by Curran Associates, Inc. (2016)

For permission requests, please contact the Institute of Physics at the address below.

Institute of Physics
Dirac House, Temple Back
Bristol BS1 6BE UK

Phone: 44 1 17 929 7481
Fax: 44 1 17 920 0979

techtracking@iop.org

Additional copies of this publication are available from:

Curran Associates, Inc.
57 Morehouse Lane
Red Hook, NY 12571 USA
Phone: 845-758-0400
Fax: 845-758-2633
Email: curran@proceedings.com
Web: www.proceedings.com

Table of contents

Volume 757

27th Micromechanics and Microsystems Europe Workshop
28–30 August 2016, Cork, Ireland

Accepted papers received: 21 September 2016
Published online: 11 November 2016

Preface

Papers

Introduction to Proceedings of MME 2016

This volume includes selected papers from MME 2016 – the 27th Micromechanics and Microsystems Europe Workshop. MME dates all the way back to 1989, when 25 participants gathered to hear about trends in the new and emerging field of microsystems technology in Twente, The Netherlands. In 2001, MME was held in Cork for the first time, and, fifteen years on from that gathering, we were delighted to welcome MME back to Ireland once more.

60 high quality submissions were received from authors affiliated to 19 countries, and a selection of those accepted for the conference are collected here. The works span all aspects of MEMS and microsystems technology, including design and fabrication, simulation and modelling, experimental evaluation, and commercial translation.

We acknowledge the important contributions made by our four invited speakers, who discussed exciting recent developments and their future visions for microsystems technology.

Our conference Gala Dinner was held at The Old Midleton Distillery, home of the world-famous Jameson Irish whiskey, where delegates took a guided tour along the old distillery trail through mills, maltings, stillhouses, warehouses and kilns – and, of course, participated in an Irish whiskey tasting tutorial!

The Tyndall National Institute and Cork Institute of Technology have received generous support from a number of agencies and industrial partners, including Science Foundation Ireland, Fáilte Ireland, and the Institute of Physics. We sincerely thank them for their assistance with the hosting of MME 2016.

Thanks also to Programme Committee, our Local Organizing Committee for their assistance with event logistics, and to the management and staff of the CIT Cork School of Music.

We hope that you enjoy this Proceedings, and look forward to seeing you in Uppsala, Sweden, for MME 2017.

Yours sincerely,

Dr. Conor O'Mahony and Dr. Martin Hill

Local Co-Chairs – MME 2016

MME 2016 gratefully acknowledges the support of the following partners:

sfi
Fondúreacht Eolaíochta Éireann
Dá bhfuil romhainn
Science Foundation Ireland
For what's next

Kurt J. Lesker® Company

Polytec

scienion
ENABLING LIFE SCIENCE

IntelliSense

cork CONVENTION BUREAU

Fāilte Ireland
National Tourism Development Authority

meetinireland.com

inspire

Zwick / Roell
Testing Systems

KEYSIGHT TECHNOLOGIES

IOP
Institute of Physics

Journal of Physics: Conference Series **757** (2016) 011001 doi:10.1088/1742-6596/757/1/011001

Organisation

Local Co-chairs

Dr. Conor O'Mahony, Tyndall National Institute, University College Cork, Cork, Ireland
conor.omahony@tyndall.ie

Dr. Martin Hill, Cork Institute of Technology, Cork, Ireland
martin.hill@cit.ie

Local Organising Committee

Dr. Ruth Houlihan, Tyndall National Institute, Cork, Ireland

Dr. Oskar Olszewski, Tyndall National Institute, Cork, Ireland

Ms. Suzanne O'Callaghan, Tyndall National Institute, Cork, Ireland

Mr. Andrea Bocchino, Tyndall National Institute, Cork, Ireland

Ms. Francesca Cavo, Tyndall National Institute, Cork, Ireland

International Steering Committee

Leon Abelmann, KIST Europe, Germany & University of Twente, Netherlands (Chair)

Harish Bhaskaran, Oxford University, UK

Jose Higino Correia, DEI, University of Minho, Guimarães, Portugal

Peter Enoksson, Chalmers University of Technology, Sweden

Sami Franssila, Aalto University, Finland

Daniel Garcia Sanchez, UPMC, France

Martin Hill, Cork Institute of Technology, Cork, Ireland

Jacopo Iannacci, Fondazione Bruno Kessler, Italy

Hüseyin Kizil, Istanbul Technical University, Turkey

Per Ohlckers, Vestfold University College, Norway

Robert Puers, KU Leuven, Belgium

José Luis Sánchez-Rojas, University of Castilla–La Mancha, Spain

Ulrich Schmid, Vienna University of Technology, Austria

Uwe Schnakenberg, IWE-RWTH-Aachen, Aachen, Germany

Herbert Shea, EPFL, Switzerland

Program Committee

Conor O'Mahony, Tyndall National Institute, Ireland

Martin Hill, Cork Institute of Technology, Cork, Ireland

Leon Abelmann, KIST Europe, Germany

José Luis Sánchez-Rojas, University of Castilla–La Mancha, Spain

Maria Tenje, Uppsala University, Sweden

Journal of Physics: Conference Series **757** (2016) 011001 doi:10.1088/1742-6596/757/1/011001

Participants beside a Jameson whiskey copper still at the Old Midleton Distillery in advance of the MME 2016 Gala Dinner. (Photo courtesy of Piotr Mackowiak).

Winners of the MME 2016 Best Paper Awards. Left to right: Dr. Martin Hill (Cork Institute of Technology - Co-Chair); Tijmen Hageman (KIST Europe, Germany); Zahra Khaji (Uppsala University, Sweden); Hannes Mehner (TU Ilmenau, Netherlands) and Dr. Conor O'Mahony (Tyndall National Institute - Co-Chair). (Photo courtesy of Piotr Mackowiak).

Journal of Physics: Conference Series **757** (2016) 011002 doi:10.1088/1742-6596/757/1/011002

Peer review statement

All papers published in this volume of *Journal of Physics: Conference Series* have been peer reviewed through processes administered by the proceedings Editors. Reviews were conducted by expert referees to the professional and scientific standards expected of a proceedings journal published by IOP Publishing.

A thin silicon thermoelectric nanowire characterization platform (TNCP) equipped with nanoporous electrodes for electrical contact formation

S. Hoda Moosavi[1*], Michael Kroener[1], Maxi Frei[2], Fabian Frick[1], Sven Kerzenmacher[2], and Peter Woias[1]

[1]Laboratory for Design of Microsystems, Department of Microsystems Engineering – IMTEK, Freiburg, Germany, [2]Laboratory for MEMS Applications, Department of Microsystems Engineering – IMTEK, Freiburg, Germany

[*]Email: hoda.moosavi@imtek.uni-freiburg.de

Abstract. We report on the fabrication of a silicon platform for the thermoelectric and structural characterization of single nanowires, equipped with nanoporous electrodes. Controlled wafer thinning to a thickness of 160 µm results in platform chips, which can be inserted into Transmission Electron Microscopes (TEM) for the nanowire's structural composition analysis. Our fabrication approach comprises the Bosch process (ICP), and "dicing before grinding" techniques to achieve this small thickness. To study the idea of developing a "plug-and-measure" platform, we have developed a novel approach for self-adhesion between a contact electrode and a nanowire by nanoporous electrodes. Due to the increased surface-to-volume ratio and increased van-der-Walls forces nanowires stick firmly to the electrodes for a good thermal and electrical connection. This innovative technique does also avoids, in best case, separate steps for contact formation.

1. Introduction

The day-by-day increasing demand for energy sources, fossil fuels, and other scarce resources reveals the indispensability of development of "clean" energy supplies. Among most of the energy-consuming technical processes waste heat is eventually dissipated into the ambient and not used further. Thermoelectric energy harvesting, which uses this else lost energy to power small embedded systems, offers a chance to thereby improve energy efficiency. Therefore, lots of effort has been put into the search for finding better thermoelectric materials [1–3]. Due to enhanced scaling effects, thermoelectric nanostructures are predicted to be more efficient compared to bulk material [4]. This, however, must be examined precisely for nanostructures, such as nanowires. The presented platform is designed to measure these properties and its functionality is proven by a thermoelectric nanowire assembled to it by means of dielectrophoresis phenomena [3]. Yet, nanowire assembly alone is not a guarantee to get a proper electrical contact between electrodes and the nanowire. Different techniques are available to form such an electrical contact; these are focused ion beam deposition (FIB) [1], as well as using shadow masks [3]. Both techniques are time-intensive approaches and require special facilities to be performed. Another wafer-scale, assembly technique is based on covering the electrodes with low temperature solders [5]. That technique, however, relies on a complicated cleanroom process. Here we suggest improving the contact by making the platform electrodes nanoporous. The porous platinum electrodes are fabricated by a special electro deposition technique [6,7]. As a result of the increased surface-to-volume ratio the stickiness between the nanowire and the electrode is increased. This can be taken as a proof of concept, indicating that surface treatment investigations in this field are promising.

2. Thermoelectric Nanowire Characterization Platform (TNCP)

Several platforms for thermoelectrical nanowire characterization have been developed [8–10]. However, a comprehensive investigation of the specimen's thermoelectric behaviour also relies on studying its chemical composition and growth direction. We have therefore developed the first platform enabling thermoelectric measurement in addition to structural analysis via Transmission Electron Microscopy (TEM) [3,5,11,12]. TEM compatibility requirements include a suspended specimen to avoid electron scattering, and also a TNCP chip diameter less than 3 mm, and a thickness of less than 200 μm.

2.1. TNCP design

Our final TNCP design, a silicon chip of 2 mm × 2 mm × 160 μm, includes a membrane structure to assure a proper thermal insulation during measurements, as well as low electron scattering in the TEM. It also is equipped with measurement electrodes and micro heaters for the Seebeck measurements. The TNCP design is shown in Figure 1.

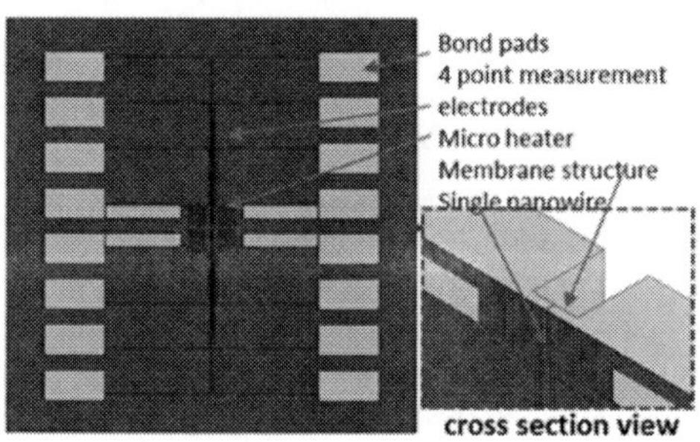

Figure 1. The TNCP design.

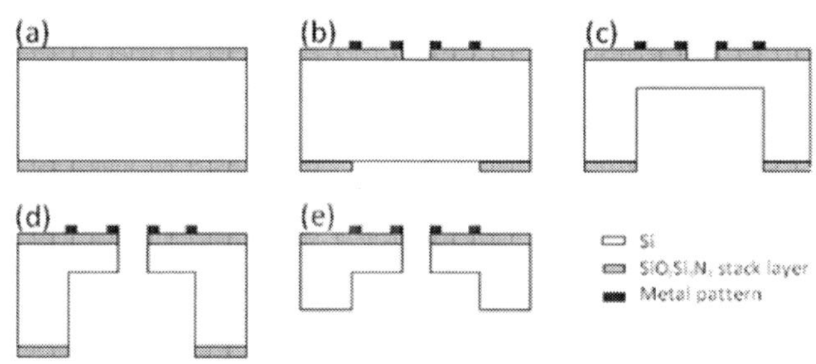

Figure 2. TNCP Fabrication processes.

2.2. TNCP fabrication

The fabrication process starts with a 380 μm thick silicon wafer, coated with 300 nm SiO_2/Si_3N_4 (Fig. 2-a). The wafer's frontside is processed with Pt sputtering, lithography and IBE to generate the metallization structure, followed by a backside and frontside etching at the mask layer via lithography and RIE process in succession (Fig. 2-b). STS-ICP is performed to form the 360 μm deep backside trenches (Fig. 2-c). In another step, an ICP process is utilized to open the 20 μm frontside trenches

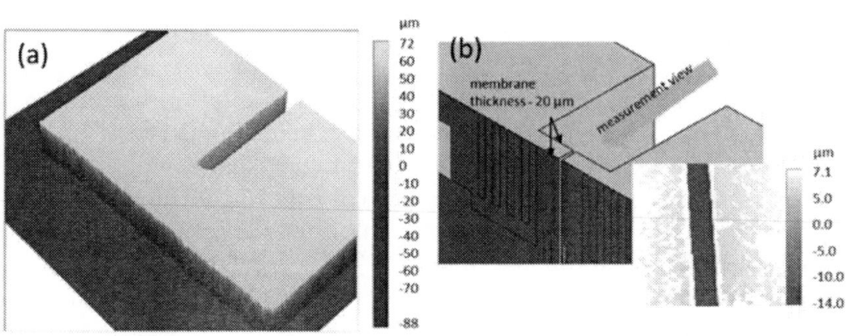

Figure 3. Veeco [wyko NT 9100] white light profilometer results, (a): scanning the TNCP backside indicates the chip's proper thickness of 160 μm. (b): profilometer image from the TNCP's backside trench. The membrane thickness is (appr.) 20 μm.

and to generate the membrane structure (Fig. 2-d). However, it is very probable that as soon as the trench is opened, the plasma penetrates to the membrane's backside and causes underetching. Protective resist spin coating is not an option due to the large aspect ratio of the opening in the backside. This problem was solved by modifying the ICP process. The wafers exposed to the plasma in ICP are Helium cooled from the backside. By reducing the maximum Helium flow rate from 25 to 16 SCCM, as soon as a through hole is generated in the wafer, the Helium leak rate increases and the process is terminated automatically. With this process modification the expected membrane thickness is reached (Fig. 3-b), without the requirement of expensive SOI wafers. Afterwards the wafer is partially diced to a depth of 160 μm. 220 μm rear grinding is performed at DISCO Hi-TEC Europe GmbH which leads to the diced 160 μm TNCP platforms (Fig. 2-e, Fig. 3,-a).

3. Nanoporous Platinum

It has been shown that a Bi_2Te_3 nanowire assembled onto the TNCP shows a contact resistance in the order of GΩ [3], indicating that no electrical contact with the underlying platinum layer is established. If the nanowire would be automatically contacted after the assembly, without the need for further processing, this would highly increase the platform's ease-of-use. Our idea is to contact the nanowire by maximizing the area between the nanowire and the electrode surface, making use of Van-der-Waals forces. We also assume that current will find more ways of flowing through thin non-conductive layers on the nanowire surface (e.g. oxide) which otherwise prohibit electrical contact. The larger contact surface in our approach is achieved by producing nanoporous platinum electrodes.

3.1. Fabrication

Our strategy for fabrication of porous platinum relies on the galvanic co-deposition of Cu and Pt, and subsequent electrochemical dealloying of the non-noble Cu component. Deposition and dissolution times have strong influence on the structure and stability of the layers, with high ratios of dissolution-to-deposition time leading to less dendritic structures. The optimized time for deposition and dissolution is found to be 9 s and 4.5 s, respectively [6]. The deposition and dealloying potentials are -0.5 V and 0.7 V, respectively. Using these parameters, we performed several deposition cycles to find the suitable number for our application. As a result, already one cycle gives rise to the expected porous structure, as shown in Figure 4-a.

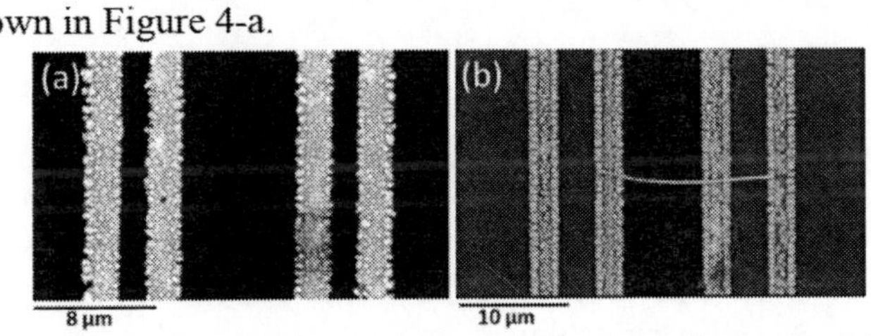

Figure 4. (a): SEM-picture of the porous electrodes of TNCP. (b): A nanowire assembled on the realized nanoporous Pt electrodes by dielectrophoresis phenomena.

3.2. Measurement

The nanowire assembly to the platform is performed using dielectrophoresis technique [3], as shown in Figure 4-b. As a result of increased surface-to-volume ratio the stickiness between the nanowire and the electrode is increased. A good clue for that is shown in Figure 5-a. The nanowire breakage over the gap happened during droplet evaporation in a dielectrophoresis assembly. The force from the droplet pressure change during the evaporation was higher than the mechanical strength of the nanowire and fractured it in the gap. Yet, the remaining parts of the nanowire on the electrode were intact and unmoved. This observation demonstrates that the stickiness between the nanoporous electrodes and the nanowire is enhanced due to an increased contact surface area. In the electrical measurements, a two point measurement on an assembled Bi_2Te_3 nanowire yields a resistance value of 0.92 MΩ ± 0.06 MΩ (Fig. 5-b). This magnitude in comparison to the GΩ values obtained on unprocessed TNCPs reported by [3] shows an enhancement. Although no reliable electric contact is achieved (which would be approximately 2000 Ω [12]), still the surface treatment for automatic nanowire assembly to the electrodes deserves further investigation.

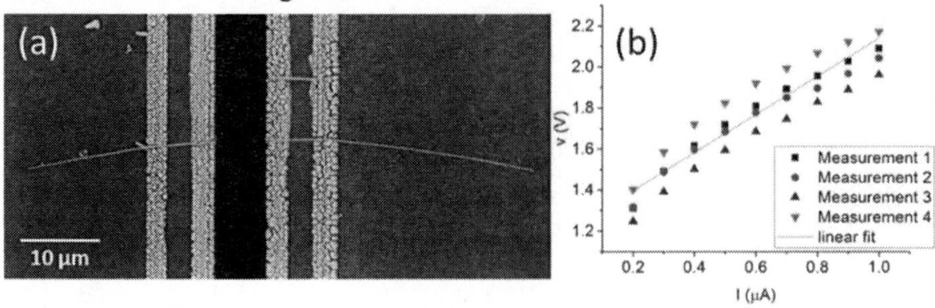

Figure 5. (a): The electrode and nanowire adhesion enhancement. (b): I-V curve of a Bi_2Te_3 nanowire, on a porous electrode.

4. Conclusion and outlook

We have successfully fabricated a TEM compatible chip for nanowire characterization. The fabrication is based on the dicing before grinding technique. We have presented the electrodeposition parameters to achieve a nanoporous micro electrode. The nanowires assembled to such a platform stick properly to the electrodes as a result of the enhanced common area. The electrical measurement showed a reduced resistance between nanowire and the nanoporous electrode, compared to planar electrodes. To sum up, although no reliable electrical measurement is achieved, still this proof of concept indicates that surface treatment investigation in this field can be promising. For future experiments, it would be desirable to get the nanowires into even closer physical contact with the electrode, for example by letting the nanowire to sink into and be surrounded by a nanowire grass grown on the electrodes.

Acknowledgment

This study was supported by the German Research Association DFG (SPP 1386). Also the authors are grateful to Dr. Uwe Pelz for precious discussions. Also we thank Dr. Ali Moazenzadeh for valuable helps for the chips wire bonding.

References

[1] D. Kojda, R. Mitdank, et al. 2014 *Semicond. Sci. Technol.* **29** 124006. doi:10.1088/0268-1242/29/12/124006
[2] D. Kojda, R. Mitdank, et al., 2015 *Phys. Rev.* B **91** 1–13. doi:10.1103/B.91.024302
[3] Z. Wang, S.H. Moosavi, et al, 2015 in: O. Eibl, K. Nielsch, N. Peranio, F. Voelklein (Eds.), *Thermoelectric Bi_2Te_3 Nanomater.*, first, Wiley
[4] L.D. Hicks, M.S. Dresselhaus, 1993 *Phys. Rev.* B **47** 16631–16634. doi:10.1103/B.47.16631
[5] S.H. Moosavi, M. Kroener and P. Woias 2016 *Sens. Actuators A: Phys., http://dx.doi.org/10.1016/j.sna.2016.05.006*
[6] C. Köhler, A. Kloke, et al. 2013 *J. Power Sources*. 242 255–263. doi:10.1016/j.jpowsour.2013.05.035
[7] A. Kloke, C. Köhler, et al., 2012 *Adv. Mater.* **24** 2916–2921. doi:10.1002/adma.201200806
[8] S. Karg, P. Mensch, et al., 2013 *J. Electron. Mater.* **42** 2409–2414. doi:10.1007/s11664-012-2409-7
[9] L. Shi, D. Li, et al., 2003 *J. Heat Transfer*. 125 881. doi:10.1115/1.1597619
[10] F. Voelklein, M.C. Schmitt, et al. 2011 in: A. Hashim (Ed.), *Nanowires - Implementations Applications.*, doi:10.5772/19656
[11] S.H. Moosavi, M. Kroener and P. Woias 2015 *Procedia Eng.* 120 210–214. doi:10.1016/j.proeng.08.612
[12] Z. Wang, S.S. Adhikari, et al., 2013 *IEEE 26th Int. Conf. Micro Electro Mech. Syst.*, pp. 508–511. doi:10.1109/MEMSYS.2013.6474290

Electrochemical microsensor system for cancer research on photodynamic therapy *in vitro*

J Marzioch[1], J Kieninger[1], J A Sandvik[2], E O Pettersen[2], Q Peng[3] and G Urban[1]

[1]University of Freiburg-IMTEK, Georges-Köhler-Allee 103, 79110 Freiburg, Germany

[2]University of Oslo, P.O.Box 1072 Blindern, 0316 Oslo, Norway

[3]The Norwegian Radium Hospital, Ullernchauseen 70, 0310 Oslo, Norway

Julia.marzioch@imtek.uni-freiburg.de

Abstract. An electrochemical microsensor system to investigate photodynamic therapy of cancer cells *in vitro* was developed and applied to monitor the cellular respiration during and after photodynamic therapy. The redox activity and therefore influence of the photodynamic drug on the sensor performance was investigated by electrochemical characterization. It was shown, that appropriate operation conditions avoid cross-sensitivity of the sensors to the drug itself. The presented system features a cell culture chamber equipped with microsensors and a laser source to photodynamically treat the cells while simultaneous monitoring of metabolic parameter *in situ*. Additionally, the optical setup allows to read back fluorescence signals from the photosensitizer itself or other marker molecules parallel to the microsensor readings.

1. Introduction

Photodynamic Therapy (PDT) is a two-step cancer treatment using a photosensitizing drug followed by local irradiation with visible light. PDT is a clinically established method to treat bladder, colon or skin cancer benefiting from the relatively small stress for the patient [1]. In the case of an insufficient treatment additional therapies, e.g. radiation therapy, are still applicable. Although PDT is applied clinically, the biochemical processes during and after the therapy are only partly understood [2]. Nowadays, cell models for the therapy are evaluated by single point (end point) measurements only, due to the lack of an appropriate *in vitro* monitoring platform to trace metabolic effects continuously. Thus transient effects, which are supposed to occur during the application in patients, will be overseen in the classical cell model. The proposed electrochemical sensor system enables the monitoring of changes in metabolism during and after the treatment in cell culture. Due to the integration of the sensor onto a transparent glass chip the parallel optical and electrochemical monitoring of the cells is enabled.

In this work, the photosensitizer drugs are characterized electrochemically with respect to possible interference with the metabolic monitoring. Afterward the system's performance is demonstrated monitoring oxygen concentration during and after photodynamic therapy in breast cancer cell culture. Recent developments brought up several microsystems enabling PDT *in vitro* along with optical readout [6-9]. In contrast, the presented system focuses on the combination of electrochemical sensors together with optical access, both also possible during the therapy itself.

2. Electrochemical Microsensor System

A platform to monitor the metabolic effects of cancer cells during and after PDT was developed. The system is based on a transparent sensor chip using electrochemical sensors [3, 4], integrated into an optical setup [5]. In this work, a chronoamperometric oxygen sensor was chosen as an example to demonstrate the feasibility of continuous measurement of cellular respiration. The transparency of the sensor chip enabled the optical inspection of the cells by an inverted microscope.

The sensor chip was embedded in a microsensor system. This arrangement allowed the cultivation of cells in a small chamber directly on the sensor chip, see figure 1 and 2.

Figure 1: Electrochemical sensor with cell culture area (in transparent plastic cylinder).

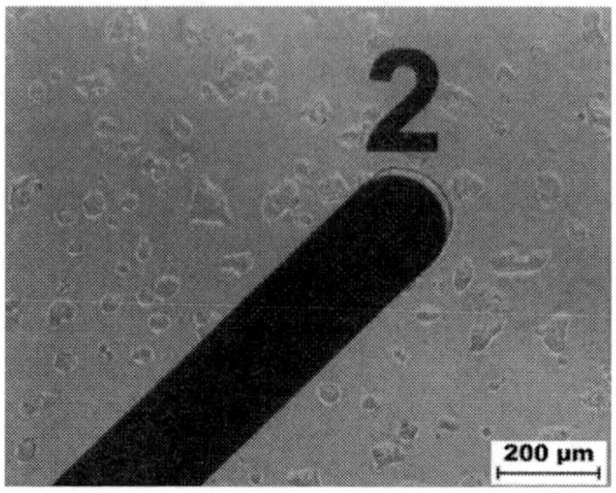

Figure 2: T-47D breast cancer cells 24 h after seeding on sensor chip next to the working electrode with connection line.

Figure 3: Setup for excitation of cell cultures with laser light for PDT featuring simultaneous detection of emitted fluorescence light and monitoring of metabolic parameters with electrochemical sensors. The light from the laser source (1) is deflected at a dichroic mirror (2) through the objective (3) to the cell culture chip (4). Fluorescence emission from the cell culture chamber passes through the dichroic mirror (2) and is detected by the CMOS-camera (5). The laser intensity is monitored by a detector (6).

The microsensor system was integrated into a setup together with a laser source for the photodynamic therapy and a fluorescence microscope, see figure 3. The beam from the diode laser (473 nm) was reflected at a dichroic mirror and reached the microsensor system through an objective lens. A CMOS camera at the end of the optical path allowed to detect fluorescence light emitted from the photosensitizer during the treatment. Additionally, this feature opened the possibility to use a fluorescence marker after the treatment.

3. Results

3.1. Electrochemical Investigation

An electrochemical investigation was performed to investigate the processes of the photosensitizer PpIX itself. PpIX was dissolved following an established protocol by using a solution named "Gomer" consisting of a mixture of perchloric acid, distilled water and methanol. The Gomer solution in a typical PpIX concentration of 70 µM results in a methanol concentration of 0.5 M. Since this is a considerable high concentration of methanol a cyclic voltammetric measurement, in the range of -0.6 V to 1.2 V was performed to investigate possible effects. The measurement was compared to a typical phosphate buffered saline (PBS) curve, see figure 4. It shows a prominent peak at 0.3 V, which is characteristic for the oxidation of methanol. The measurement was repeated with a 70 µM PpIX solution in PBS, once under normal oxygen atmosphere and once after the solution was purged with Nitrogen. The additional significant contribution due to the methanol oxidation was not visible anymore, see figure 5. In cyclic voltammetry experiments, no significant dependency of the redox behavior of the photosensitizer regarding the oxygen concentration can be observed.

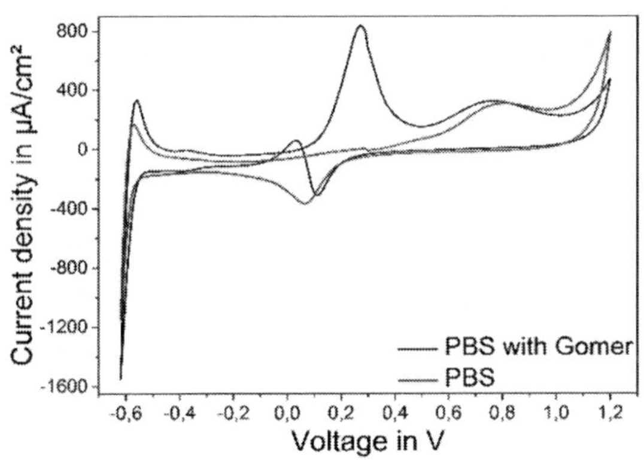

Figure 4: Cyclic voltammetric measurement of Gomer in PBS

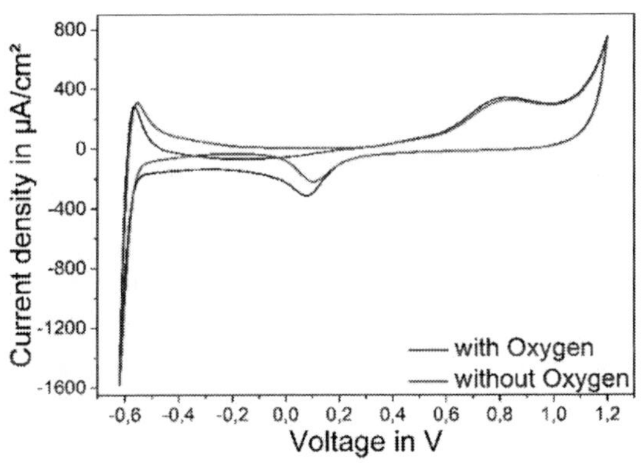

Figure 5: Cyclic voltammetric measurement of 70 µM PpIX in PBS with and without oxygen

3.2. Cell Measurements

T-47D breast cancer cells were successfully cultured on the microsensor system and incubated with the photosensitizer PpIX for PDT, see figure 3. The cellular oxygen concentration as indicator for cellular respiration was monitored before and after the therapy. The oxygen measurements were performed using a 3-step chronoamperometric protocol. The protocol runs with a waiting time of 50 s between each 10 s measurement. From every cycle the least 4 s are averaged to form a data point. In order to interpret the results of the cell measurements a calibration of the current density compared to the oxygen concentration was done. The chip was measured under cell environmental conditions at a temperature of 37 °C with oxygen concentrations between 0.5 and 20.95 %, which correspond to a range between 4.9 and 198 µM of dissolved oxygen. The resulting calibration curve of the oxygen sensor with a slope of -0.7µA/cm²/µM is shown in figure 6. A linear dependency between current density and oxygen concentration was found. The electrochemical measurements revealed the increase in the oxygen concentration due to reduced cell respiration and cell death after the treatment, see figure 7. The photodynamic dose was chosen to demonstrate an incomplete treatment. Approximately 10 h after the treatment a slow decrease in oxygen tension became visible. Here, it was possible to observe *in situ* the increase in respiration caused by cell repopulation. The sudden increase and subsequent decrease at approximately 65 h is caused by the medium exchange and the resulting disturbance of the oxygen diffusion gradient dropping towards the cells.

These results demonstrate the importance of continuous metabolic monitoring as phenomena like the repopulation effect would be overseen using single point measurements only.

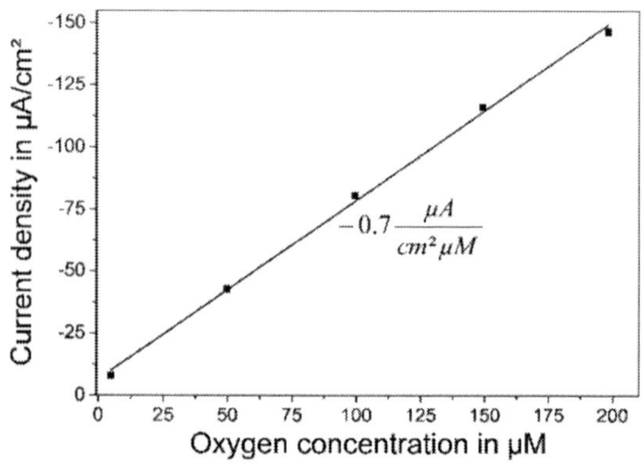

Figure 6: Oxygen sensor calibration in cell culture medium at 37 °C

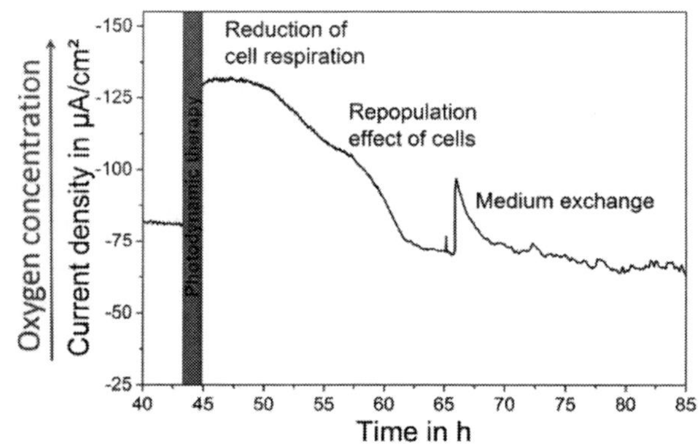

Figure 7: Pericellular oxygen measurement before and after photodynamic therapy on T-47D breast cancer cells

4. Conclusion and Outlook

The presented results show the applicability of the microsensors system to monitor oxygen concentration during PDT in a cell culture during and after treatment. It was demonstrated by the electrochemical characterization that the presence of the photosensitizer drug itself does not lead to any significant interference with sensor readings. A protocol for efficient culturing T-47D breast cancer cells on the sensor chip has been established in order to provide an on-chip treatment. The results suggests that the cellular respiration can be used as an indicator for cell death. Additionally, the system enables the monitoring of possible and undesired repopulation effects after the treatment. It became evident, that the continuous metabolic monitoring reveals effects, which would be overseen using single point measurements only.

Future work will focus on the combination of sensors for different metabolic parameter, in a first step by the integration of biosensors for glucose and lactate.

References

[1] Dolmans D.E.J.G.J, Fukumura D and Jain R.K 2003 TIMELINE: Photodynamictherapy for cancer *Nat Rev Cancer* **3** 380-7

[2] Shahzidi S et al. 2011 Simultaneously targeting mitochondria and endoplasmic reticulum by photodynamic therapy induces apoptosis in human lymphoma cells *Photochem Photobio* **10** 1773-82

[3] Kieninger J, Aravindalochanan K, Sandvik J.A, Pettersen E.O and Urban G.A 2014 Pericellular oxygen monitoring with integrated sensor chips for reproducible cell culture experiments *Cell Prolif* **47** 180-8

[4] Pettersen E.O et al. 2015 Targeting tumour hypoxia to prevent cancer metastasis. From biology, biosensing and technology to drug development: the METOXIA consortium *J Enzyme Inhib Med Chem* **30** 689-721.

[5] Marzioch J, Kieninger J, Sandvik J.A, Pettersen E.O, Peng Q and Urban G 2015 Photodynamic therapy-in vitro investigation using an electrochemical microsensor system *Procedia Eng* **120** 468-71
[6] Jedrych E, Pawlicka Z, Chudy M, Dybko A and Brzozka Z 2011 Evaluation of photodynamic therapy (PDT) procedures using microfluidic system *Analytica Chimica Acta* **683** 149–55
[7] Cho S et al. 2013 Droplet-based microfluidic platform for high-throughput, multi-parameter screening of photosensitizer activity *Anal Chem* **85** 8866–72
[8] Yang Y, Yang X, Zou J, Jia C, Hu Y, Du H and Wang H 2015 Evaluation of photodynamic therapy efficiency using an in vitro three-dimensional microfluidic breast cancer tissue model *Lab Chip* **15** 735–44
[9] Jastrzebska E, Bulka M, Rybicka N and Zukowski K 2015 Analysis of the efficiency of photodynamic therapy using a microsystem for mono-, co- and mixed cultures *Sensors and Actuators B: Chemical* **221** 1356–65

Crystalline growth of AlN thin films by atomic layer deposition

S Sadeghpour, F Ceyssens and R Puers
ESAT-MICAS, KU Leuven, Leuven, Belgium
E-mail: sina.sadeghpour@esat.kuleuven.be

Abstract. Aluminum nitride (AlN) thin film was grown by plasma enhanced atomic layer deposition using trimethylaluminum and ammonia precursors. A method was found to have crystalline thin film AlN with almost zero thickness variation and a truly one layer deposition of atoms per each cycle of the process. The growth rate saturated at $\sim$ 1 Å/cycle, and the thickness was proportional to the number of reaction cycles. The preferred crystal orientation, uniformity of the nucleation and the surface roughness of the grown AlN were investigated. X-ray diffraction (XRD), atomic focused microscopy (AFM) and scanning electron microscopy (SEM) were carried out to analyze the crystallinity and properties of the films.

1. Introduction

Aluminium nitride (AlN) is a piezoelectric material with high band gap ($\sim$6.2 eV) [1], high thermal conductivity [2], high electric resistance ($10^{13}\Omega$ cm), as well as low expansion at high temperatures [3]. It is a good candidate for many applications such as surface acoustic wave (SAW), high frequency resonators for electronic and sensing applications. The advantage of AlN with respect to other piezoelectric materials is its possibility to be deposited in a CMOS compatible process. Plasma enhanced atomic layer deposition (PE-ALD) is one of the more advanced methods of thin film growing at low temperature, between 150 to 300°C. ALD is a type of chemical vapor deposition based on sequential, surface-adsorption limited reactions. In ALD, gaseous precursors and oxidisers are applied to the substrate in separate pulses and every pulse at most a complete monolayer is adsorbed. Therefore, in each pulse a single layer of atoms is deposited.

ALD deposition of AlN has already been demonstrated based on using trimethylaluminum (TMA) or aluminum chloride (AlCl4) as the aluminum source [4, 5]. The use of TMA as the aluminum precursor results in higher deposition rates and lower doses to achieve saturation [6], which means more uniform nucleation on the surface of the substrate in the first cycle of the deposition. This will causes a higher crystallinity when growing AlN.

In this paper, we report the crystalline growth of AlN thin films by PE-ALD using TMA and ammonia (NH_3) as the aluminum and nitrogen source, respectively. Different parameters such as temperature, pressure, pulse time and the effect of substrate were investigated. Also, the effect of annealing on the crystallinity after deposition of the AlN thin film is presented. Although the characteristics of ALD-deposited AlN were already investigated [7–9], to our knowledge the combination of parameters for growth of crystalline AlN thin films by ALD discuessed here has not been extensively studied before.

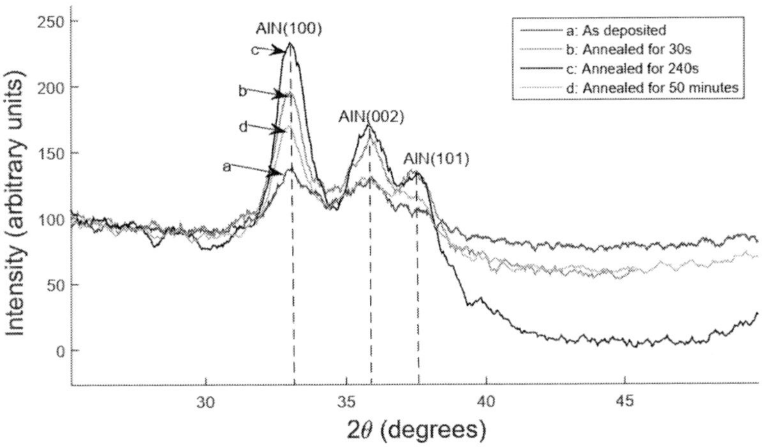

Figure 1. XRD pattern of AlN grown by the first recipe, as-deposited and annealed at 900°C for different duration, shows polycrystalline hexagonal structure.

2. Experimental details

(100) Si wafers were cleaned using Piranha and Hydrofluoric acid sequentially. The reason that plasma is used in the process is to reduce the reaction temperature of ammonia, which is around 550°C, to less than 300°C in order to prevent from the decomposition of TMA, which occurs above 300°C [10].

Several deposition experiments were done by our SENTECH PE-ALD machine, which can be divided in two different recipes. The second recipe is the improved version and will be discussed in section 3. Furthermore, the effect of the substrate and bottom electrodes on crystallization were investigated by using Si(100) and glass substrates and sputtered Ti on Si. Actually, it was shown that Ti works promising as an electrode for crystalline AlN growing [11].

The deposition rate of the grown AlN by PE-ALD were calculated by dividing its thickness to the number of deposition cycles. The thickness of the grown AlN layer was measured by reflectometry in at least ten points of each sample and was confirmed by cross section scanning electron microscopy (SEM) of the deposited AlN. Then the thickness variation was calculated by comparing the measured thickness in different points on the sample. The crystallinity and surface roughness of the grown film was investigated by X-ray diffraction (XRD) measurement and atomic force microscopy (AFM), respectively.

3. Results and discussion

3.1. First recipe

For the first experiment, the AlN was deposited on a 3″ Si(100) substrate at 300°C using nitrogen as carrier and purging gas, and TMA and ammonia as aluminum and nitrogen source, respectively. The power of the plasma was set to 200 W and the flow rates of the ammonia and nitrogen were set to 40 and 50 sccm, respectively. Each cycle of the deposition process consists of a 30ms TMA pulse, purging the reactor with nitrogen for 4 seconds, exposure to NH_3 plasma for 15 seconds, and purging again for 8 seconds. To keep the plasma chamber clean of TMA during the TMA pulse, the shutter between plasma and reactor chamber was opened only through NH_3 plasma exposure. This cycle was repeated 2000 times to reach to a reasonable thickness in order to make the XRD measurement possible.

The averaged measured thickness is equal to 84 nm, which means that the grow rate is equal

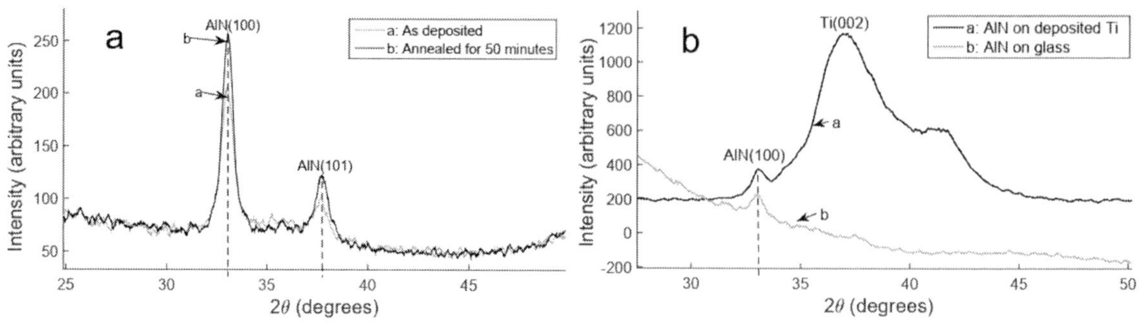

Figure 2. XRD pattern of AlN grown by the second recipe, as-deposited and annealed at 900°C on (a)Si and on (b)Ti and glass substrate.

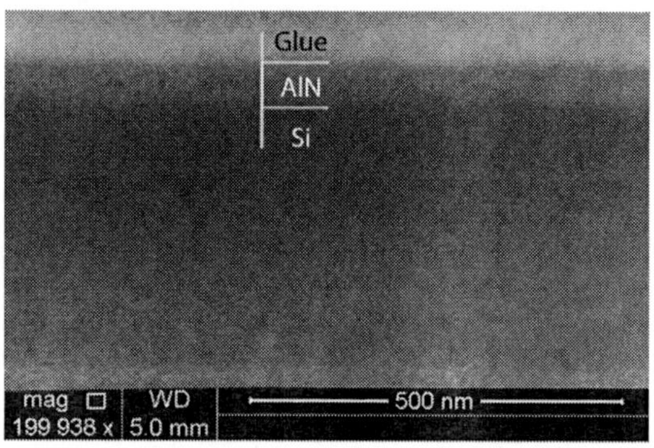

Figure 3. Cross-sectional SEM-image of a 100nm AlN film on Si(100), deposited by the second recipe

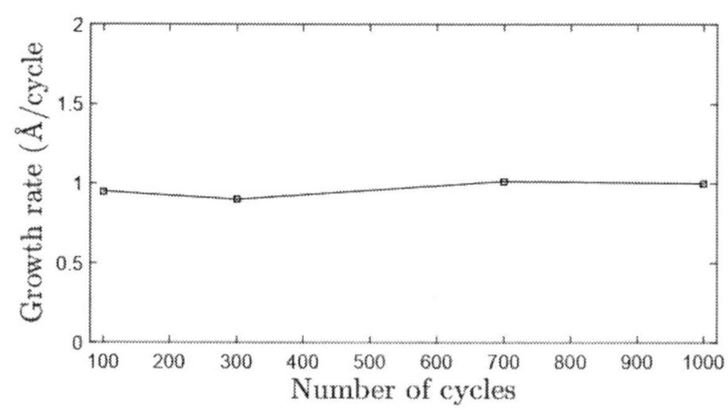

Figure 4. Growth rate of AlN film as function of cycles' number, deposited by the second recipe.

to 0.42 Å/cycle. This is even less than 1 Å/cycle, which means that the deposition is not truly one layer of atoms per cycle. It is reasonable to suppose that the incomplete reaction of NH_x to methyl groups of TMA as well as nonuniform nucleation on the surface of the substrate in the early cycles are the two main reasons of the low growth rate. This nonuniform growing leads to random nano crystallization, which was proven by XRD measurement of the grown AlN. Figure 1 shows the existence of two (100) and (002) crystal orientations of AlN in equal density. The figure also shows the beneficial effect of annealing at 900°C for varying periods. Longer annealing time doesn't have a significant effect on the crystallization. A short annealing time is shown to be superior to longer times, which is confirming the benefits of rapid thermal annealing (RTA) claimed in literature [12]. The thickness variation by this recipe is above 10 percents.

3.2. Second recipe

In order to eliminate the problem of nonuniform nucleation, after the cleaning process with HF and Piranha, all wafers were put again in Piranha for 10 minutes to cover the whole surface with OH groups. A uniformly hydroxilized starting surface appeared critical for obtaining strongly crystalline layers. This can be understood out of the need for hydroxyl groups to react with TMA during the ALD process, a reaction in which the hydrogen of the hydroxyl group is replaced

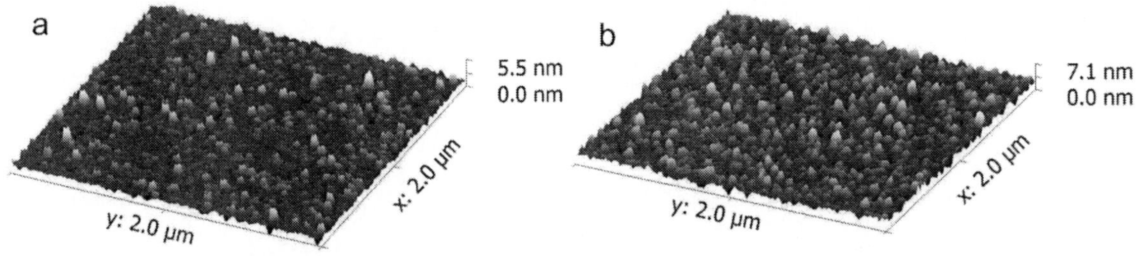

Figure 5. Surface morphology of the deposited AlN by PE-ALD on (a) Si and (b) sputtered Ti,which was measured by AFM.

with Al. Furthuremore, during the ALD reaction cycles, the TMA molecules were trapped in the reactor chamber for a few seconds by closing the throttle after the pulse to increase the incubation period. This process modification aimed at causing a higher probability of nucleation on surface of the substrate. Also, the exposure time of the substrate to NH_3 plasma was increased to obtain a sufficient removal of methyl groups of deposited TMA molecules, which leads to less carbon contamination of the deposited AlN [13]. The temperature was decreased to 280°C in order to prevent the decomposition of TMA and obtain a self limiting growth of AlN.

The second recipe was repeated for 1000 cycles and each cycle includes 2×(40ms TMA pulse with closed throttle to trap the atoms, waiting for 10 seconds while the throttle is kept close, purging for 10 seconds with fully open throttle), 20 seconds of NH_3 plasma exposure, and again purging for 10 seconds. The power and pressure of the plasma in this process were kept at 200W and 70Pa, respectively. The thickness of the grown AlN was 100nm, which demonstrates a growth of 1 Å/cycle. This is clear evidence of strong nucleation of deposition of a full atom layer per cycle. Figure 3 shows the cross-section image of the grown AlN film, which has been obtained by focused ion beam (FIB) etching and cross-section SEM in the same time. As shown in figure 4 the growth rate is almost constant with respect to the number of cycles. The thickness variation by this recipe is less than 1 percent.

Figure 2 shows the XRD pattern and the effect of annealing of the grown AlN on a Si(100) substrate, on borosilicate glass, and on sputtered Ti on Si. It is shown that the (100) orientation of AlN has a peak significantly higher than the previous recipe and also, the substrate does not have a remarkable effect on the preferred orientation of the grown AlN, but has an effect on the quality of the film and surface morphology. Figure 5 proves this fact by showing the AFM result of the deposited AlN on Si and Ti. The absolute value of the surface roughness of 100nm thick AlN deposited on Si and Ti is 0.35 nm and 0.71 nm, respectively.

In our experiments, the plasma power lower than 100W and plasma pressure higher than 100Pa lead to insufficient removal of methyl groups, causing deposition of low quality porous AlN. Figure 6 shows the XRD pattern of the grown AlN on Si(100) by the second recipe, but with plasma pressure and power equal to 150 Pa and 100W respectively. As shown, no crystalline phase was observed from the XRD pattern.

In each cycle of the ALD process one layer of either Al or N atoms is deposited, which means that the deposition units are atoms. As shown in figure 1 and 2, the dominant crystal orientation of the grown AlN thin film is (100); it is in contrast with [14], in which the preferred crystal orientation is (002) whenever the growth unit is atoms instead of Al-N dimer.

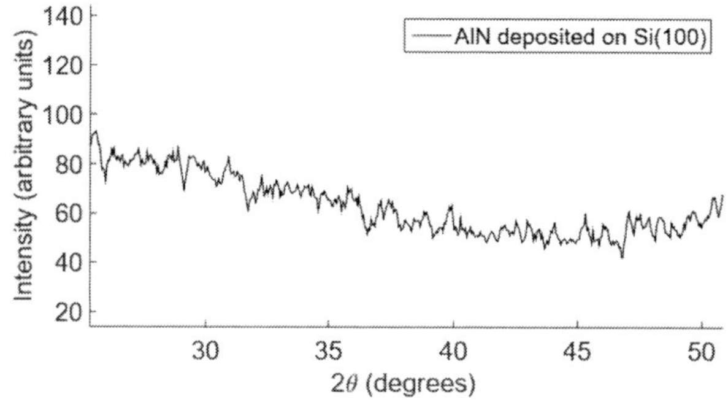

Figure 6. XRD pattern of AlN grown by the second recipe on Si(100), at plasma pressure and power equal to 150 Pa and 100 W, respectievly.

4. Conclusions

In this study, we have deposited crystalline AlN thin films at 280°C with zero thickness variation and constant growth rate with respect to the number of reaction cycles. It was found that the incubation period and dose of the TMA precursor strongly influence the crystallinity. Also, in the improved recipe a higher exposure time to the NH_3 plasma is used to increase the removing of methyl groups and improve the crystallization of the grown film. Furthermore, it was shown that the substrate type (Si(100), glass or Ti on Si(100)) does not have significant effect on the preferred crystal orientation, but has effect on surface roughness and grain size. The most important output of this study is the observation that crystalline growth of AlN thin film by PE-ALD is mostly dependent to the uniform nucleation of atoms at the primary cycles of the process, which we enhanced by ensuring a hydroxylated starting surface and by increasing the incubation times.

Acknowledgements

This research was made possible by Frederik Ceyssens research fellowship from FWO-Flanders and has received funding from the European Union's Seventh Framework Programme (FP7/2007-2013) / ERC grant agreement n 340931, Thalys.

References

[1] Motamedi P and Cadien K 2015 *J. Cryst. Growth.* **421** 45-52
[2] Goerke S et al 2015 *Appl. Surf. Sci.* **338** 35-41
[3] Figge S, Krncke H, Hommel D and Epelbaum B 2009 *Appl. Phys. Lett.* **94** 101915
[4] Perros A, Hakola H, Sajavaara T, Huhtio T and Lipsanen H 2013 *J. Phys. D: Appl. Phys.* **46** 505502
[5] Lee Y, Kang S 2004 *Thin Solid Films* **446** 227231
[6] Ozgit C, Donmez I, Alevli M and Biyikli B 2011 *J. Cryst. Growth.* **335** 51
[7] Van Bui H, Wiggers F B, Gupta A, Nguyen M D, Aarnink A A, de Jong M P and Kovalgin A Y 2015 *Journal of Vacuum Science & Technology A* **33** 01A111
[8] Leskel M, Ritala M 2002 *Thin solid films* **409** 138-146
[9] Alevli M, Ozgit C, Donmez I, Biyikli N 2011 *J. Cryst. Growth.* **335** 51-57
[10] Puurunen R, Lindblad M, Root A and Krause A 2001 *Phys. Chem. Chem. Phys.* **3** 1093-1102
[11] Doll J, Petzold B, Ninan B, Mullapudi R and Pruitt B 2010 *J. Micromech. Microeng.* **20**
[12] Cao D, Cheng X, Xie Y, Zheng Li, Wang Z, Yu X, Wang J, Shenc D and Yua Y 2015 *RSC Adv.* B **5** 37881-37886
[13] Bosund M, Sajavaara T, Latinen M, Huhtio T and Lipsanen H 2011 *Applied Surface Science* **257** 7827-7830
[14] Ishihara M, Li S J, Yumoto H, Akashi K, Ide Y 1998 *Thin Solid Films* **316** 152-157

HF-Release of Sacrificial Layers in CMOS-integrated MOEMS structures

S Döring, M Friedrichs, W Pufe, M Schulze

Fraunhofer Institute for Photonic Microsystems (IPMS), Maria-Reiche-Str. 2, 01109 Dresden, Germany

Sebastian.doering@ipms.fraunhofer.de

Abstract. In this paper we will present details of the release process of SiO_2 sacrificial layers we use within a multi-level MOEMS process developed by IPMS. Using such sacrificial layers gain a lot of benefits necessary for the production of high-end MOEMS devices like high surface quality and great surface planarity. However the HF-release of the sacrificial layer can be connected with specific issues. We present, which mechanisms are involved in the release process and how knowing them, can be the key for an optimized performance of the device. More-over we will present how to protect the CMOS backplane of our devices from unwanted HF attack during the release.

1. Introduction

The IPMS develops various types of Spatial light modulators (SLM), a kind of MOEMS [1], [2]. Applications for SLMs are high speed pattern generators for DUV lithography, for mask inspection, for wave front correction in adaptive optics and for projection systems. Each of these applications requires special material properties and structures especially for the parts of actuator suspension and mirrors and IPMS offers a wide range of possibilities to serve these purposes. SLMs with only small deflections can be fabricated using small air gaps and thicker spring. If higher deflection is required, the gap has to be increased. However in such case softer and thus mostly thinner hinges are needed to counteract the weaker electrostatic force at a given maximal drive voltage of the CMOS. Such MOEMS often use a reinforcement level additionally to the hinge and to the mirror level to fulfil the requirements on the device. A scheme and an exemplary image of such a three-level MOEMS are shown in figure 1 and 2.

The fabrication of most of our SLM is done on customized CMOS backplanes. For a successful integration of the MOEMS structure on the CMOS the use of SiO_2 as sacrificial layer is most promising. To obtain full movability of our mirrors, sacrificial layers, which provide low surface roughness and good homogeneity of thickness, are necessary. For our preparation process we use undoped SiO_2 as sacrificial layer material, which is deposited by plasma-enhanced chemical vapour deposition (PECVD) using Tetraethylorthosilicate (TEOS) as precursor material. A subsequent treatment by chemical-mechanical polishing (CMP) is performed to define the final actuator gap height and obtain a smooth and plain surface for the subsequent mirror deposition. The final release is performed by using gas-phase etch with hydrogen fluoride (HF) [3]. Most of the metal structures contain Aluminium based alloys, which are used to prepare soft hinges, reinforcements and high reflective mirrors. Details about the used materials and the preparation process can be found in [4].

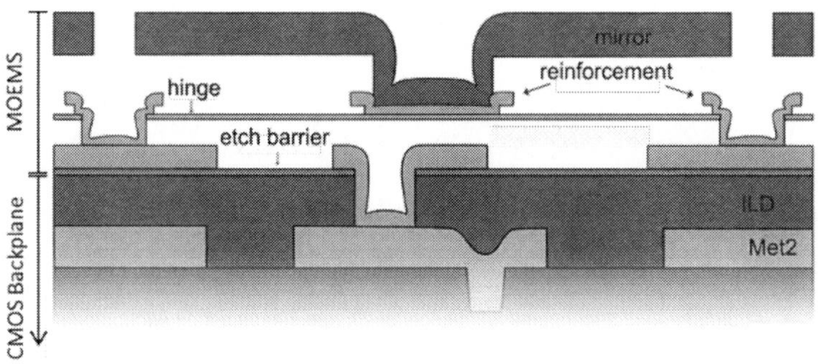

Figure 1. Schematic cross-section image of three-level MOEMS structure after the HF-release.

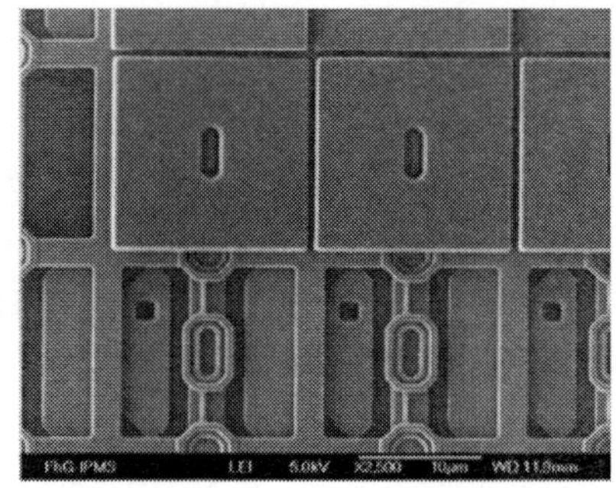

Figure 2. Top view SEM image with mirror and hinge layers of single pixels.

2. HF-Release kinetics

2.1. Release mechanisms

A major topic one has to take into account using sacrificial layers is the etch kinetics of the release process. It is essential to know how fast the etching is done and which mechanisms lead to the specific etching fronts. From this information it is possible to adjust the duration of etching to the given MOEMS structures and the sacrificial volume, so an optimal processing refine of the device can be provided. Appearance of under-exposing to HF would lead to remaining oxide. In such a case e.g. if these remains connect movable structures, mirror deflection to its full extend would fail. On the other hand over-exposure can cause shortcomings, too. Especially metal surfaces are sensitive to corrosion by either the HF itself or water vapour occurring during the reaction chain of the etching. Keeping the release time on a low limit would also minimize risk and amount of corrosion of the open metal parts in the structure.

Unfortunately the movement of the etch front is not isotropic into the sacrificial volume over the full time of the release. We could observe three different mechanisms, which determine the actual shape of the etch front at a given time for a specific MOEMS design. The different mechanisms of etch front movement are shown in figure 3.

For our devices, the etch process starts in the slits between the mirrors. At the beginning the etch front moves with an isotropic manner into the SiO_2 forming the shape shown in subfigure 3A. However shortly after beginning, two anisotropic mechanisms become important and disturb the isotropic shape. The first one is the so called "topology based" (figure 3B) and is caused by the structure placed under the etched SiO_2, especially by filled vias. Long time before the etching front hits this structure, the etching velocity into the vertical direction increases. The reason for this behaviour is not fully understood but we assume that it is correlated with the behaviour during via filling. Since the growth of the oxide layer is not uniform from the bottom and the sidewalls, during the CVD several zones of growth exist and their fronts will meet at a certain point. In the worst case this can lead to voids inside the via but most probably provides areas with deviating density or stoichiometry, which not only exist inside the via but also above and slightly aside them, due to a low reduction of the different properties. Such disturbed areas show different etching sensitivity compared to the surrounding material on top of unstructured parts of the MOEMS. One example of this mechanism is shown in figure 4. One can see that the shape of the underlying hinge structure is patterned into the SiO_2, even though there is still an oxide layer with a thickness of approximately 200 nm on top of the spring.

While the first anisotropic mechanism is dominant in the vertical direction, the second one, which is called the "interface layer etch", is more prominent in the horizontal one. As shown in the layout in figure 3C, there is an oxide/oxide interface within the sacrificial layer. During the preparation of the device, the bottom oxide layer acts as stop for the etching of the hinge layer. The etching process influences the surface of this bottom layer in several ways although the exact reaction flow is still not fully understood. Our etching gas is a mixture of Cl_2 and BCl_3. Usually resist reacts with Chlorine in molecular as in elemental form to vinyl-like polymers, which are deposited on the surface. Usually, they should be removed during a subsequent cleaning, but residues are possible. A second process is a two-step reaction of the SiO_2 itself. Firstly it is reduced to Si by BCl_3 forming BOCl and finally reacts with Chlorine to $SiCl_4$ [5]. From all this reactions it can be assumed, but is not yet proven, that in the interface of two sacrificial layers there are phases of polymers, BOCl, Si or $SiCl_4$, which can change the interface properties of both layers or react with HF or later reaction products during the release and thus change the release kinetics at interfaces. The consequence is that along the interface of the two oxide layers, the release is faster than it is inside of each layer. The etching front of the interface layer etch has typically a wedge-like shape as shown in figure 3C, which – like the topology based etch – deviates strongly from the isotropic behaviour. In consequence one has to investigate the actual kinetics for each individual design to provide the best release process since the ratio of surface areas to sacrificial volume as well as the structure shape is very individual for different MOEMS designs.

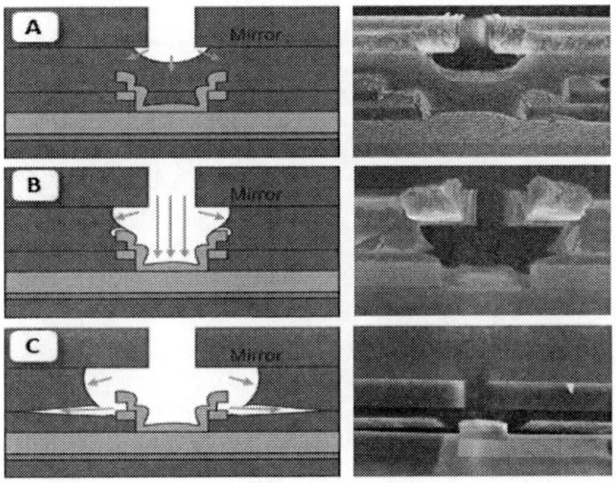

Figure 3. Schematic (left) and representative SEM images (right) of the release processes. (A) Isotropic Etch, (B) topology based etch, (C) interface layers etch.

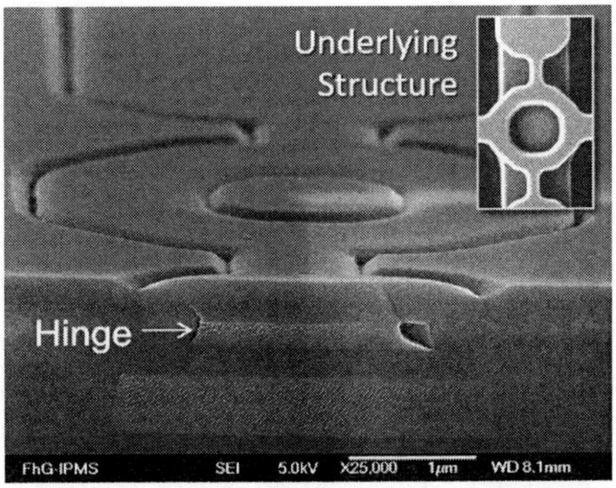

Figure 4. SEM image taken in oblique view showing the topology based etch on top of the hinge structures without deposited mirrors (inset).

The actual kinetics comes into account for the etching of the mirror array of the device. However, a big part of the device outside the mirror array has no significant topology but offers a high open area. Etching of the sacrificial layer in this area therefore takes more time. Even if the mirrors are completely released and fully movable, it is possible, that SiO_2 is still remaining on the device as it is shown schematically in figure 5. This remaining SiO_2 can possibly cause charging effects, which could have a negative influence on the operation of the device.

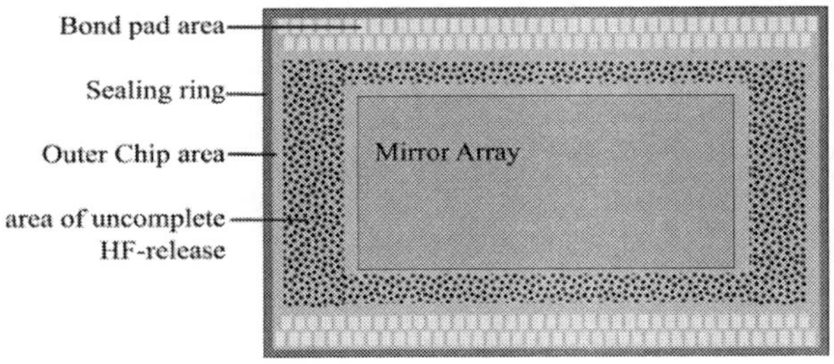

Figure 5. Schematic top view on an SLM device. In contrast to the bond pad and the mirror array area, the outer chip area (black dotted area) does not have any surface topology and etching rate is lower. As a consequence SiO_2 can partly remain on the surface.

2.2. Residues
Another issue, which becomes important with the release kinetics, are residues from the sacrificial layer, which remain after the release process. Their shape can be string-like as shown in figure 6 or also tend to be canvas-like depending on the actual release kinetics. Often the residues contain carbon, probably from the TEOS, together with fluorine, oxygen and maybe boron and small amounts of other elements as can be measured using EDX as seen in figure 7. Since they are not solved in HF or later plasma treatments, they may stay in the device even for later operation. At this point, they may affect general performance or functionality in terms of restriction of single mirrors. We assume, that they agglomerate at the border, at which two etching fronts meet in the later release process. By changing the shape of the structure under the mirror one can influence the velocity and shape as well as possible numbers and direction of movement for etching fronts at a certain point. A very popular method to decrease the HF release duration is to use additional holes in the mirror. Having an optimized number and arrangement of these release holes can significantly change the release kinetics and consequently the position, where possible residues occur. Even though IPMS has developed several remedies to solve the residue issue afterwards the release process, avoiding it by pushing residues from crucial to non-crucial positions is the most promising one and provides a significant increase of the performance of the SLM device.

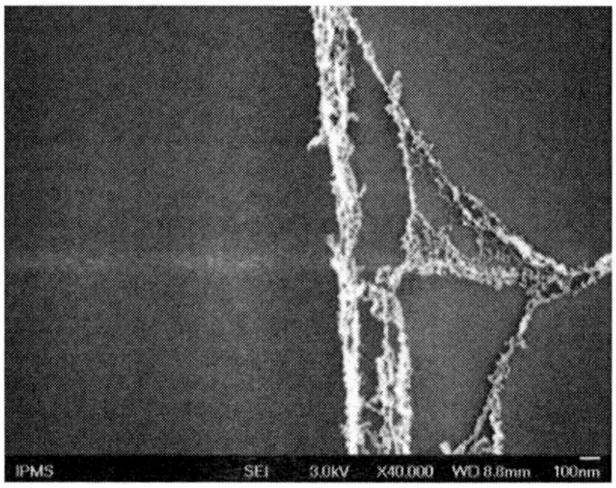

Figure 6. SEM-image of a typical residue on the mirror backside taken after the HF release.

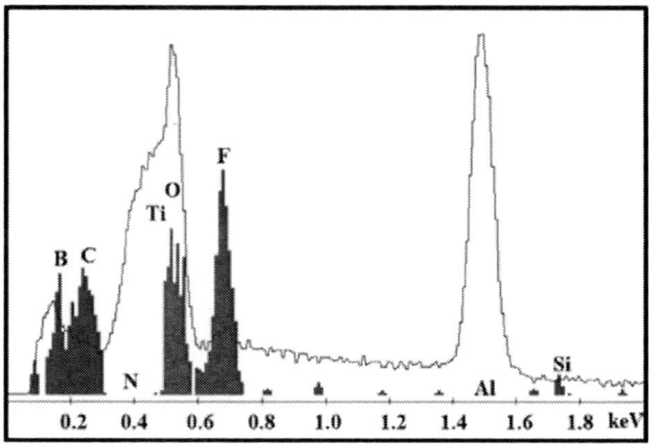

Figure 7. Analysis of a mirror backside affected by residues by energy dispersive X-ray spectroscopy. The full spectrum is shown as black line, while the red bars are calculated as the difference between an affected and an unaffected area.

2.3. Protection of the CMOS
Although the use of SiO_2 as sacrificial layer material has many benefits for the MOEMS part of the device, it can be unfavourable for the CMOS part. Since SiO_2 is also used for insulating layers inside the CMOS, these layers have to be protected during the release process. Therefore an effective barrier layer between the top electrode layer and the inter-layer dielectrics is necessary. From different possible materials, which are HF inert and electrical insulating [6], [7], we chose Al_2O_3 as best suited for our needs. Since it showed less defects and better coverage, Al_2O_3 prepared by atomic layer deposition (ALD) is superior to sputtered one. Devices prepared with an etch stop from ALD-Al_2O_3 did not show any hint of etching attack to the CMOS from MOEMS side. As shown in figure 8, the USG underneath the Al_2O_3-layer still exists without visible damage, while the air gap is fully released.

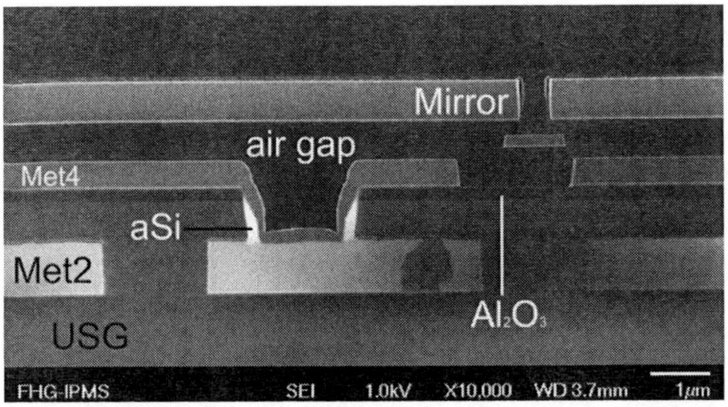

Figure 8. Cross-section of a MOEMS device prepared by ion polishing. For improvement of the mechanical stability the air gap, the mirror slits and the space above the mirrors is filled by epoxy glue. Since the contrast between USG and Al_2O_3 was not sufficient in the original image, the Al_2O_3 layer was slightly recoloured manually .

3. Summary

We presented details of how SiO_2 is used as sacrificial layer material in our fabrication process for multi-level MOEMS devices. After a brief introduction to the layout and the used preparation techniques, the HF release kinetics was investigated. It was shown, that isotropic etching is only one mechanism at the very beginning of the release process, while the later process is dominated by two anisotropic mechanisms caused by topology and interfaces. However the release kinetics has a significant impact on the occurrence and position of residues, still present under the mirror after the HF release. Since they can affect the movability of the mirrors, it is proposed to use and optimize the etch kinetics with respect to each individual MOEMS design for a better performance of the device. As a second goal we presented effective methods to protect the CMOS backplane during the release by using an ALD-Al_2O_3 etching barrier underneath the address electrode and a double-seal ring made of metal on the edges of the diced chip.

Acknowledgments

We thankfully acknowledge the contributions of S Mirabella to this work.

References

[1] Ljungblad U-B, Askebjer P, Karlin T, Sandstrom T and Sjoeberg H 2005 *Proc.of SPIE* (San Jose, CA) vol 5721 p 43-52
[2] Schmidt J-U, *et al.* 2010 *Proc. of SPIE* (Brussels) 77162L
[3] Jang W I, Choi C A, Lee M L, Jun C H and Kim Y T 2002 *J. Micromech. Microeng.* **12** 297-306
[4] Mirabella S, Friedrichs M and Schulze M 2014 *Micromech. and Microsys. Eur. Workshop MME* (Istanbul) vol 24
[5] Garozzo G, Beffumo F and La Magna A 2005 *IEEE Trans. Semicond. Manufact.* **18** 355-8
[6] Yang W S and Kang S W 2006 *Thin Solid Films* **500** 231-6
[7] Bakke T, Schmidt J-U, Friedrichs M and Völker B 2005 *Micromech. and Microsys. Eur. Workshop MME* (Göteborg) vol 16

Journal of Physics: Conference Series **757** (2016) 012005 doi:10.1088/1742-6596/757/1/012005

Three step deep reactive ion etch for high density trench etching

B Lips and R Puers

Department ESAT-MICAS, KU Leuven, Kasteelpark Arenberg 10, 3000 Leuven, BE

E-mail: `bram.lips@esat.kuleuven.be`, `robert.puers@esat.kuleuven.be`

Abstract. A three step Deep Reactive Ion Etch (DRIE) process is developed to etch trenches of 10 µm wide to a depth of 130 µm into silicon with an etch rate of 2.5 µm min^{-1}. The aim of this process is to obtain sidewalls with an angle close to 90°. The process allows the etching of multiple trenches with high aspect ratios that are closely placed together. A three step approach is used as opposed to the more conventional two step approach in an attempt to improve the etching selectivity with respect to the masking material. By doing so, a simple AZ6632 positive photoresist could be used instead of the more commonly used metal masks which are harder to remove afterwards. In order to develop this process, four parameters, which are the bias power, processing pressure, step times and number of cycles, are evaluated an optimized on a PlasmaPro 300 Cobra DRIE tool from Oxford Plasma Technology.

1. Introduction

Deep reactive ion etching (DRIE) is used to create deep, steep sided features in silicon wafers with aspect ratios (etch depth/feature width) beyond 10:1. The technique is developed and licensed by Robert Bosch GmbH [1] and relies on the repeated alternation of isotropic silicon etching and passivation steps to obtain anisotropic profiles. F-based plasmas such as SF_6 are generally used for fast isotropic silicon etching. whereas F-based inhibitor plasmas such as C_4F_8 are generally used to obtain directionality. Figure 1 shows the general process flow. Classically, the selective removal of the fluorcarbon and the isotropic etching is combined into one step: the etching step. However as indicated by others [2], a higher etch selectivity with respect to the masking material can be obtained by subdividing the etch step in two parts. The first part is then optimized towards selective removal of the fluorcarbon film deposited at the bottom of the trench, whereas the second part is optimized towards fast isotropic etching. This gives rise to a so called three step DRIE process. Such a process is a big advantage when used in combination with CMOS processing. Because of the higher etch selectivity with respect to the masking material, photoresists can now be used as masking material even for through wafer trenches and holes. This in contrast to the more conventional materials used for that, such as oxides or metals, which are harder to remove afterwards without damaging useful oxides or metals in the CMOS stack underneath.

The main considerations when developing a DRIE process include side-wall verticality, bottom roughness and etching rate of both silicon and masking material. All of these are affected by process parameters such as plasma power, pressures, gas flows, bias power, cycle times and substrate cooling. This is a wide set of processing parameters that are all interconnected and

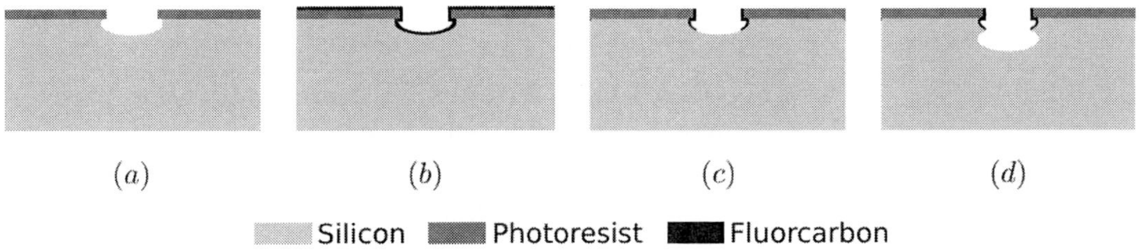

Figure 1. DRIE process. (*a*)Isotropic silicon etch. (*b*)Fluorcarbon film deposition. (*c*)Selective fluorcarbon film removal. (*d*)Isotropic Silicon etch.

thus need careful optimization for each specific etching job. This paper will mainly focus on four of these parameters, which are bias power, processing pressure, step times and number of cycles. All the experiments are conducted on a PlasmaPro 300 Cobra DRIE tool from Oxford Plasma Technology.

2. Experimental details

For the experiments, 500 µm thick <100>boron doped Si-wafers with a resistivity of 10-20 Ω cm are used. These are spincoated with the positive photoresist AZ6632 at a speed of 2000 rpm for 30 s. The soft bake is conducted at 120 °C for 4.5 min. Before developing in an 4:1 diluted 351-solution, the resist is UV-exposed with a dose of 100 mJ cm^{-2}. The development is finished with a hard bake of 4.5 min at 125 °C. This is done in order to improve the etch resistivity of the resist during the DRIE processing. The actual patterns are parallel grooves with a length of 5000 µm, a width of 10 µm and a pitch of 18 µm. Grooves with equal length but smaller widths and pitches are also patterned and etched alongside but are not further discussed in this paper.

In order to save costs and reduce the etch mask variability of different experiments, only small pieces (~4 cm^2) of the patterned wafer are processed. This is done through manual diamond cleaving. However, a carrier substrate is still needed to properly clamp and cool the pieces in the DRIE tool. The carrier also functions as a protection for the bottom electrode. In this work SiOx was used as carrier and reused across the different experiments. Fomblin®-oil or thermal grease was used to temporarily bond the two.

A three step process provided by the DRIE tool manufacturer was used as a starting point. The effect of parameter changes on the output will be discussed next. After DRIE processing, samples were again cleaved with a diamond to inspect the cross section with a FEI XL30 secondary ion emission microscope. Photoresist thicknesses before and after processing were measured with a Dektak®XL stylus profilometer.

3. Results and discussion

The initial parameters of the three step process are given in table 1. Due to the addition of the extra breakthrough step as compared to the more conventional two step DRIE process, it is now possible to alter process parameters after fluorcarbon removal at the bottom of the trench. Main parameters altered here are: SF_6-flow, pressure and bias power. After 100 cycles, the process of table 1 resulted in a 68 µm deep trench with a Si to mask etch selectivity of 52:1. The profile is severely negative tapered with a sidewall angle of 91.9°. This limits the usability of the process if one wants to etch trenches with a small pitch, especially if the trenches needs to be very deep. Therefore the effect of several parameters is studied in order to straighten the trench profile without compromising on the etch selectivity in the existing three step process.

Table 1. Three step process parameters and their initial values.

Step	Time (s)	SF_6 (sccm)	C_4F_8 (sccm)	Pressure (mbar)	Plasma power (W)	Bias power (W)
Deposition	2	10	200	0.033	2500	0
Breakthrough	2	200	10	0.040	2000	35
Etch	2	400	10	0.093	2500	0

3.1. Bias power

The first parameter under study is the bias power. According to [3], sidewall angles become more positive tapered by lowering the bias power in the etch step of a two step process. As can be seen in figure 2, similar results are obtained in this paper with the three step process by lowering the bias power during the breakthrough. An interesting side effect is the increase of the etch selectivity with respect to the resist mask. Which confirms that resist mask erosion is mainly caused by ionic impact. Switching off the bias power after fluorcarbon removal, as is done in a three step process, is thus a viable route to enhance the Si to resist etch selectivity.

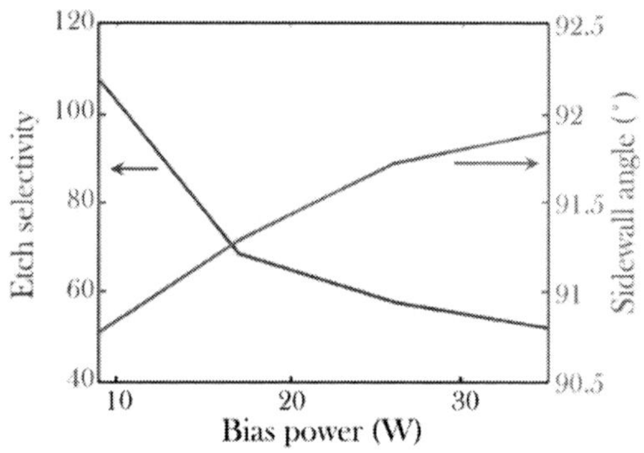

Figure 2. Effect of bias power during the breakthrough on the Si to resist etch selectivity and sidewall angle.

One drawback however of decreasing the breakthrough power is the reduced overall etch rate, which dropped from 6.8 µm min^{-1} at 35 W to 5.9 µm min^{-1} at 9 W. This is due to the reduced contribution of the breakthrough step to the actual silicon etching. Similar results should thus be found by keeping the bias power constant and reducing the breakthrough step time. Unfortunately a 2 s step time is already close to the minimal switching time of the machine, which is limited by the response time of the mass flow controllers. Another limiting factor in this experiment was the accuracy of the automatic matching unit. If bias power was lowered to 4 W, matching became impossible, resulting in total power reflection and almost no silicon etching.

3.2. Breakthrough pressure

With bias power adjusted to 9 W the effect of breakthrough pressure is investigated. Lowering the pressure from 40 µbar to 30 µbar further improved the sidewall angle to 90.4°. This is most probably due to the longer mean free path of SF_6-ions which in term leads to a lower ion dispersion in the dark region between the substrate and the plasma [2]. Thus causing less etching of the sidewall polymer. However both selectivity and overall etch rate dropped to 84.6:1 and 5.5 µm s^{-1} respectively. Due to the limited pump capacity and the chosen flow rates, the minimal attainable pressure is 25 µbar. Yet this pressure is never experimented with since the pressure in the subsequent etching step, which is 3.7 times higher, was already difficult to reach in the 2 s time slot.

3.3. Deposition time
A natural parameter to tune in order to change the sidewall angle, is the deposition step time. The longer this step lasts, the more polymer is deposited and the harder it becomes to remove it from the sidewalls in the subsequent step. As a consequence more positive tapered slopes are obtained [4]. Doubling the deposition time to 4 s resulted in a perfect sidwall slope of 90° and a 1.4 times better conservation of the mask material. But due to overpassivation, the etch rate severely dropped to 2.25 $\mu m\,s^{-1}$. As stated before, the sidewall angle of deep trenches is the main focus of this research. Therefore further optimization of the deposition time is omitted at this point. But it is important to mention that reducing the deposition time to overcome overpassivation will lead to higher overall etch rates, without affecting both sidewall angle and mask removal rate. Testing the process developed so far to check if the observations still hold with more cycles is now more interesting in achieving deep straight walled trenches.

3.4. Number of cycles
By doubling the number of cycles to 200 and keeping all the other parameters the same, trenches were, as expected, twice as deep. However the sidewall angle slightly increased to 90.4°. This is due to an effect called imaging force. As explained in [2] and [5] imaging force is the deflection of ions entering a trench towards the sidewalls due to the negative potential of these conducting silicon sidewalls with respect to the plasma glow. As trenches become deeper, more deflection will occur which in term leads to the observed profile. As the number of cycles is increased

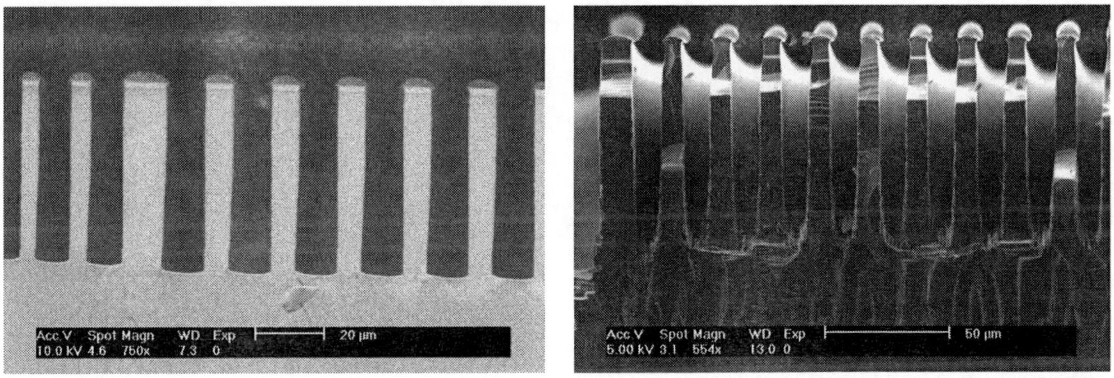

(*a*) (*b*)

Figure 3. Etching result after (*a*)200 cycles and (*b*)400 cycles.

even further to 400 cycles the observed profile changes completely. As can be seen in figure 3: sidewalls become positive tapered with an angle of 89.1° and grassing is observed at the bottom of the trench. The formation of the grass and the positive sloped sidewalls are due to an imbalance in polymer removal and deposition [6]. Too much polymer is deposited with respect to breakthrough and etch step. The reason why it only appears in deeper trenches is because the polymer removal at the bottom becomes less effective at higher aspect ratios. This is a consequence of imaging forces and dispersion of ions due to collisions in the plasma glow [2], basically the same mechanisms that are thought to cause the aspect ratio dependent etching (ARDE) [2].

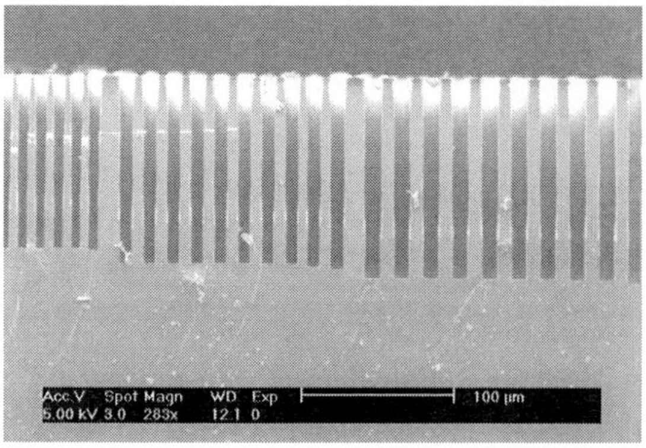

Figure 4. Result after 400 cycles with 9 W bias power in the first half and 14 W in the second half. Trench widths from left to right are 6, 8 and 10 µm.

To prevent grassing from occurring, more directional ions during the breakthrough are thus needed. Since pressure is already minimized, bias power should therefore increase. However an increase in bias power also leads to faster mask removal and worse selectivity as indicated in figure 2. Because the higher bias power is only needed at high aspect ratios one can first etch trenches with a low bias power and later on switch to a higher bias power. The result of this approach, where the first 200 cycles are etched at 9 W and the subsequent 200 cycles at 14 W, is shown in figure 4. These trenches have sidewall angles of 89.7° and are 130 µm deep. The sidewall damage seen at 80 % of the trench depth is most probably due to mask erosion as explained in [7].

Conclusion
By analysing the effect of bias power, breakthrough pressure, deposition time and number of cycles on etch characteristics in a three step DRIE process. A recipe is developed to etch 130 µm deep trenches in silicon with a width of 10 µm. The sidewalls are almost vertical with an angle of 89.7°, which is a strict requirement for high density deep trench etching. Excellent silicon to photoresist etch selectivity of 50:1 is achieved due to limited usage of high energy ions during breakthrough. This obviates the use of metal or oxide masks for through wafer etching applications such as: comb drives, springs and through wafer vias. Which is a big advantage when combined with CMOS processing. Currently the proposed 400 cycle process with an etch rate of 2.45 $\mu m\,min^{-1}$ is used to etch mechanical structures with thicknesses starting from 100 µm.

Acknowledgement
This research was made possible by the Hercules Foundation for heavy equipment (AKUL 034 and ZW1115), and received funding from the European Research Council under the European Union's Seventh Framework Programme (FP7/2007-2013) / ERC grant agreement n° 340931.

References
[1] Robert Bosch GmbH 1994 *Pat. 4,855,017 and 4,784,720 (USA) and 4241045C1* Germany
[2] Jansen H V, De Boer M J, Unnikrishnan S, Louwerse M C and Elwenspoek M C 2009 *J. Micormech. Microeng.* **19** 033001-41
[3] Ayón A A, Braff R, Lin C C, Sawin H H and Schmidt M A 1999 *J. Electrochem. Soc.* **146** 339-49
[4] Abdolvand R and Ayazi F 2008 *Sens. Actuators* A **144** 109-16
[5] Elwenspoek M and Jansen H V 2004 *Silicon Micromachining* (Cambridge: Cambridge University Press)
[6] Ayón A A, Chen S, Lohner A, Spearing S M, Sawin H H and Schmidt M A 1998 *Proc. Material Research Society (Cambridge)* vol 546 Heuer A H and Jacobs S J (Cambridge: Material Research Society) p51
[7] Meng L and Yan J 2015 *J. Micormech. Microeng.* **25** 035024-32

Asymmetric resonance frequency analysis of in-plane electrothermal silicon cantilevers for nanoparticle sensors

Maik Bertke[1,2,*], **Gerry Hamdana**[1,2], **Wenze Wu**[1,2], **Markus Marks**[1], **Hutomo Suryo Wasisto**[1,2] **and Erwin Peiner**[1,2]

[1] Institute of Semiconductor Technology (IHT), Braunschweig University of Technology, Hans-Sommer-Straße 66, D-38106 Braunschweig, Germany
[2] Laboratory for Emerging Nanometrology (LENA), Langer Kamp 6a, D-38106 Braunschweig, Germany

*E-mail: M.Bertke@tu-bs.de

Abstract. The asymmetric resonance frequency analysis of silicon cantilevers for a low-cost wearable airborne nanoparticle detector (Cantor) is described in this paper. The cantilevers, which are operated in the fundamental in-plane resonance mode, are used as a mass-sensitive microbalance. They are manufactured out of bulk silicon, containing a full piezoresistive Wheatstone bridge and an integrated thermal heater for reading the measurement output signal and stimulating the in-plane excitation, respectively. To optimize the sensor performance, cantilevers with different cantilever geometries are designed, fabricated and characterized. Besides the resonance frequency, the quality factor (Q) of the resonance curve has a high influence concerning the sensor sensitivity. Because of an asymmetric resonance behaviour, a novel fitting function and method to extract the Q is created, different from that of the simple harmonic oscillator (SHO). For testing the sensor in a long-term frequency analysis, a phase-locked loop (PLL) circuit is employed, yielding a frequency stability of up to 0.753 Hz at an Allan variance of 3.77×10^{-6}. This proposed asymmetric resonance frequency analysis method is expected to be further used in the process development of the next-generation Cantor.

1. Introduction

Since the last few years, more and more types of nanoparticles (NPs) have been used in different consumer goods, resulting in an increasing awareness towards releasing these NPs in air. Because of their small size, possessed hazard, and presence in both indoor and outdoor environments, they can become a huge danger or risk when they are exposed to human body via either direct deposition on the skin or inhalation through respiratory systems [1]. Therefore, there have been many studies concerning the characteristics and health effects of different types of airborne NPs in the environmental communities as well as the strategies to manage their exposure [2].

To ensure a safety of the workers who directly handle and interact with NPs during the manufacturing of the NP-based devices or products, it is necessary to equip them with a low-cost wearable direct-reading tool for continuously monitoring the airborne NP mass concentration levels [3]. Therefore, owing to their detection simplicity and cost-effective batch fabrication, various types of resonant micro/nanoelectromechanical systems (M/NEMS) were developed as NP mass-sensitive sensors by several world-leading research groups to meet the demands (e.g., microfluidic film bulk acoustic resonator (FBAR) [4], electrothermal silicon cantilever [5], filter-fiber nanoresonator [6], and vertical silicon nanowire resonator [7]). The NP measurement principle is based on the monitoring of the resonance frequency shifts of the M/NEMS caused by the NPs additionally deposited on their surfaces. In our second generation of the pocket-sized *can*tilever-based airborne NP detec*tor* (Cantor-2), the in-plane electrothermal piezoresistive silicon cantilever is connected to a phase-locked loop (PLL) circuit to realize a real-time tracking system of the resonance frequency [8, 9]. However, the phenomenon of the asymmetric resonance frequency signals produced by the employed cantilevers has not been analyzed in details. Thus, in this work, several designs of the electrothermal cantilever resonators are investigated in terms of their resonance characteristics, which will then be used as a basis for further improvement of the Cantor. Moreover, their frequency stability is also measured to define the Allan variance of the device in a long-term test.

2. Sensor preparations

The electrothermal silicon cantilever resonators with different free-end geometries (Figures 1(a)-(d)) were simulated using the finite element modelling (FEM) tool of COMSOL Multiphysics 4.3b to determine their figure of merit (i.e., $FoM = m_c/(f_0 \times A_c \times Q)$ [10]) as listed in Table 1. In this case, m_c, f_0, and A_c denote the mass of the cantilever, the resonance frequency, and the area of the collection surface, respectively. It should be noted that all these cantilevers have a total length of 1000 μm and a beam spring width of 170 μm. By exciting the heating resistor on the cantilever with a DC bias of 5 V, the temperature (T) increases from 293.15 K to 306.1 K (i.e., $\Delta T = 12.95$ K) resulting in a cantilever bending (Figure 2(a)). The calculated maximum stress obtained from FEM is 3.63×10^7 N/m² [9].

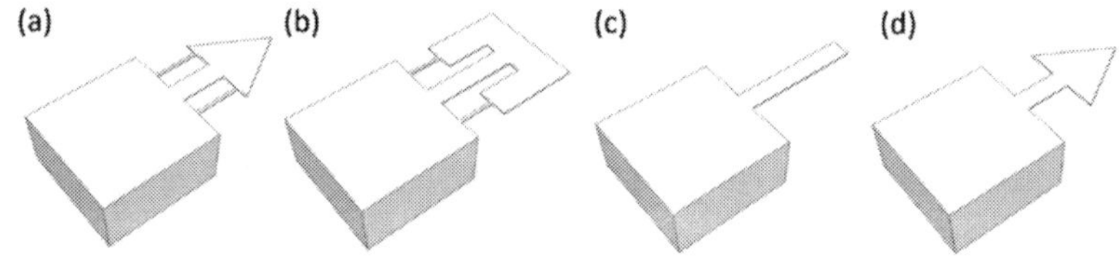

Figure 1. Designs of (a) a triangular cantilever with strivings (C_{ts}), (b) a rectangular cantilever with strivings (C_{rs}), (c) a rectangular cantilever (C_s), and (d) a triangular cantilever (C_t).

Table 1. FEM results of four cantilever resonators with different free-ends.

Cantilever	f_0 [kHz]	A_c [mm²]	m_c [μg]	Q	FoM
C_{ts}	351,380	0.248	7.80	9497	9.4×10^{-9}
C_{rs}	166,228	0.570	17.92	5541	3.4×10^{-8}
C_s	217,150	0.139	4.37	4343	3.3×10^{-8}
C_t	185,329	0.248	7.80	3497	4.9×10^{-8}

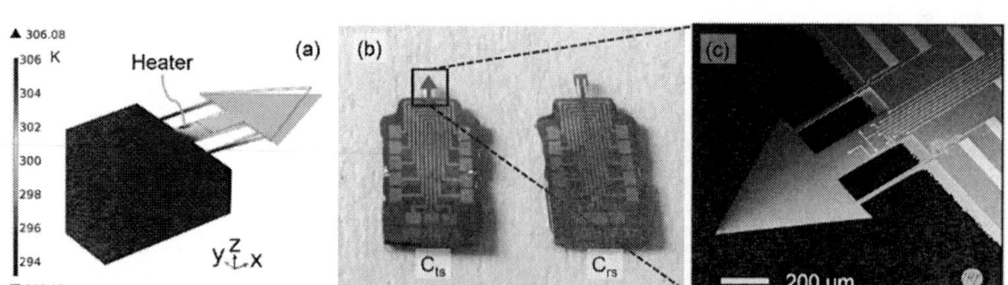

Figure 2. (a) 3-D FEM result, (b) photograph, and (c) scanning electron micrograph of one of the fabricated sensors (i.e., the triangular cantilever with strivings C_{ts}).

Subsequently, these in-plane cantilevers were fabricated using bulk silicon wafers and bulk micromachining techniques, leading to lower cost of materials and fabrication processes, respectively (Figures 2(b) and (c)). The detailed fabrication steps have been described previously, which mainly employ photolithography, dopant diffusions using borofilm 100 and phosphorosilica (Emulsitone), thermal oxidation, metal evaporation, and inductively coupled plasma (ICP) cryogenic dry etching processes [9].

3. Resonance frequency analysis

To analyze the resonance behaviors of the fabricated cantilevers (Figures 3(a)-(d)), a frequency sweep was carried out using the measurement setup described in [10]. The main relevant parameters of the measured resonance curves are the f_0, the Q and the relative amplitude of the peak, which can be directly extracted from a fitting procedure. Because of the asymmetry of the resonance curve, a fitting function in relation to the Fano resonance is developed replacing the commonly used simple harmonic oscillator (SHO) fitting function [11]. The Fano resonance is based on a system of two coupled vibrations where one is forced to its oscillation directly and their superposition will result in an asymmetric resonance behavior (Figure 4) [12].

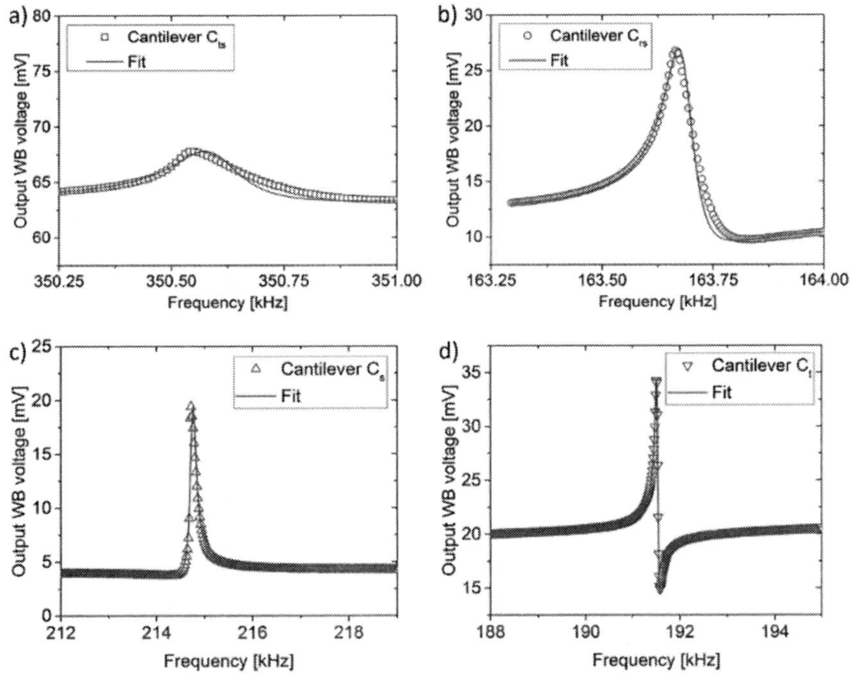

Figure 3. Measured resonance curves of cantilevers and their corresponding fittings of (a) C_{ts}, (b) C_{rs}, (c) C_s, and (d) C_t.

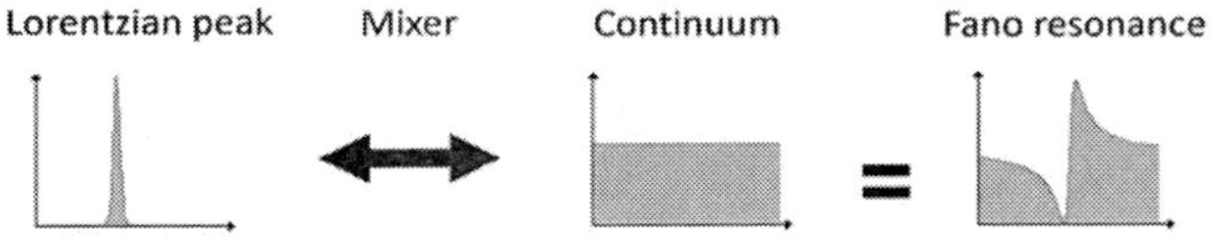

Figure 4. Illustration of a Fano resonance as a mixed combination of a discrete state and a continuous background.

Journal of Physics: Conference Series **757** (2016) 012006 doi:10.1088/1742-6596/757/1/012006

The Fano resonance frequency response can be described implicitly by [12]:

$$\frac{q^2-1}{f^2+1}+\frac{2qf}{f^2+1}+1=\frac{(f+q)^2}{f^2+1}, \tag{1}$$

where q and f are the asymmetry factor and the frequency, respectively. The frequency sweep response σ around the f_0 is then given by:

$$\sigma=\frac{\left(\frac{f-f_0}{g}+q\right)^2}{\left(\frac{f-f_0}{g}\right)^2+1}\times H+\sigma_0+t\times f \tag{2}$$

with line width and gain parameters g and H, respectively, and an offset comprising a constant (σ_0) and linearly varying term ($t \times f$). The proposed Q is calculated by combining two established methods (i.e., $Q_{q=0} = f_0/2g[(2)^{0.5}-1]^{0.5}$ and $Q_{q=1} = (2f_0^2+2g^2)/(4g\times f_0)$ [13]) with their newly introduced correction factors of $(1-|q|^{0.5})$ and $|q|^{0.5}$ as:

$$Q=\frac{f_0}{2g\sqrt{\sqrt{2}-1}}\left(1-\sqrt{|q|}\right)+\frac{2f_0^2+2g^2}{4g\times f_0}\sqrt{|q|}\ \text{ for } q=\pm 0\ldots 1. \tag{3}$$

A simulation of an electrical RLC circuit using MATLAB is used as a reference to validate Equation (3) replicating the different resonance curves. In this case, an excitation is generated to bring the circuit to resonance followed by a cut off at steady state. The yielded decaying response amplitude can then be fitted over the time. The damping of the oscillation defines the quality factor of the system [13,14]. Figures 5(a)-(c) depict the resonance curves with different asymmetry levels calculated using Equation (2). For the validation, the frequency responses of the fitted RLC circuit are added. The relative deviation between the calculated and simulated f_0 and Q are listed in Table 2. Small differences between these two curves (Figure 5(a)) causes the deviation of f_0 and Q at $q = 1$. We find that inaccuracies of the Equation (3) at $q = 0.5$ and $q = 0.25$ are still comparable with that of $q = 1$ (i.e., $\Delta Q < 0.7$), proving its reliability.

Table 2. The relative deviation between the calculated and simulated validation results of the f_0 and Q in terms of the quality of developed Equation (3).

Asymmetry factor	$\Delta f_0/f_0$	$\Delta Q/Q$
$q = 1$	3.03×10^{-4}	6.8×10^{-3}
$q = 0.5$	2.01×10^{-4}	13.6×10^{-3}
$q = 0.25$	4.01×10^{-4}	4.2×10^{-3}

Table 3. Comparison of the f_0 and Q obtained from measurement and FEM.

Cantilever	$f_{0\_measured}$ [Hz]	$Q_{\_measured}$	f_{0_FEM} [Hz]	$Q_{\_FEM}$
C_{ts}	350,595 ± 2	2401 ± 48	351,380	9497
C_{rs}	163,687 ± 1	2491 ± 57	166,228	5541
C_s	214,755 ± 4	2087 ± 95	217,150	4343
C_t	191,530 ± 1	3242 ± 43	185,329	3497

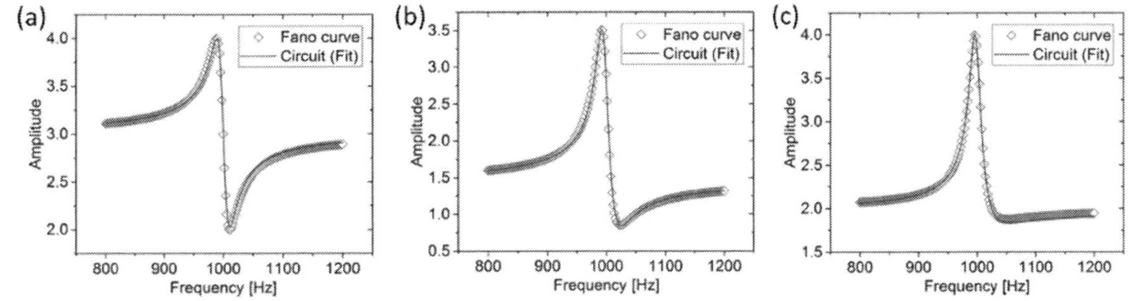

Figure 5. Fitting analysis of the Fano resonance and the RLC circuit curves for verifying the quality of the developed Equation (3) for (a) $q = 1$, (b) $q = 0.5$ and (c) $q = 0.25$.

Furthermore, Table 3 shows good agreement of the f_0 obtained from FEM and the fitting of the measurements with four different cantilever sensors using Equations (2) and (3). However, the Q values of the designed cantilevers with strivings (C_{ts} and C_{rs}) are much lower than expected, which can be due to the simplification of the 3-D model geometries used in the FEM.

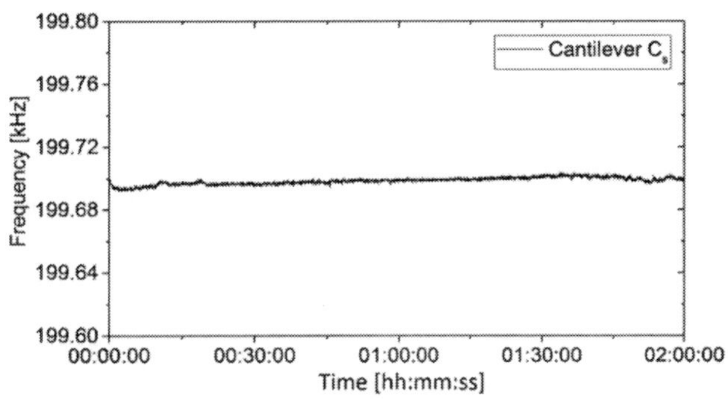

Figure 6. Measurement result of the long-term stability of a cantilever resonator with PLL-based tracking electronics.

Moreover, in the long-term stability tests using a homebuilt PLL circuit [15], the fabricated cantilevers without strivings (C_s and C_t) have demonstrated a frequency stability of up to 0.753 Hz at an Allan variance of 3.77×10^{-6} (Figure 6). Although the proposed sensors have shown promising results in terms of the simulation and measurement (i.e., high $Q > 2000$), some geometry modifications as well as fabrication parameter adjustments are still required to enhance the performance of the devices. Besides that, the described asymmetric resonance frequency fitting method (i.e., based on Fano resonance method) is expected to be employed supporting the optimization of the next-generation Cantors to precisely estimate their resonance characteristics (i.e., f_0 and Q).

4. Conclusions

Electrothermal silicon piezoresistive cantilever sensors with different free-end geometries have been analyzed in terms of their asymmetric resonance behaviors. A novel fitting model based on the Fano resonance approach has been developed to precisely extract the values of the resonance frequency and quality factor (Q) of the devices, replacing the simple harmonic oscillator (SHO) fitting model. Moreover, despite the promising results from long-term stability tests, further experiments of the cantilever resonators in airborne nanoparticle (NP) exposure are still needed to justify the resonance characteristics of the devices in the conditions of before and after being polluted with NPs.

References

[1] Pini M, Salieri B, Ferrari A M, Nowack B and Hischier R 2016 Human health characterization factors of nano-TiO_2 for indoor and outdoor environments *Int. J. Life Cycle Assess* 1-11 http://dx.doi.org/10.1007/s11367-016-1115-8.

[2] Díaz-Soler B M, Martínez-Aires M D and López-Alonso M 2016 Emerging risk in the construction industry: Recommendations for managing exposure to nanomaterials *DYNA* **83**(196), 48-54, http://dx.doi.org/10.15446/dyna.v83n196.56608.

[3] Kuhlbusch T A J, Asbach C, Fissan H, Göhler D and Stintz M 2011 Nanoparticle exposure at nanotechnology workplaces: A review *Particle and Fibre Toxicology* **8**(22) 1-18 http://dx.doi.org/10.1186/1743-8977-8-22.

[4] Paprotny I, Doering F, Solomon P A, White R M and Gundel L A 2013 Microfabricated air-microfluidic sensor for personal monitoring of airborne particulate matter: Design, fabrication, and experimental results *Sens. Actuators A Phys.* **201** 506–16 http://dx.doi.org/10.1016/j.sna.2012.12.026.

[5] Beardslee L A, Addous A M, Heinrich S, Josse F, Dufour I and Brand O 2010 Thermal excitation and piezoresistive detection of cantilever in-plane resonance modes for sensing applications *J. Microelectromech.* **19** 1015–17 http://dx.doi.org/10.1109/JMEMS.2010.2052093.

[6] Schmid S, Kurek M, Adolphsen J Q and Boisen A 2013 Real-time single airborne nanoparticle detection with nanomechanical resonant filter-fiber *Scientific Reports* **3** 1288 http://dx.doi.org/10.1038/srep01288.

[7] Wasisto H S, Merzsch S, Stranz A, Waag A, Uhde E, Salthammer T and Peiner E 2013 Femtogram aerosol nanoparticle mass sensing utilising vertical silicon nanowire resonators *Micro Nano Lett.* **8**(10) 554–8 http://dx.doi.org/10.1049/mnl.2013.0208.

[8] Wasisto H S, Uhde E and Peiner E 2016 Enhanced performance of pocket-sized nanoparticle exposure monitor for healthy indoor environment *Build. Environ.* **95** 13–20 http://dx.doi.org/10.1016/j.buildenv.2015.09.013.

[9] Wasisto H S, Merzsch S, Uhde E, Waag A and Peiner E 2015 Handheld personal airborne nanoparticle detector based on microelectromechanical silicon resonant cantilever *Microelectron. Eng.* **145** 96–103 http://dx.doi.org/10.1016/j.mee.2015.03.037.

[10] Wasisto H S, Wu W, Uhde E, Waag A and Peiner E 2015 Electrothermal piezoresistive cantilever resonators for personal measurements of nanoparticles in workplace exposure *Proc. SPIE 9517, Smart Sensors, Actuators, and MEMS VII, and Cyber Physical Systems* **95170**B http://dx.doi.org/10.1117/12.2180151.

[11] Wasisto H S, Merzsch S, Waag A, Uhde E, Salthammer T and Peiner E 2013 Airborne engineered nanoparticle mass sensor based on a silicon resonant cantilever *Sens. Actuator B Chem.* **180** 77–89 http://dx.doi.org/10.1016/j.snb.2012.04.003.

[12] Miroshnichenko A E, Flach S and Kivshar Y S 2010 Fano resonances in nanoscale structures *Rev. Mod. Phys.* **82** 2257–98 http://dx.doi.org/10.1103/RevModPhys.82.2257.

[13] Manzaneque T, Hernando H and Rodríguez-Aragón L 2010 Analysis of the quality factor of AlN-actuated micro-resonators in air and liquid *Microsyst. Technol.* **16** 837-45 http://dx.doi.org/10.1007/s00542-009-1003-2.

[14] Toledo J, Manzaneque T, Hernando-García J and Vázquez J 2014 Application of quartz tuning forks and extensional microresonators for viscosity and density measurements in oil/fuel mixtures *Microsyst Technol.* **20**(4) 945–53 http://dx.doi.org/10.1007/s00542-014-2095-x.

[15] Wasisto H S, Zhang Q, Merzsch S, Waag A and Peiner E 2014 A phase-locked loop frequency tracking system for portable microelectromechanical piezoresistive cantilever mass sensors *Microsyst. Technol.* **20**(4), 559–69 http://dx.doi.org/10.1007/s00542-013-1991-9.

Design and Simulation of Bistable Microsystem with Frequency-up conversion effect for Electrostatic Energy Harvesting

Bogdan Vysotskyi[1,2], Fabien Parrain[1], Elie Lefeuvre[1], Xavier Leroux[1], Denis Aubry[2], Philippe Gaucher[2]

[1] IEF, University Paris XI - Paris-Sud, Orsay, France.

[2] MSSMat, CentraleSupélec, Châtenay-Malabry, France.

bogdan.vysotskyi@u-psud.fr

Abstract. This work is dedicated for the study of energy harvesters implemented in form of microelectromechanical systems (MEMS) used to harvest ambient vibrations for powering standalone electronic devices. The previewed application is to power a leadless pacemaker with mechanical energy of the heartbeat, which requires the amount of power typically more than 1µW. The target of the presented article is to combine the effect of bistability and non-linear coupling by electrostatic effect in order to achieve the high value of bandwidth at the low frequency under the low accelerations. Such system is expected to bring high power density performance. This study is performed mostly by numerical simulation.

For the present day energy harvesting is viewed as a promising and innovative alternative for the batteries in the task of powering a wide variety of standalone devices. The examples could be found in different fields: starting from the infrastructure monitors and sensor networks up to biomedical applications.

Conception and design of the energy harvesting device that will be able to deliver >1µW of power under the low frequency excitation (20-30Hz) with acceleration of 1g is in the scope of this work. Such parameters are not very far from the values of the human heartbeat, so the presented device could serve as a starting point in development of the mechanism capable of powering the pacemaker [1]. Numerous examples of the energy harvesters operating with such input parameters had already been proposed, but generally they are not compatible with biomedical applications due to the size [2] and use of magnetic elements [3]. However, introducing the non-linearity in electrostatic energy harvester could produce a dramatic effect on its performance [4].

In the bistable system, the switching between stable positions depends only on the value of the external applied force [5], and it could be used to access the low frequencies keeping the bandwidth high for the external excitation.

It had been clearly demonstrated that maximum power that can be converted by an inertial power generator is proportional to the operational frequency of the transducer [6]. So, the frequency-up conversion is a promising strategy to increase the output power value.

Design of the device and the simulation of its performance is presented in this paper.

1. Design of the Frequency-up conversion System

Presented energy harvester using bistable electrostatic frequency-up conversion system consists of two suspended masses with bistable and linear springs, variable capacity attached to linear oscillator and two pair of electrodes: one to impose bias between to oscillators and separately another one to apply voltage on variable capacity. Oscillator with bistable springs gets excited from the vibrating medium - it serves as a seismic mass, then comes into contact with linear oscillator which leads to electrostatic pull-in followed by accompanied movement and, finally, pull-out. As a result, energy is transmitted from a seismic mass to the transducer that is attached to linear oscillator. Thus, independent oscillation after pull-out is a source of produced electrical energy.

Bistability of an oscillator is achieved by using a double curved beam spring geometry. The switching forces are chosen to be 1.3mN and 0.5mN for the double curved beam mechanism. For the variable capacitance part, a gap closing finger geometry was implemented as it is seems to be an optimal solution to maximize the amplitude of capacitance variation keeping the limited maximum displacement. Stoppers are placed in order to prevent the electrostatic stitching.

The schematics of the device is shown on the Fig. 1.

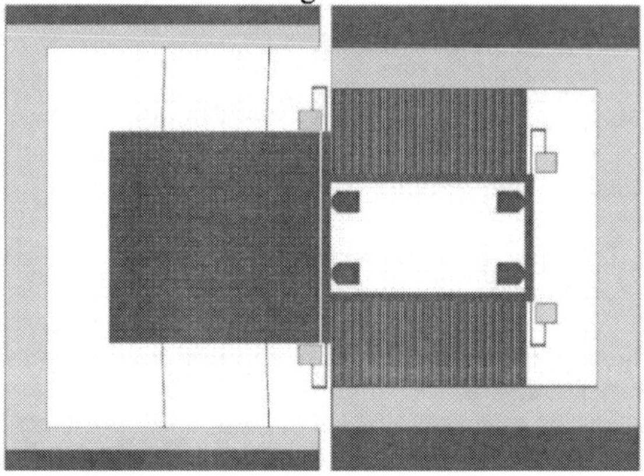

Figure 1. Bistable electrostatic energy harvester with frequency up-conversion. Mechanical oscillator with bistable springs is presented on the left side, linear resonator with comb fingers is on the right side. Orange represents silicon, yellow – gold contacts.

Device dimensions are 1.5cm x 1cm x 85μm. Total device volume is 0.013 cm^3, which is much lower than limiting value of 1cm^3, so it is compatible with previewed application. Summary of device parameters is presented on Table 1.

Parameter	Value
Curved beam length, mm	2
Capacitive finger length, mm	2
Seismic mass, g	0.22
Stopper proximity distance, μm	1
Inter-finger distance, μm	20
Linear oscillator resonant frequency, Hz	4000
Cmax, pF	102
Cmin, pF	10

Table 1. Summary of designed parameters.

2. Modelling the Dynamics

To simulate the highly non-linear dynamics of the coupled two oscillators system with frequency up-conversion the equivalent circuit approach is used. Using electrical circuit simulation is a simple way to confine the multiphysics phenomena in electrical equations, with easy way to include a variety of non-linear effects (springs softening and hardening) [7]. Equivalent circuit had been implemented in LTspice IV. Its electrical parameters correspond unambiguously to the basic mechanical properties of both resonators (such as spring stiffness, damping and mass), and, thus, the same evolution equations are solved [8]. Electrostatic contact between two movable parts was simulated as for the ideal capacitor. Schematics of equivalent circuit is shown on the Fig. 2.

The non-linearity of bistable mechanism was simulated by adding a restoring force of double curved beam in the oscillator circuit. The non-linear stiffness coefficients had been obtained from the finite element model developed in ANSYS©. This model allows simulating the time evolution of the displacement of both movable masses in the frequency up-conversion system.

Biasing voltage between two oscillators was set to 1.2V. The displacement of seismic mass and linear resonator with excitation of 1g at 25Hz is shown on the Fig. 3.

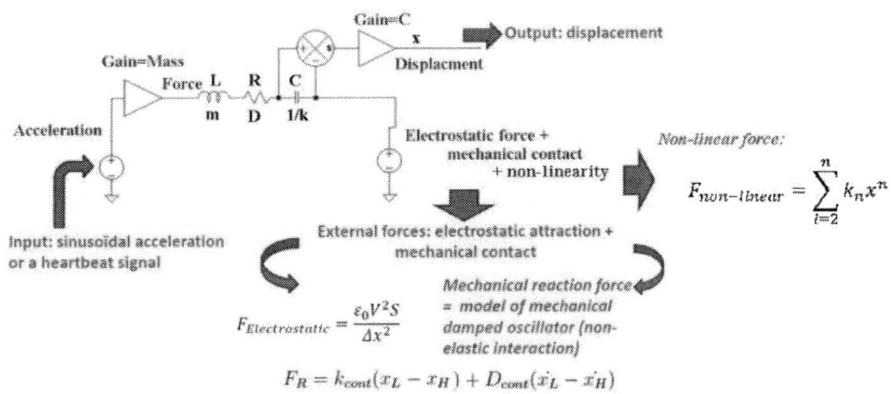

Figure 2. Equivalent circuit of bistable oscillator used to simulate the displacement evolution.

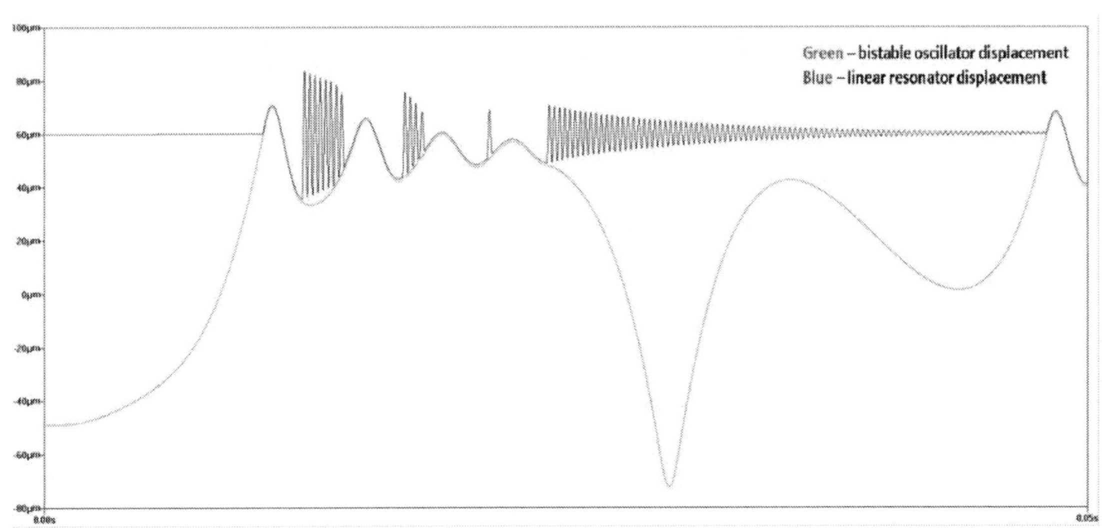

Figure 3. Spice simulation that indicates electrostatic frequency up-conversion in the system of two mechanical oscillators. Displacement of the bistable resonator (green) and linear resonator (blue).

The notion of bandwidth for non-linear systems still remains ambiguous. Here we will use this term in order to describe the frequencies at which the bistable mechanism could be switched. Keeping the same 1g acceleration, the sweep and selective frequencies analysis had been performed. Both revealed that with given parameters seismic mass could be excited at frequency range of 1-100Hz, which could be explained by highly non-linear nature of bistable mechanism, which is mostly sensitive to applied inertial force, but much less to the frequency of excitation.

3. Energy Harvesting Simulations

To evaluate the power output of the energy harvester, a simple model of the interface circuit with biasing capacitor, diode rectifier and energy storage capacitor had been chosen [9]. Results of simulated power output as a function of biasing capacitor voltage and store capacitor voltage are shown on the Table 2.

V bias / V store	Harvested power, μW
8V/4V	0.93
10V/5V	1.5
12V/6V	2.4

Table 2. Simulated power output as a function of biasing voltage.

Maximum biasing voltage had been chosen to impose slightly smaller attraction force in transducer in comparison to restoration force of the linear spring. Thus, with the highest simulated biasing voltage applied to the transducer, the power density could reach the value of 185μW/ cm^3.

4. Device Fabrication Procedure

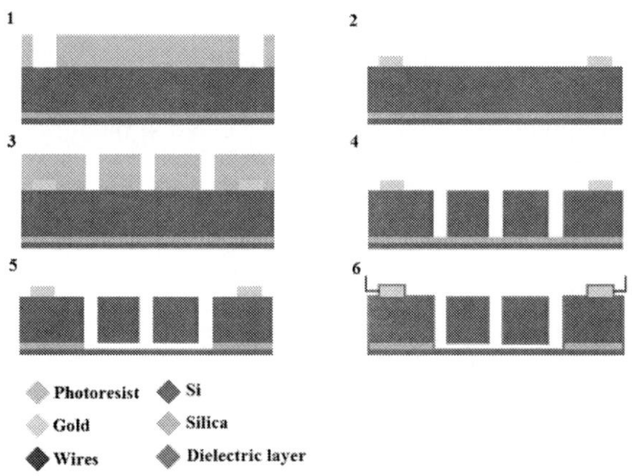

Figure 4. Schematic representation of electrostatic frequency up-conversion system fabrication process (not for scale).

The system discussed before had not been fabricated yet, however fabrication technique had already been developed and tested on prototypes. Device is previewed to be created on 4-inch SOI wafer of 85μm device thickness layer with classical Si technologies that are easily accessible for industrial implementation. The offered process flow is shown on the Fig. 4.
First, the Ti35ES photoresist is deposited by spin-coating, which is followed by UV lithography and development (Fig. 4, 1). Next, gold layer thin film is evaporated on the wafer and lift-off process is performed (Fig. 4, 2). After, next step of S1828 photoresist spin-coating and UV lithography is made (Fig. 4, 3) and device is shaped with deep reactive ion etching (DRIE) of Si layer (Fig. 4, 4). Finally,

the MEMS is liberated with HF vapour etching of buried oxide layer (Fig. 4, 5). The wires are bonded to gold electrodes, and atomic layer deposition of thin dielectric layer is done (Fig. 4, 6).

5. Conclusion

In the present article, a design of the bistable electrostatic energy harvester with frequency-up conversion mechanism is offered and its performance is simulated. As stated by the results obtained, seismic mass with bistable springs is expected to be sensible to the excitations at low frequencies with the broad band and at relatively low amplitude (of the order of 1g). An equivalent circuit had been used in order to simulate the run of both mechanical and electrical parts of the energy harvester. Simulated harvester output is exceeding the value of 1μW. The power output is highly dependable on the voltage applied to the variable capacitor. Energy harvester is previewed to be produced in form of MEMS device with classical clean room technologies on SOI wafer. An optical and electrical characterization set-ups are created and tested on trial devices – so, both dynamics of the oscillators and final power output could be measured.
However, the main future challenge for the bistable springs based system is the gravitational offset that interferes in the device dynamics if it is not positioned horizontally. Up to now, it is considered to be a major issue not only for the presented system, but for a number of MEMS in general and it is a target for the future development.

Acknowledgements

This work is supported by the “IDI 2015” project funded by the IDEX Paris-Saclay, ANR-11-IDEX-0003-02.

References

[1] M. Deterre, "Toward an energy harvester for leadless pacemakers", Ph.D thesis, Universite Paris Sud - Paris XI, France, 2013.
[2] M.A.Halim, J.Y.Park, "Piezoceramic based wideband energy harvester usingimpact-enhanced dynamic magnifier for low frequency vibration", Ceramics International, 41-1, 2015.
[3] K. Ashraf, M. H. Md. Khir, J. O. Dennis and Z. Baharudin, "A wideband, frequency up-converting bounded vibration energy harvester for a low-frequency environment", Smart Materials and Structures, 22, 2013.
[4] Y. Lu, F. Cottone, S. Boisseau, F. Marty, D. Galayko, and P. Basset, "A nonlinear MEMS electrostatic kinetic energy harvester for human-powered biomedical devices", Applied Physics Letters, 107, 2015.
[5] J. Qiu, J. H. Lang, A. H. Slocum, "A Curved-Beam Bistable Mechanism", Journal of Microelectromechanical Systems, 13, No. 2, 2004.
[6] R. Guillemet, P. Basset, D. Galayko, T. Bourouina, "Design optimization of an out-of-plane gap-closing electrostatic Vibration Energy Harvester (VEH) with a limitation on the output voltage", Analog Integr Circ Sig Process, Springer, 2011.
[7] D.S. Nguyen, E. Halvorsen, "Analysis of vibration energy harvesters utilizing a variety of nonlinear springs", Proceedings of the PowerMEMS, 2010.
[8] S. Risquez, M. Woytasik, J. Wei, F. Parrain, E. Lefeuvre, "Design of a 3D Multilayer Out-of-plane Overlap Electrostatic Energy Harvesting MEMS for medical implant applications", Proceedings of DTIP, pp. 5-9, 2015.
[9] E. Lefeuvre, S. Risquez, J. Wei, M. Woytasik and F. Parrain, "Self-Biased Inductor-less Interface Circuit for Electret-Free Electrostatic Energy Harvesters", Proceedings of the PowerMEMS, 557 , 2014.

A micropump driven by electrochemically produced short-lived bubbles

I V Uvarov[1], S S Lemekhov[1], A E Melenev[1] and V B Svetovoy[1,2]

[1]Yaroslavl Branch of the Institute of Physics and Technology RAS, 150007, Universitetskaya 21, Yaroslavl, Russia
[2]MESA+ Institute for Nanotechnology, University of Twente, PO 217, Enschede 7500 AE, The Netherlands

i.v.uvarov@bk.ru

Abstract. A new working principle for electrochemical micropump with the gas termination time as short as 100 microseconds is presented. It is based on water electrolysis with a fast change of voltage polarity. A simple electrochemical micropump is designed to demonstrate this pumping principle. The device consists of a working chamber with metallic electrodes, inlet and outlet diffusers, and channels for liquid. The chamber and the channels are filled with the electrolyte that plays a role of the pumped liquid. The pump was tested in different regimes. One of these regimes related to formation and termination of short-lived microbubbles is especially promising. Long time stability of the electrodes is demonstrated.

1. Introduction

Microfluidic devices are widely used in chemical, biological and medical applications including microreactors, implantable drug delivery systems, micro total analysis systems (μTAS), lab-on-a-chip [1]. Micropump is an essential component of these devices responsible for generation of the fluid movement. Several actuation mechanisms are known: pneumatic, piezoelectric, electromagnetic, and others [2]. Electrochemical actuation is based on the reversible electrochemical reactions that enable gas bubble expansion and contraction by electrolysis in the aqueous electrolyte solution. In comparison with the other mechanisms, the electrochemical actuation provides large driving force, accurate flow control, low power consumption, low heat generation and compliance with the lab-on-a-chip technology. The main drawback is a long response time due to slow recombination of the gas inside of the working chamber [3]. Recently a short-time electrolysis using alternating polarity voltage pulses was demonstrated, which allows fast generation and termination of the gas in the chamber [4]. Here we describe operation of a simple electrochemical micropump in the new pumping regime based on a short-time electrolysis of alternating polarity.

2. Design, operation principle, and fabrication of the micropump

Design of the pump is schematically shown in figure 1. The pump contains a chamber with a diameter of 500 μm and metallic electrodes at its bottom. Inlet and outlet diffusers are connected to the chamber. The length of each diffuser is 200 μm, the neck width is 20 μm, and the diffuser angle is 9.5°. Channels connected to the diffusers having a width of 250 μm. The pump has a planar configuration, all its elements have a rectangular cross-section with the same height of 6 μm. Since the pump is used only for demonstration purpose the working fluid of the pump is the electrolyte, which is 1 M solution of Na_2SO_4 in distilled water.

The pump operates in a cyclic mode. Each cycle consists of pumping and supplying phases. In the pumping phase a series of short (~10 μs) voltage pulses of alternating polarity is applied to the electrodes. These pulses produce hydrogen and oxygen due to water electrolysis but all the gas is collected in nanobubbles [4]. The pressure increases in the chamber and pushes the fluid through the diffusers. More liquid is going via the outlet diffuser because it has smaller flow resistance. In the supplying phase the pulses are switched off and the gas in nanobubbles recombine into water in 100 μs or so releasing the pressure. This time more liquid is going via the inlet diffuser. Thus, the liquid is pumped from the inlet to the outlet. Then the cycle is repeated. The pump has no movable parts such as passive check valves. Flow direction is determined by the geometrical design of the diffusers.

The pump was constructed from two parts: the glass substrate with deposited metallic electrodes, and the polydimethylsiloxane (PDMS) structure containing channels for electrolyte. These parts were fabricated separately and then bonded together. Electrodes were deposited on the glass substrate by magnetron sputtering. Two types of electrodes were used: one is 100 nm thick Pt layer with an adhesion layer (Ti, 10 nm thick) and the other one is Ti/Al layers with a thickness of 100/500 nm on the same adhesion layer. PDMS structure was fabricated by pouring the liquid PDMS pre-polymer onto the silicon mold, degassing in vacuum and curing at the elevated temperature. To achieve a strong bond between the glass substrate and the PDMS structure, oxygen plasma treatment was used. Exposed surfaces were brought into contact immediately after the treatment. The electrodes and the chamber were aligned manually.

The fabricated pump with close-up view of the working chamber is shown in figure 2. For clarity channels are filled with water colored by food dye. The electrodes have interdigitated structure to increase the efficiency of the electrolytic bubble generation and pumping. The diameter of the overall electrode area is 450 μm which is close to the diameter of the working chamber. This is done in order to saturate the entire volume of the chamber evenly with the gas bubbles.

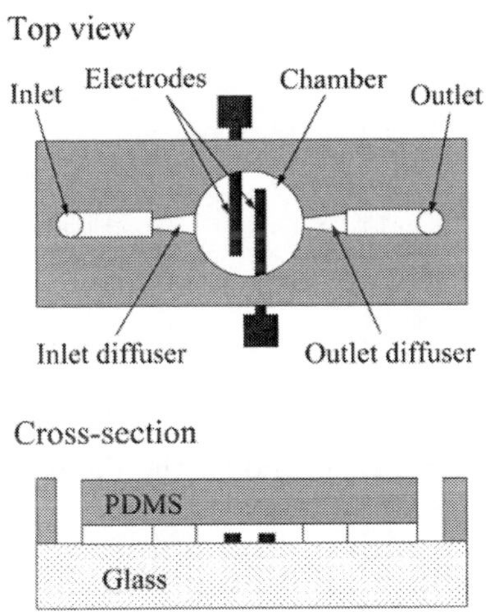

Figure 1. Design of the micropump.

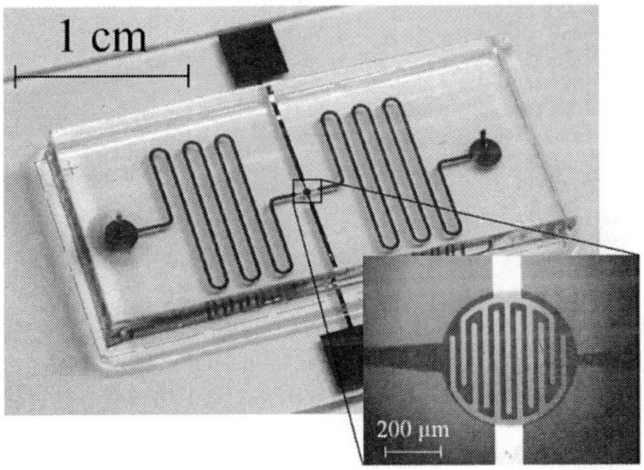

Figure 2. Fabricated micropump.

3. Testing of the device

3.1. Single polarity mode

Standard operation mode of the electrochemical pump is the single polarity mode. In the pumping phase a voltage pulse is applied to one of the electrodes while the other electrode is grounded. Hydrogen is produced above the negative electrode and oxygen is produced above the positive one. The hydrogen and oxygen bubbles are spatially separated. The gases do not mix and have no chance to react. In the supplying phase the voltage is removed and the bubbles recombine into water during a rather long time.

The micropump with platinum electrodes was tested in the single polarity mode. In the pumping phase the voltage pulse with an amplitude of 5 V and duration of 1 ms was applied to the sample. The pulses were generated by Keysight 33512B waveform generator and amplified 5 times using a homemade amplifier. A picture of the working chamber made in the moment $t = 200$ ms is shown in figure 3. The gas bubbles are clear visible in the chamber. The bubbles disappear completely in about 5 s restricting the duration of the supplying phase. Gas termination restricts the operation frequency of the pump by a value of 0.2 Hz.

3.2. Alternating polarity mode

Recently a short-time electrolysis by alternating polarity voltage pulses was demonstrated, which allows fast generation and termination of the gas in the chamber [4]. This regime can be used for fast pumping of liquids. The voltage applied to the working electrode (the other one is grounded) is shown in figure 4. In the pumping phase it is a series of N rectangular pulses of alternating polarity with duration t_0 and amplitude U. These pulses produce hydrogen and oxygen above the same electrode. For short pulses ($t_0 \sim 10$ µs) all the gas is collected in nanobubbles, which are invisible because they do not scatter light [4]. Due to gas production the pressure increases in the chamber and pushes the fluid through the diffusers. In the supplying phase the pulses are switched off. Unlike the single polarity mode, hydrogen and oxygen molecules are not spatially separated and form not only H_2 and O_2 bubbles, but also bubbles containing mixtures of H_2 and O_2 [5]. The latter disappear in the reaction, which is ignited in nanobubbles spontaneously. The mechanism of the reaction is still not completely clear but the reaction happens with the assistance of the gas-liquid interface [5]. When the pulses are switched off the pressure is reduced in about 100 µs. Short recombination time makes it possible to reduce significantly the duration of the supplying phase.

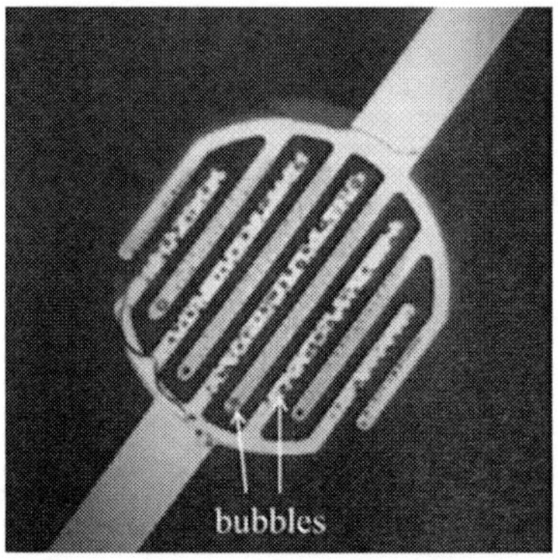

Figure 3. Working chamber in the single polarity mode.

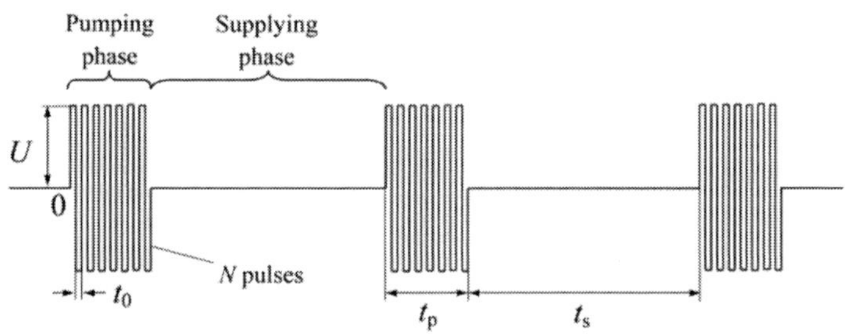

Figure 4. Voltage applied to the electrode in the alternating polarity mode.

The pump with platinum electrodes was tested in the alternating polarity mode. A series of $N = 6000$ pulses was applied to the electrodes. The pulses were repeated with a frequency of $f = 100$ kHz ($t_0 = 5$ µs) at the amplitude $U = 12.5$ V. Duration of the pumping phase was $t_p = Nt_0 = 30$ ms. For this test duration of the supplying phase was chosen to be $t_s = 220$ ms to make the process observable with the naked eye but much shorter t_s can be used. Total duration of the pumping cycle was $t_p + t_s = 250$ ms.

The expected flow rate produced by the micropump can be estimated as follows. Total number of gas molecules M produced during the pumping phase is

$$M = \frac{3I_F t_p}{4e} \approx 1.4\times10^{15},$$

where e is the absolute value of the electron charge and $I_F \approx 10$ mA is the amplitude of the Faraday current. All these molecules are densely packed in nanobubbles with a size of $2r \sim 100$ nm. Pressure in these bubbles is dominated by a high Laplace pressure, $P \approx 2\gamma/r \approx 28$ bar, where $\gamma \approx 0.072$ J/m^2 is the surface tension of water. Significant part of the produced molecules recombines back into water without producing a useful work. As a conservative estimate one can take that only 10% of gas molecules [5] takes part in the pumping. Assuming also that the design of the diffusers provides a medium level of rectification $\varepsilon = 0.1$ [7] we come to the estimate of the flow rate

$$R \sim 0.1\varepsilon\, f_{\rm p} \frac{MkT}{P} \approx 5 \text{ nl/min},$$

where $f_p = (t_p+t_s)^{-1} = 4$ Hz is the frequency of pumping and the thermal energy kT originates from the equation of state. This value is three orders of magnitude smaller than the flow rate $R \sim 10$ μl/min in micropumps with much larger volume of the chamber [3, 8, 9]. The pump described in this paper allows much higher pumping frequency of about 1 kHz or, probably, even more. The use of higher f_p can compensate for the low pumping rate.

The flow rate was determined experimentally observing the movement of the meniscus in the channel. The result $R_{\rm exp} \approx 0.15$ nl/min is significantly smaller than was predicted. It happens because PDMS is a very soft material with Young's modulus as low as $E = 1.3$ MPa. Since the meniscus is moving the overpressure in liquid P_l is estimated somewhat larger than the Laplace pressure produced by the meniscus $P_l \geq \gamma/h = 12$ kPa, where $h = 6$ μm is the height of the channel. FEM modelling of the pump has shown that for this pressure the maximal deflection of the upper wall will be of $d_0 = 2.2$ μm. With this deflection the increment of the chamber volume $\Delta V \approx 0.14$ nl is close to the volume of gas produced in one cycle $\delta V = 0.1MkT/P \approx 0.21$ nl. Therefore, instead of pumping a significant part of the produced gas is used to inflate the chamber. It is clear that such a soft material cannot be used for fabrication of the chamber and channels. It is planned to use glass instead of PDMS to make the chamber stiff.

An important feature of the alternating polarity mode is a gradual wear of the electrodes. The working chamber after several hundred cycles is shown in figure 5. A strong damage of Pt electrodes was observed, fingers were thinned and blackened. To all appearance the damage was caused by a significant energy release in the close proximity of the electrode surface as the result of the combustion reaction between hydrogen and oxygen in nanobubbles [6]. Damage of the electrodes was accompanied by the decrease in the Faraday current and the decrease in the flow rate. Therefore, the choice of electrode material is essential to ensure the reliability of the pump. Almost no damage was observed for the aluminum electrodes covered by 100 nm of titanium. These electrodes provided smaller Faraday current than the platinum electrodes due to formation of titanium oxide on the surface during the electrochemical process. The pump with Ti/Al electrodes needed higher voltage amplitude or longer series of pulses in order to obtain the same flow rate as the pump with Pt electrodes.

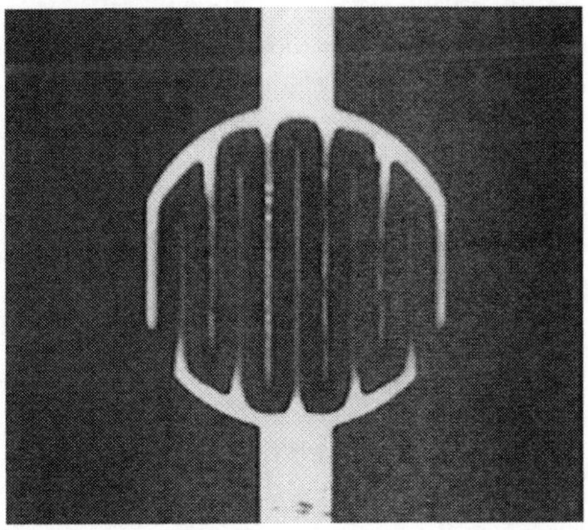

Figure 5. Working chamber in the alternating polarity mode.

Journal of Physics: Conference Series **757** (2016) 012008 doi:10.1088/1742-6596/757/1/012008

3.3. Alternating polarity mode: long pumping phase

Extreme phenomena were observed when a relatively long series of pulses was applied to the device. The number of pulses was increased to $N = 20000$ so that the duration of the pumping phase was $t_p = 100$ ms. Duration of the supplying phase was chosen to be $t_s = 900$ ms, and the total duration of the cycle was $t_p + t_s = 1$ s. The driving pulses have the same parameters: $U = 12.5$ V, $f = 100$ kHz. The experiment has been done for the device with Ti/Al electrodes.

The chamber at the beginning of the pumping phase is shown in figure 6(a). Several bubbles are visible inside of the chamber, which appear during the filling of the device with the electrolyte and are not related to the electrochemical process. The end of the pumping phase is shown in figure 6(b). A bubble of about 100 μm in diameter emerges in the middle of the chamber, it is marked by the red circle. This bubble is not visible in the frame before and in the frame after. Since its image is not sharp we can conclude that the bubble exists much less than 40 ms (video was made at 25 fps). At the end of the pumping phase the liquid is sharply pushed out of the chamber through the diffusers. The chamber returns to its initial state in less than 40 ms as figure 6(c) shows.

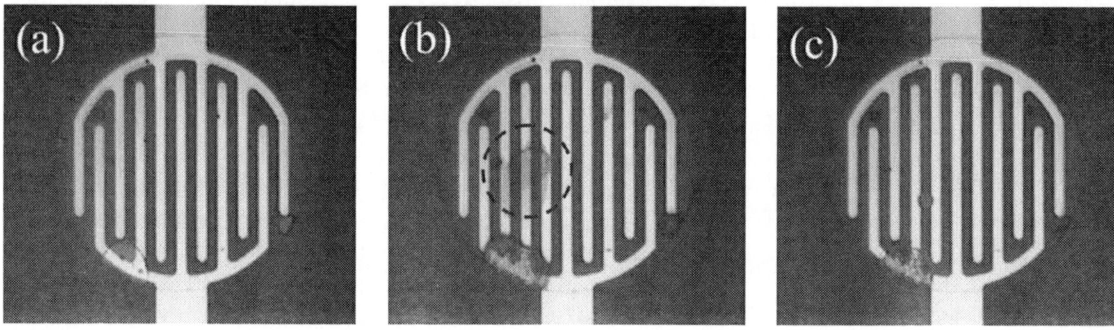

Figure 6. Chamber of the micropump. (a) The beginning of the pumping phase. (b) The end of the pumping phase. (c) 40 ms after the frame (b).

The observations can be explained as follows. A relatively long series of the voltage pulses provides a high concentration of the H_2 and O_2 nanobubbles inside of the chamber. When the concentration is high enough, nanobubbles merge into a microscopic bubble, which can be vaguely seen in the optical microscope. The combustion reaction ignites spontaneously in this microbubble producing a significant pressure jump that pumps the liquid.

Appearance of short-lived microbubbles produced by short voltage pulses of alternating polarity was demonstrated by our group previously. Bubbles with a typical size of 10 μm were observed in a closed microchamber with dimensions of 100×100×5 μm^3 [10]. They lived just a few microseconds and were accompanied by pressure jumps of 0.1-1 bar. The time scale of the events was too short to observe visually the dynamics of the bubbles. To all appearance similar processes happen in the chamber of the micropump. High concentration of H_2 and O_2 nanobubbles is produced by the electrochemical process; these bubbles merge and form a microbubble containing stoichiometric mixture of gases; the reaction in the microbubbles is ignited spontaneously releasing significant energy that drives the pump. The process was shot by an ordinary video camera with a frame rate of 25 fps and the images were not synchronized with the driving signal. The time resolution of the camera is 40 ms but the time scale of the processes in the chamber is expected to be on the time scale from micro- to milliseconds. To follow the dynamics of the bubbles we plan to use a fast video camera.

The alternating polarity mode using the long pumping phase allows to achieve a relatively high flow rate. The meniscus in the channel is moving at least 100 μm per cycle. For the pumping frequency $f_p = 1$ Hz we obtain a flow rate of $R = 9$ nl/min, which is almost two orders of magnitude higher than the flow rate observed in the regime without formation of the short-lived bubbles. More detailed investigation needed for the pumping regime with short-lived bubbles.

4. Conclusions

A new pumping principle based on water electrolysis with a fast change of voltage polarity was demonstrated using a simplest valveless micropump. In comparison with normal single polarity mode, the alternating polarity mode allowed to achieve higher operation frequency of the pump due to a very short gas termination time. The observed flow rate in this regime was lower than expected. This is because the chamber of the micropump was made of PDMS, which is deformed during pumping leading to increase of the chamber volume. To obtain the maximum flow rate the chamber of the micropump has to be fabricated from a more stiff material. Fast degradation of the platinum electrodes was also observed. In contrast, almost no damage was observed for the aluminum electrodes covered by a titanium layer. At some parameters of the driving pulses short-lived microbubbles flicker in the chamber. This pumping regime is accompanied by a significant increase in the pumping rate even for the chamber made of PDMS and worth of more detailed investigation.

Acknowledgments

This work is supported by the Russian Science Foundation (Grant No. 15-19-20003).

References

[1] Whitesides G M 2006 *Nature* **442** 368-73
[2] Au A K, Lai H, Utela B R and Folch A 2011 *Micromachines* **2** 179-220
[3] Yi Y, Buttner U, Carreno A A A, Conchouso D and Foulds I G 2015 *J. Micromech. Microeng.* **25** 105011
[4] Svetovoy V B, Sanders R G P, Ma K and Elwenspoek M C 2014 *Sci. Rep.* **4** 4296
[5] Svetovoy V, Postnikov A, Uvarov I, Sanders R and Krijnen G 2016 *Energies* **9** 94
[6] Svetovoy V B, Sanders R G P, Lammerink T S J and Elwenspoek M C 2011 *Phys. Rev.* E **84** 035302(R)
[7] Singhal V, Garimella S V and Murthy J Y 2004 *Sens. Actuators* A **113** 226-35
[8] Lui C, Stelick S, Cady N and Batt C 2010 *Lab Chip* **10** 74-9
[9] Li P-Y, Sheybani R, Gutierrez C A, Kuo J T W and Meng E 2010 *J. Microelectromech. Syst.* **19** 215-28
[10] Postnikov A V, Uvarov I V, Prokaznikov A V and Svetovoy V B 2016 *Appl. Phys. Lett.* **108** 121604

Experimental study of PDMS mechanical properties for the optimization of polymer based flexible pressure micro-sensors

T H N Dinh, E Martincic, E Dufour-Gergam and P-Y Joubert

Institut d'Electronique Fondamentale (IEF), CNRS, Université Paris Sud, Université Paris-Saclay, Orsay, France

thi-hong-nhung.dinh@u-psud.fr

Abstract. This paper reports on the optimization of flexible PDMS-based normal pressure capacitive micro-sensors dedicated to wearable applications. The deformation under a normal force of PDMS thin films of thicknesses ranging from 40 μm to 10 mm is firstly experimentally studied. This study points out that for capacitive micro-sensors using bulky PDMS thin films as deformable dielectric material, the sensitivity to an applied normal load can be optimized thanks to an adequate choice of the so-called form ratio of the involved PDMS thin film. Indeed, for capacitive micro-sensors exhibiting 9 mm^2 electrodes, the capacitance change under a 6 N load can be adjusted from a few percent up to over 35% according to the choice of the load-free thickness of the used PDMS film. These results have been validated thanks to electromechanical characterizations carried out on two flexible PDMS based capacitive normal pressure micro-sensor samples fabricated with two different thicknesses. The obtained results open the way to the enhanced design of PDMS based pressure sensors dedicated to wearable and medical applications. Further works will extend this study to a wider range of sensor dimensions, and using numerical modelling.

1. Introduction

Thanks to progresses made in the fabrication of flexible electronics and micro-sensors, wearable physiological measurement devices are more and more developed for ambulatory and self-monitoring purposes [1]. Among them, devices using polymer based capacitive micro-sensors are very popular since they gather advantageous features, such as being sensitive to various physiological quantities as well as being easily wearable. This kind of sensors is basically constituted of metallic electrodes deposited on - or imbedded in - dielectric polymer films. Besides, among the existing polymers Polydimethylsiloxane (PDMS) is a primary reference material, hence suitable for biocompatible and wearable use [2]. Furthermore, PDMS being a dielectric material [3], it is appropriate to be used in capacitive sensing applications. Finally, PDMS films are known to be highly deformable, allowing flexible sensors to be produced. Therefore they are particularly well suited to be part of wearable capacitive force or pressure sensors. Indeed, associated to metallic electrodes, the deformation of the PDMS films under pressure induces capacitance changes between electrodes that can be read out for pressure measurement. For that reason, PDMS-based capacitive sensors are found in many medical applications involving pressure monitoring, such as tactile sensing [4], gesture or gait analysis [5]. Since PDMS films are deformable under shear and normal forces, three-axis pressure capacitive micro-sensors array can be developed with appropriate electrode design [6]. However, it has been shown that the thickness of a PDMS film as well as the surface on which the load is applied can strongly affect its deformation properties and hence the resulting capacitance change [7,8]. Finally, PDMS thin films are known to be hardly compressible, with a Poisson coefficient close to 0.49 [9]. This is why capacitive normal pressure sensors constituted of bulky PDMS films are often reported as lacking sensitivity to normal pressure [6,10]. Hence authors have proposed structured PDMS films to enhance the sensor sensitivity to normal pressure [10]. However the fabrication processes associated to these designs are more complex and can result in higher manufacturing costs. In addition, the obtained structure makes it difficult to foresee or adjust the sensor sensitivity and measurement range. In this paper, the mechanical behavior of bulky PDMS film samples under normal pressure is experimentally investigated. This study aims at analyzing and optimizing the sensitivity of bulky flexible capacitive normal pressure sensors. In addition, PDMS-based sensor samples are fabricated and electromechanically characterized. The measured sensitivity is discussed comparatively to the previously determined mechanical properties of the used PDMS films. Finally, optimization rules for the design of enhanced capacitive normal pressure flexible micro-sensors are highlighted.

2. Sensors design

2.1. *Basic principle of the sensor*

For the sake of simplicity, a simple parallel plate capacitive normal pressure micro-sensor is considered in this study. The side length of the considered micro-sensor is of a few millimeters, as required e.g. for wearable plantar pressure monitoring applications. The sensor is constituted of two square-plate copper electrodes facing one another. The electrodes are separated by a deformable dielectric film made of PDMS. The sensor is deposited on a Kapton substrate, used to provide both flexibility and mechanical robustness to the fabricated device. In load-free conditions, the capacitance of such sensor is given by [11]:

$$C = (2.47b + 1)\frac{\varepsilon_0 \varepsilon_r}{d} L^2 \tag{1}$$

where ε_0 and ε_r are the permittivity of vacuum and the relative permittivity of PDMS, respectively; d is the distance between the electrodes and L is the electrode side-length. The capacitance of the sensor is altered when pressure is applied, since the distance d between the two electrodes is changed. Reading out the capacitance change gives information about the applied pressure.

2.2. *Evaluation of mechanical behavior of PDMS*

In order to quantitatively estimate the mechanical properties of PDMS films under normal pressure, PDMS film samples of various thicknesses were realized. The films are of a Sylgard®184 type. They feature an oligomer/curing agent ratio of 10/1 by weight, and they are cured at 75°C during 1 hour. The fabricated film samples have thicknesses ranging from 40 µm to 9.8 mm.

During fabrication, the thickness of the film samples is controlled by the speed of spin, during the spin coating [12]. After fabrication, the film samples are mechanically characterized using a CSM Instruments micro-indentation tester. The tester is implemented with cylindrical indentation tips of radius $r = 1.7$ and 2.5 mm. Each film sample is positioned under the indentation tip thanks to a 3 axis motorized table featuring a 0.25 µm position resolution. The indentation tip, normal to the film surface is driven into the sample by applying a load of 6 N, while the position of the tip is measured. Therefore the set-up allows the PDMS thickness change under load to be evaluated as a function of the applied load. Based on Eq. 1, it is possible to foresee the capacitance change of a PDMS based capacitive pressure sensor, ΔC, with the given applied load, using:

$$\Delta C_{\%} = 100\frac{\Delta d}{d - \Delta d} \tag{2}$$

where d is the initial load-free thickness of the PDMS film and Δd is the measured PDMS thickness change under load. For PDMS film samples ranging from $d = 40$ μm to 9.8 mm, the capacitance changes expected for the two considered indentation tips are plotted in Figure 1.

One can observe that for an indentation tip of a given radius r, there exists an optimal PDMS film load-free thickness d that maximizes the capacitance change under load. In other words, the sensitivity of the designed sensor can be adjusted or optimized according to the relevant choice of the so-called form ratio ρ, defined by:

$$\rho = \frac{S}{d} \quad (3)$$

where S is the indentation surface upon which the load is applied. Based on the experiments carried out for two indentation surfaces (resulting from the two different radii), it appears (Figure 1) that the capacitance change is optimized for the same value of the form ratio $\rho_{opt} \approx 3.9$ mm ± 7% (all curve fittings have a coefficient of determination R^2 better than 98.8 %). In order to validate these conclusions, pressure sensor samples are designed with $\rho = \rho_1$ and $\rho = \rho_2$ (Figure 1). We describe how these sensors are fabricated and mechanically characterized in what follows.

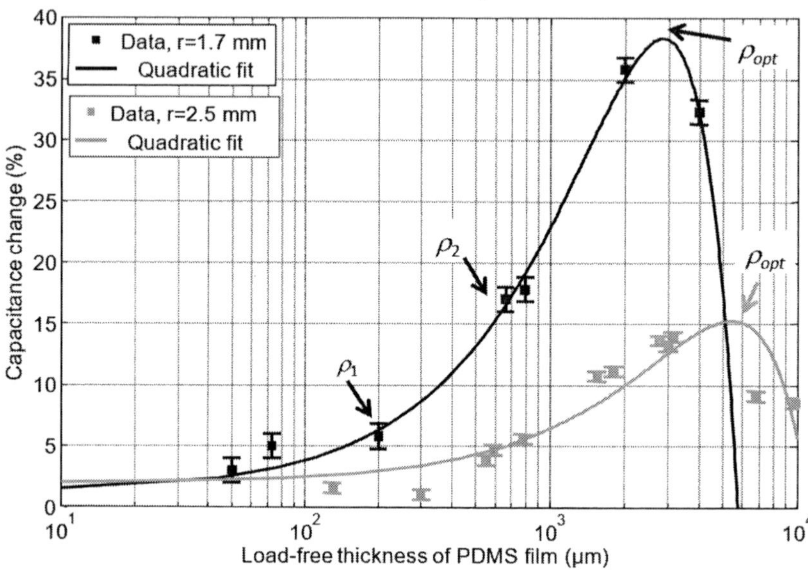

Figure 1. Estimation of sensor capacitance changes versus load free thickness of the used PDMS films, for two indentation-tip radii, and for an applied load of 6 N. Error bars are twice the standard deviation of the data estimated from 4 measurement cycles.

3. Sensor fabrication

3.1. Fabrication process

In order to fabricate PDMS based pressure sensors, it is necessary to deposit copper electrodes at each side of PDMS films. It is however difficult to do so because of the low surface energy and of the highly compliant nature of PDMS films. To bypass this drawback, the so-called "film-transfer" fabrication process [12,13] is implemented. Indeed, this fabrication process enables assemblies using heterogeneous or even incompatible materials and technologies to be considered. The film-transfer process involves 4 fabrication steps (Figure 2). In the first step, copper top electrodes are realized on a donor wafer covered by a low adhesion layer and copper bottom electrodes are fabricated on a target wafer using the micro-molding process (step 1). Then, the fabricated copper lines are covered with PDMS (step 2). It should be noted that a gap corresponding to the position of the conductor pad on the target substrate is intended for further electrical connection purposes. The third step aims at aligning and bonding the two wafers using an uncured PDMS layer as glue. Finally, after removing the donor wafer, a capacitor with a PDMS dielectric layer is obtained (step 4). The PDMS films used to fabricate the sensors are identical to the ones used for characterization purposes.

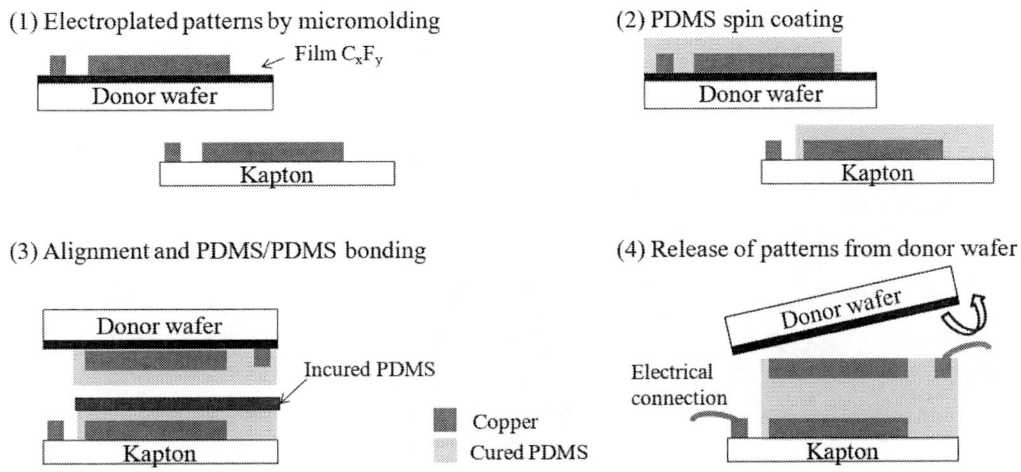

Figure 2. The four-step fabrication process of the sensors

3.2. *Fabricated devices*

In this work, sensor samples are fabricated on 10 cm diameter Kapton substrates. The fabrication process enables a large number of sensors together with their connection tracks to be fabricated on the same substrate with a transfer yield of 100%. In what follows, two types of sensor are considered for electromechanical characterizations. They are of same size (side-length L = 3 mm) but exhibits different PDMS film thicknesses, as presented in Table 1. The connection tracks are 100 μm wide. An example of a realized sensor is shown in Figure 3. The load-free capacitances of the fabricated sensors are measured by means of a HP1492A impedance analyzer operating at 1MHz. For the eighteen fabricated sensor samples, the measured values are close to the expected ones, which were evaluated using Eq. 1 in which ε_r is assumed to be equal to 2.68 in PDMS [3]. The observed discrepancies between experimental and expected values can be attributed to the possible inaccurate estimation of thickness d, to a possible misalignment of electrodes during fabrication, as well as to the approximations used to build the model of Eq. 1 [11]. The variability between the measured capacitances of sensors of same type can also be attributed to the slight inhomogeneity of the PDMS thickness over the whole film, which was evaluated to be less than 5% according to our thickness measurements.

Table 1. Sensor sample features and characterization results

Sensor type	d	ρ	Load-free Capacitance (pF)			ΔC max
	(mm)	(mm)	Expected	Experiment	Discrepancy	(at 6 N)
1	0.20	45	1.23	1.27±0.06	3.2 %	6%
2	0.66	13.6	0.49	0.48±0.02	2 %	17%

3.3. *Electromechanical characterization*

In addition, electromechanical characterizations were carried out on two sensor samples. In order to estimate their capacitance change under normal pressure, a dedicated electromechanical test bench was set up (Figure 3). The set-up includes an indenter tip associated to a spring loaded commercial force gauge, fixed on a PC controlled 3-axis robot arm. The load applied to the capacitive sensors is

Journal of Physics: Conference Series **757** (2016) 012009 doi:10.1088/1742-6596/757/1/012009

monitored using the force gauge, and the capacitance changes are measured using the same impedance analyzer, as a function of the applied load. The capacitance changes obtained in the 0-6 N force range for sensors of type 1 and 2 are plotted in Figure 4. Each measurement point results from a set of 20 different measurements implemented during 20 load-and-release cycles.

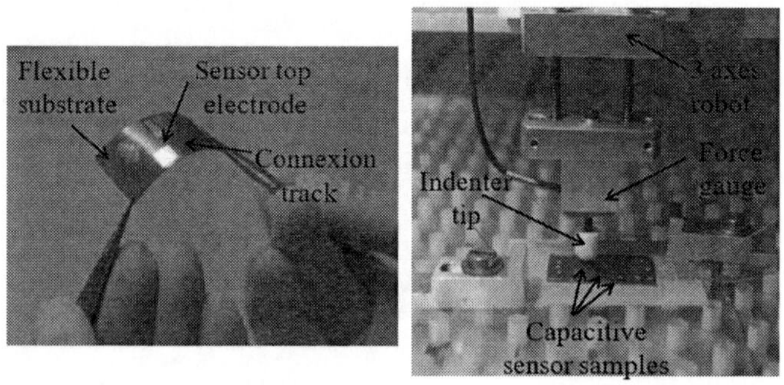

Figure 3. Capacitive pressure sensor sample and electromechanical characterization set-up.

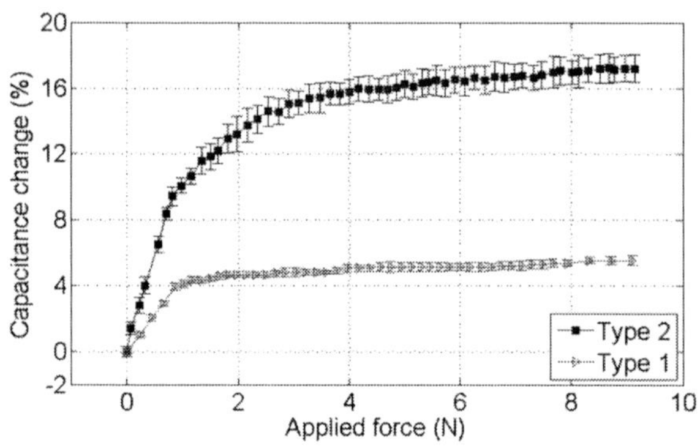

Figure 4. Capacitance changes of sensor samples of type 1 and 2 (Table 1) versus applied force. Error bars are twice the standard deviation of the data, estimated from 20 measurements.

It may be noted that the maximum capacitance change reaches less than 6 % for type 1 sensor, which is designed with a load-free PDMS thickness d = 200µm, and a form ratio ρ_1 = 45 mm. For the same load, the capacitance change is 17% for the type 2 sensor, which is designed with ρ_2 = 13.6 mm. Since ρ_2 is closer to the optimal form ratio ρ_{opt} than ρ_1, the type 2 sensor is, as expected, more sensitive than the type 1 sensor (Figure 1 and Figure 4). For these sensor dimensions, the maximum capacitance change under 6N load would have been in the order of 37% for a load-free thickness of PDMS d = 3mm (Figure 1). The capacitance change is then lowered if the used PDMS film is of a greater thickness. These results confirm that the sensor sensitivity to normal stress can be optimized according to the form ratio ρ of the designed sensor.

4. Conclusion

In conclusion, PDMS thin films have been fabricated and their mechanical properties under normal load have been experimentally studied. Thanks to this study it was established that the capacitance changes of micro-sensors constituted of bulky PDMS thin films and metallic electrodes can be optimized according to their form ratio. Indeed for a 3 mm side length micro-sensor, the capacitance change under a 6 N load can vary from a few percents up to more than 35%, if the load-free thickness of the film is adequately chosen. To validate these results, PDMS based normal pressure micro-sensor samples have been fabricated using a transfer of film fabrication process, and electromechanically characterized. Based on the characterization of two different sensor samples, it has been experimentally confirmed that the form ratio of the used PDMS film enables the sensitivity of the sensor to be optimized for a given side length of the sensor. These results should be validated thanks to an extended study involving a wider range of sensor dimensions, either via an experimental study or by means of numerical computations. Nevertheless, the results presented in this study open the way to the design of enhanced flexible pressure sensors dedicated to wearable applications.

Acknowledgments
This work has benefited from the financial support of the LabeX LaSIPS (ANR-10-LABX-0040-LaSIPS) managed by the French National Research Agency under the "Investissements d'avenir" program (n°ANR-11-IDEX-0003-02). The authors wish to thank technical staff of the clean room facilities of the "Centrale de Technologie Universitaire (MINERVE-CTU)" for its valuable help during the experimental tests.

References

[1] Patel S, Park H, Bonato P, Chan L and Rodgers M 2012 A review of wearable sensors and systems with application in rehabilitation. *J. Neuroeng. Rehabil.* **9** 21

[2] Marois Y and Be M C 2001 Studies of Primary Reference Materials Low-Density Polyethylene and Polydimethylsiloxane : A Review *J. Biomed. Mater. Res.* **58** 467–77

[3] Tiercelin N, Coquet P, Sauleau, R, Senez V and Fujita H 2006 Polydimethylsiloxane membranes for millimeter-wave planar ultra flexible antennas *J. Micromechanics Microengineering* **16** 2389

[4] Tiwana M I, Redmond S J and Lovell N H 2012 A review of tactile sensing technologies with applications in biomedical engineering *Sensors Actuators A Phys.* **179** 17–31

[5] Lei K, Lee K-F and Lee M-Y 2014 A flexible PDMS capacitive tactile sensor with adjustable measurement range for plantar pressure measurement *Microsyst. Technol.* **20** 1351–8

[6] Surapaneni R, Guo Q, Xie Y, Young D J and Mastrangelo C H 2013 A three-axis high-resolution capacitive tactile imager system based on floating comb electrodes *J. Micromechanics Microengineering* **23** 75004

[7] Yang F 2003 Thickness effect on the indentation of an elastic layer *Mater. Sci. Eng. A* **358** 226–32

[8] Miao L, Jianren S, Ying S, Christopher B and Quanfang C 2009 Thickness-dependent mechanical properties of polydimethylsiloxane membranes *J. Micromechanics Microengineering* **19** 35028

[9] Johnston I D, McCluskey D K, Tan C K L and Tracey M C 2014 Mechanical characterization of bulk Sylgard 184 for microfluidics and microengineering *J. Micromechanics Microengineering* **24** 035017

[10] Mannsfeld S C B, Tee B C-K, Stoltenberg R M, Chen C V H-H, Barman S, Muir B V O, Sokolov A N, Reese C and Bao Z 2010 Highly sensitive flexible pressure sensors with microstructured rubber dielectric layers. *Nat. Mater.* **9** 859–64

[11] H. Nishiyama M N 1994 Form and Capacitance of Parallel-Plate Capacitors *IEEE Trans. components, Packag. Manuf. Technol. part A* **17** 477–84

[12] Dinh T H N, Martincic E, Dufour-Gergam E and Joubert P-Y 2016 Capacitive flexible pressure sensor: microfabrication process and experimental characterization *Microsyst. Technol.* 465–71

[13] T.H.N Dinh E. Martincic, Martincic E, Chhaibi A, Nazeer S, Couty M, Joubert P-Y and Dufour-Gergam E 2013 Fabrication of flexible force sensor arrays by transfer of film *24th Micromechanics and Microsystems Europe Conference (MME)* (Helsinki, Finland)

Single cell array impedance analysis in a microfluidic device

Emre Altinagac[1], Selen Taskin[2], Huseyin Kizil[2]

[1] Department of Nanoscience & Nanotechnology, Istanbul Technical University, Istanbul, Turkey.

[2] Faculty of Chemical and Metallurgical Engineering, Istanbul Technical University, Istanbul, Turkey.

altinagac@itu.edu.tr

Abstract. Impedance analysis of single cells is presented in this paper. Following the separation of a target cell type by dielectrophoresis in our previous work, this paper focuses on capturing the cells as a single array and performing impedance analysis to point out the signature difference between each cell type. Lab-on-a-chip devices having a titanium interdigitated electrode layer on a glass substrate and a PDMS microchannel are fabricated to capture each cell in a single form and perform impedance analysis. HCT116 (homosapiens colon colorectal carcin) and HEK293 (human embryonic kidney) cells are used in our experiments.

1. Introduction

Introducing a lab-on-a-chip (LoC) device with two different active regions to define single cell physical/electrical characteristics for diagnosing a disease and proposing a potential drug is our primary research goal. In the first active region of the LoC, target cells are separated by dielectrophoresis (DEP) and transferred to second active region and the cells are captured as a single array in hydrodynamic traps where the impedance measurement is carried out. The final form of the LoC has two active regions on the same body. For simplicity of our research, those regions are studied distinctly. This paper focuses on the second region and preliminary outcomes of impedance analysis of single cells are presented.

Dielectrophoresis is a well-known process to manipulate/separate a target cell type in complex populations and even capturing cells in an active electrical traps in single form. Detection of early stages of cancer diseases by cost effective LoCs is a good alternative to conventional methods where the point of care applications are strictly limited [1,5].

Impedance analysis of cancer cells is carried out by applying an AC signal to interdigitated electrode couple where a single cell is captured. Impedance measurement of interdigitated electrodes are recorded for selected frequencies and when a cell is trapped, an impedance shift occurs due to the physical/electrical conditions of an interfering biological cell [6,7]. Basic electric circuit systems are mathematically modeled depending on the electrode geometries and layer-by-layer physical structures of biological cells [6,7]. These models are used for calculation of electrical parameters of interior cell structures and can be used for further applications as in the case of maintaining higher efficiencies of DEP cell separation systems.

2. Design and Fabrication

Hydrodynamic traps for single cell analysis consist of three dimensional polydimethylsiloxane (PDMS) microchannel structures to maintain the cell within the trap area under continuous fluid flow. The idea behind the design is inspired from the study of Tan [8] where a two dimensional microfluidic channel consists of two potential paths (Path 1 and Path 2) for cells to follow under continuous fluid flow. The width and the height of the microchannel is close to the diameter of a cell to maintain the stream of cells in a linear form. Path 1 has a lower resistance and a narrow end to create a trap region. When a cell is trapped in Path 1, fluid flow is blocked or resistance gets too high for cells to pull in. Subsequent cells start to follow Path 2 until the next trap region appears. Single cell trap array is arranged by this loop of fluid flow throughout the microchannel. Due to the fabrication challenges of Tan's design consisting of 5μm x 5μm narrow necks, we present a 3D PDMS microchannel which is shown in Figure 1 with a minimum feature size of 10μm. A narrow neck in Path 1 is created in "z" direction by limiting the channel height to 5μm in this region and the rest of the channel has a 15μm height. Conventional lithography processes are used to create such structures by aligning two mask to create 3D SU-8 3010 negative photoresist as a master mold for PDMS casting.

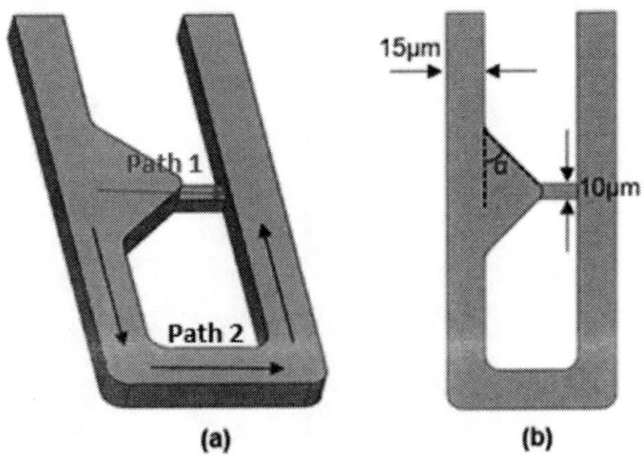

Figure 1. a) A triangular trap region is placed to Path 1 b) The flow behaviour is simulated considering three different base angles (α=30°, 45°, 60°).

Circular and triangular geometries with different dimensions and base angles are placed in Path 1 to investigate the most effective single cell trap area. The triangular trap area with a base angle of 30° has the best experimental outcome where the other geometries are most likely to trap more than one cell at a time. Distribution of fluid velocity is shown in Figure 2 and it can be seen that increasing the resistance of Path 2 by increasing the length is also a simple solution to achieve higher trapping efficiencies.

Further enhancement of preventing multiple cells trapping is done by placing a graded triangular trap area in Path 1 and can be seen in Figure 3. When a cell (1) is trapped in the acute angled region the subsequent cells (2) are cleared out of the trap area. Three different combinations are studied experimentally (α_1=30°, β_1=45°), (α_2=30°, β_2=60°) and (α_3=45°, β_3=60°).

Interdigitated electrodes with a width of 15μm and a distance of 10μm are placed under a microchannel as an array. Ti microelectrodes with a 200nm thickness are fabricated on a 400nm thick AZ 1505 photoresist coated microscope slides by DC magnetron sputtering and lift-off processes. PDMS microchannel and microelectrodes are aligned under an optical microscope and plasma-activated bonding process applied.

Journal of Physics: Conference Series **757** (2016) 012010 doi:10.1088/1742-6596/757/1/012010

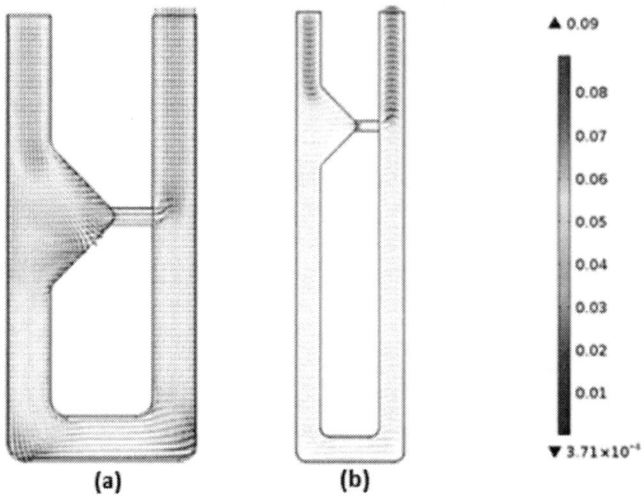

Figure 2. Fluid velocity distributions for different Path 2 lengths (α=45°). Color bar: Fluid velocity (m/s).

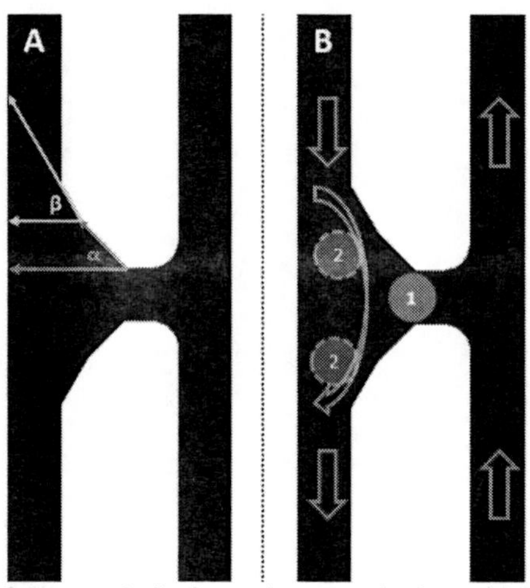

Figure 3. Schematic view of the graded triangular trap design. A) Three different combinations are studied experimentally (α_1=30°, β_1=45°), (α_2=30°, β_2=60°) and (α_3=45°, β_3=60°) B) When a cell is trapped in narrow region subsequent cells are cleared out of the trap area.

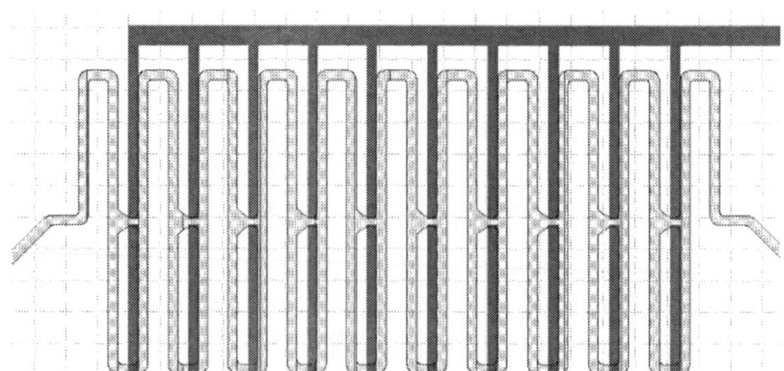

Figure 4. Schematic view of interdigitated electrodes (pink) placed under a microchannel (blue). The width of the electrodes is 15μm and the distance between each electrode couple is 10μm.

Journal of Physics: Conference Series **757** (2016) 012010 doi:10.1088/1742-6596/757/1/012010

3. Experimental Results

Various trap geometries, circular and triangular, are studied in theory and experimentally with polystyrene particles and biological cells throughout our research but the most recent experiment results are presented in this paper. HCT116 (homosapiens colon colorectal carcin) and HEK293 (human embryonic kidney) cells with different medium conductivities are used in our experiments. Conductivity of medium is adjusted by the concentration of PBS (phosphate-buffered saline) in 200mM sucrose solution (1XPBS=15,70mS/cm, 0.1XPBS=1.87mS/cm). HCT116 cells are trapped in (α_1=30°, β_1=45°) angled triangular traps with a high efficiency and shown in Figure 5. Some of the cells are deformed in narrow neck regions due to the elastic structure of biological cells and the cells in red circles are not in touch with the microelectrodes where the impedance measurement is recorded with the cells in green circles.

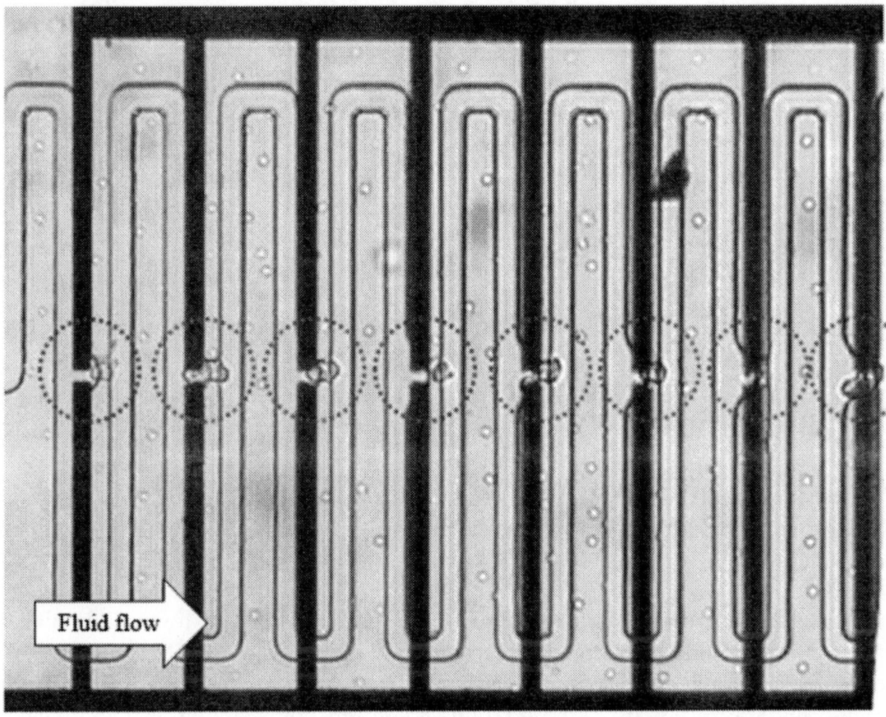

Figure 5. HCT116 cells in PBS. (α_1=30°, β_1=45°) angled triangular traps are used. The cells with no contact with the microelectrodes are shown in red circles where the cells are in contact with the microelectrodes are shown in green. Each trap area has a single HCT116 cell with a diameter of nearly 15μm.

HEK293 cells are also efficiently trapped using the same structure and the impedance shift is plotted for both cells and mediums based on the signal recorded before and after a cell trapped between electrode couples. 1V_{pp} signal with a 10-500 kHz frequency range is applied to microelectrodes and results are shown in Figure 6. It can be seen that the impedance shift varies for each cell line which can be used for further analysis or diagnosis applications. Divergence of impedance shift between cell lines are increases with decreasing the medium conductivity which makes low conductivity mediums favourable for diagnosing of cell lines from one another.

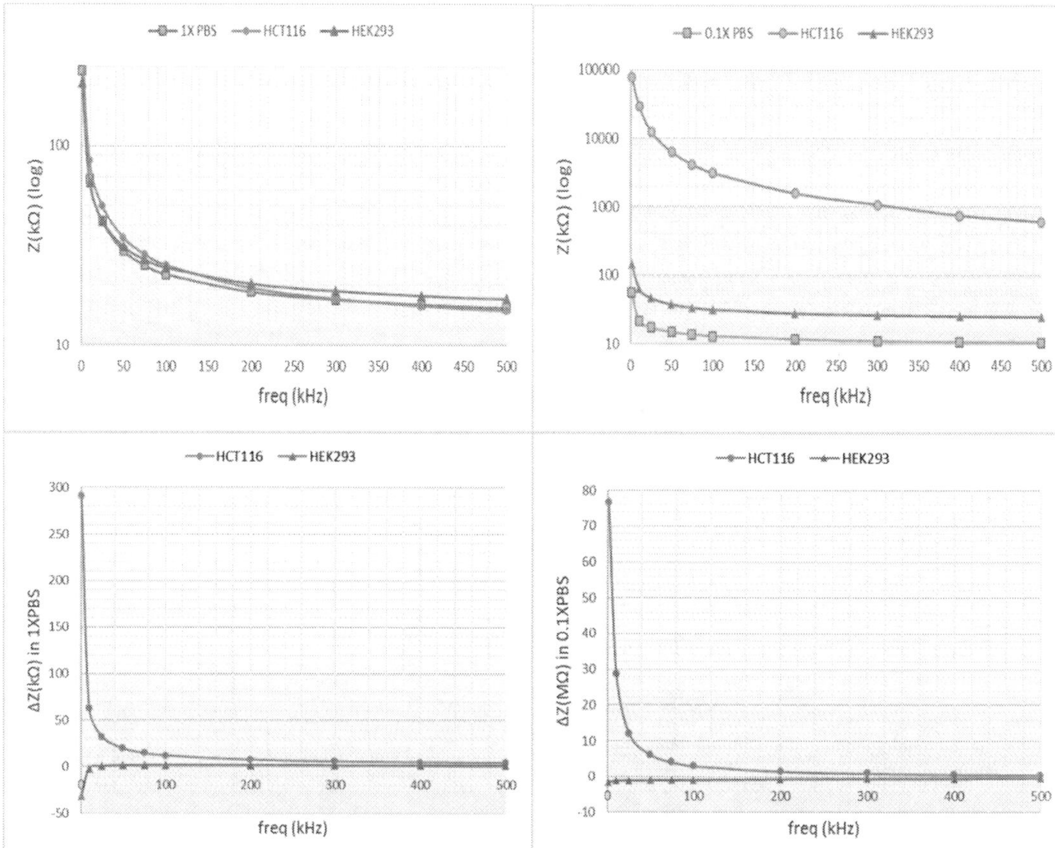

Figure 6. Impedance shift of HCT116 and HEK293 cell lines in 1XPBS (left side) and 0.1XPBS (right side) mediums. Note that ΔZ is increasing with decreasing the medium conductivity.

4. Conclusion

Our experimental results show that diagnosing of different cell lines in low conductive mediums is achievable using current single cell trap array structures. Further studies will focus on introducing the optimum medium conductivity and a frequency value for a target cell line to record the impedance shift with a minimum error. This technique will be used for calculating the physical/electrical properties of interior cell structures and the separation efficiency by DEP will be increased with gained knowledge of target cell lines. Chemical stimulation on a target cell using a specific type of drug could also be evidenced by the amount of these impedance shifts.

Acknowledgements

The authors would like to acknowledge the financial support provided by the Scientific and Technological Research Council of Turkey (TUBITAK) under Grant No. 114M802.

References

[1] J. Mateo, M. Gerlinger, D. Rodrigues, and J. S. de Bono, "The promise of circulating tumor cell analysis in cancer management", Genome Biol, 15(8), 448, 2014.

[2] N. M. Karabacak, P. S. Spuhler, F. Fachin, E. J. Lim, V. Pai, E. Ozkumur, and M. Toner, "Microfluidic, marker-free isolation of circulating tumor cells from blood samples", Nature protocols, 9(3), 694-710, 2014.

[3] C. Alix-Panabières, and K. Pantel, "Technologies for detection of circulating tumor cells: facts and vision", Lab on a Chip, 14(1), 57-62, 2014.

[4] C. Jin, S. M. McFaul, S. P. Duffy, X. Deng, P. Tavassoli, P. C. Black, and H. Ma, "Technologies for label-free separation of circulating tumor cells: from historical foundations to recent developments", Lab on a Chip, 14(1), 32-44, 2014.

[5] L. Hajba, and A. Guttman, "Circulating tumor-cell detection and capture using microfluidic devices", TrAC Trends in Analytical Chemistry, 59, 9-16, 2014.

[6] H. Park, D. Kim, and K. S. Yun, "Single-cell manipulation on microfluidic chip by dielectrophoretic actuation and impedance detection". Sensors and Actuators B: Chemical, 150(1), 167-173, 2010.

[7] Y. Demircan, A. Koyuncuoğlu, M. Erdem, E. Özgür, U. Gündüz, and H. Külah. "Label-free detection of multidrug resistance in K562 cells through isolated 3D-electrode dielectrophoresis". Electrophoresis, 36(9-10), 1149-1157, 2015

[8] . W. H. Tan and S. Takeuchi, "A trap-and-release integrated microfluidic system for dynamic microarray applications", Proceedings of the National Academy of Sciences, 104(4), 1146-1151, 2007.

Journal of Physics: Conference Series **757** (2016) 012011 doi:10.1088/1742-6596/757/1/012011

High Aspect-Ratio Neural Probes using Conventional Blade Dicing

S. B. Goncalves[1], **J. F. Ribeiro**[2], **A. F. Silva**[1], **J. H. Correia**[2]

[1] MIT Portugal Program, School of Engineering, University of Minho, Guimaraes, Portugal
[2] Department of Industrial Electronics, University of Minho, CMEMS-UMINHO, Guimaraes, Portugal

E-mail: `sgoncalves@dei.uminho.pt`

Abstract. Exploring deep neural circuits has triggered the development of long penetrating neural probes. Moreover, driven by brain displacement, the long neural probes require also a high aspect-ratio shafts design. In this paper, a simple and reproducible method of manufacturing long-shafts neural probes using blade dicing technology is presented. Results shows shafts up to 8 mm long and 200 μm wide, features competitive to the current state-of-art, being its outline simply accomplished by a single blade dicing program. Therefore, conventional blade dicing presents itself as a viable option to manufacture long neural probes.

1. Introduction

Neuroprostheses have been probing a variety of neural circuits so the scientific community can improve its knowledge about several brain diseases and disorders. Recording and stimulation of brain networks have already reached widespread clinical application. Indeed, deep brain stimulation (DBS) has proved to show beneficial effects in a variety of neurological conditions, such as depression, obsessive compulsive disorder (OCD), chronic pain, Parkinson's disease, epilepsy, essential tremor, dystonia, Tourette syndrome, blindness, deafness and spinal cord injuries [1, 2]. More recently, optogenetics has shown potential in selectively controlling dysfunctional circuits throughout optical neuromodulation, and therefore also has been proposed to assist several diseases and neuropsychiatric disorders [3].

Exploring cortical neural circuits have been a common practice in neuroscience studies. Nevertheless, the need to increase scientific insights of deeper neurons populations has triggered the development of longer penetrating devices, capable of overcoming length limitation of devices aiming for superficial brain regions. For example, the hippocampus that is frequently used as a research model for exploring both normal and pathological conditions of the nervous system, including the processes involved in memory and learning and neurodegenerative diseases [4], requires devices with shafts longer than 4 mm. Depending on the target area, probes length could require up to 10 mm long shafts in a mice-based experiment.

Accordingly, long probes designs have been reported, but mostly their approach rely on wet etching technology to form the outlines of the devices. Deep reactive ion etching (DRIE) [5–7], wet [8, 9], dry [10] and plasma etching [11, 12] technologies or a combination of these [13] has been a popular approach to define probes design. Others techniques were explored such as direct write laser (DWL) [14], microwire electrical discharge machining [15] and self-assembled

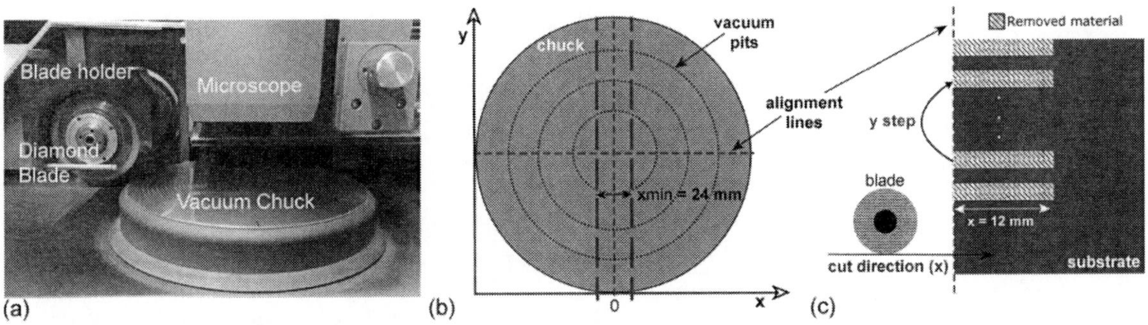

Figure 1. (a) Blade dicing setup photography. (b) Schematic of dicing cut process in the chuck, and (c) Top view detailed plan of the dicing process for producing long planar probes.

based methods [16] to achieve long designs. Nevertheless, they are complex and expensive manufacturing processes.

This paper demonstrates a fabrication approach of long and high aspect-ratio silicon probe design using standard blade dicing technology. Blade dicing has been the most widely used process in the separation of silicon wafers into individual devices both in semiconductor and micro-electro-mechanical systems (MEMS) technologies [17]. By employing this fabrication process, our approach presents some main advantages: (1) simplicity, since shaft outline is accomplish with a single technique and manufacturing runs within few blade cuts; (2) is highly scalable to wafer processes, requiring simply the repetition of cut patterns; and, (3) fabrication relies on a purely mechanical process, avoiding chemical changes in the probe bulk material. The fabrication of up to 8 mm long and 200 μm wide probe was successfully accomplished.

2. Fabrication Process

2.1. Blade Dicing

All cuts were carried out by a high precision Disco DAD 2H/6T dicing saw. Mechanical dicing process uses a dicing equipment to fully or partially cut through wafers. Fig. 1.a shows the setup used in the dicing process. Usually, a wafer is held in the chuck with the aid of suction (vacuum pits). The chuck moves at a specific and constant feed rate passing through the dicing blade in X and Y directions for device cutting (see Fig. 1.b). The feed speed or cut speed determines how fast the sample is fed towards the saw blade. The cutting tool is called the dicing blade and is mounted in the spindle of the dicing equipment. The dicing wheel (spindle) spins at a constant rotational velocity, and all experiments were carried out at 30.000 rpm spindle speed. Calibration of the blade height (Z-axis level), must be performed so the blade tip define chuck surface (promoted by electrical contact) and do not damage it at cutting time. Cutting programs for dicing silicon wafers rely mainly on three variables: speed of cut (feed speed); Y-axis step, which determines the distance between two cuts; Z-axis level, which determines the depth of cut: and X-axis step, which sets the cut length. The cut program is performed automatically, altering previous mentioned parameters depending on the fabrication goal. After completion of each sequence of cutting steps, the program repeats itself until the entire wafer is diced.

2.2. Probe manufacturing

The long probe outline is defined by a set of cuts in silicon substrate. These set of cuts and their parameters establish which is refereed as cut program. Probe manufacturing process is accomplish by a careful selection of the dicing process cut parameters. Table 1 summarizes the cut program used to produce planar shafts outline. All cuts were carried out by a DISCO Z05 blade (Z05-SD360-D1-90 52×0.08AS×40). Cutting program consisted on two different y-axis

Parameter	Value
Blade type	Z05
Blade thickness	80 μm
Cut thickness	100 μm
Cut speed	0.5 (mm s^{-1})
Z-axis calibration	0 mm (chuck surface)
X-axis step	24 mm
Y-axis steps	[300, (9×100)] μm

Table 1. Experimental dicing parameters. Square brackets represents sequential Y steps, which were repeated during the cut.

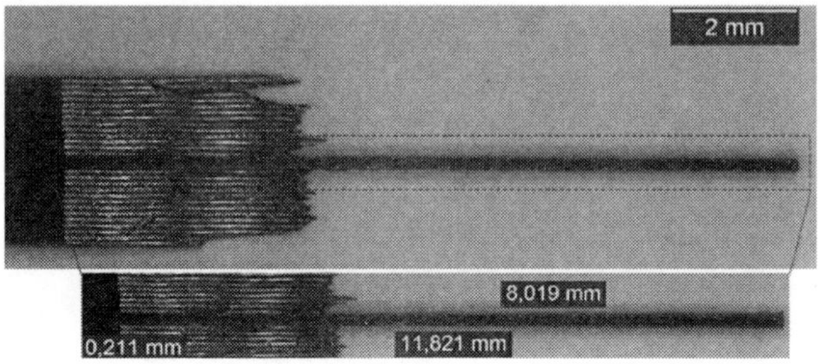

Figure 2. Dicing process resulted in 8 mm long shafts by simply employing a set of dicing parameters in a single cut program (see Table 1).

steps: the first is 300 μm and the second 100 μm. The first step determines the thickness of the shaft, while the second determines their pitch. These two steps are repeated until silicon substrate is completely diced. While probe width is set by programmed Y-axis step (300 μm), its length is accomplished by altering X-axis range of the blade. As the equipment restricts X-axis cut length to a minimum of 24 mm, using a range value inferior to its limits will inevitably result in a 24 mm long cut. Therefore, a particular modification must be made in the way substrate is positioned in the chuck surface. Substrate alignment in preparation for cut time should be made by midline of the chuck surface. This method is illustrated in Fig 1.c. Indeed, by placing the substrate at different distances from the chuck midline, it is possible to increase/decrease shafts length, in order to satisfy the application requirements.

3. Results and Discussion

The result of employing blade dicing technique in the manufacturing process of long shafts is shown in Figure 2. Indeed, in a single program comprising a set of cutting parameters (see Table 1), the outline of the probe is accomplished. Despite performing an approximately 12 mm long cut in X-axis direction, probes length is restricted to 8 mm. This is explained by the round geometry of the blade, which by the end of the cut cannot completely remove the diced substrate. Further optimization could include a combination with wet etching processes so the probe outline could be optimized to the total length of the cut. Nevertheless, an 8 mm long and 200 μm wide probe was successfully manufactured. All photos and measurements were performed with a Leica M80^{TM} stereo microscope and Leica LAS^{TM} software.

Since blade dicing is a purely mechanical process, its employment can produce undesirable mechanical vibrations, stress, and localized defects in the wafer [18]. Therefore, fabrication of challenging high aspect-ratio devices push mechanical dicing process to its limits. Manufacturing of such different designs and fragile devices represents a challenge to the dicing technology and might require modifications to the dicing process. Here an alignment adjustment to the standard operation of the dicing equipment was successfully employed. Although manual substrate alignment is performed, alignments lines in the dicing chuck allows good reproducibility of the results.

Other significant fabrication challenge consists on dicing different materials. Depending on the material, it can exhibit different behavior when diced. Typically, semiconductors like silicon, ceramics, and glasses present brittle mode cutting, which frequently induce microcracks within the material and chipping occurs at the edges [19]. This effect can be minimized by choosing proper blades and dicing parameters, such as low feed speed [20]. The blade employed is suitable for silicon substrate sawing and cuts were optimized to low speeds. As demonstrated elsewhere [21], silicon wafers diced at low speeds (< 0.5 mm s^{-1}) tend to avoid microcracks on the substrate.

4. Conclusion

The paper demonstrated the fabrication process of a long design neural probe. The manufacturing process include exclusively employing of traditional blade dicing technique. High aspect-ratio 8 mm long and 200 μm wide probe was successfully achieved by a single dicing program, which comprises a set of cut steps. The optimization of dicing parameters can broaden the application of most machining processes, allowing blade dicing technology to be employed in the production of various complex parts or devices.

Acknowledgments

S. B. Goncalves is fully supported by the FCT under grant PD/BD/105931/2014. This work is supported by the Portuguese Foundation for Science and Technology (FCT) with the reference project UID/EEA/04436/2013, by FEDER funds through the COMPETE 2020-Programa Operacional Competitividade e Internacionalizacão (POCI) with the reference project POCI-01-0145-FEDER-006941.

References

[1] Perlmutter J S and Mink J W 2006 Deep Brain Stimulation *Annu. Rev. Neurosci.* **29** 229-257.
[2] Norman R A 2007 Technology Insight: future neuroprosthetic therapies for disorders of the nervous system *Nature Clinical Practice Neurobiology* **3** 444-452.
[3] Zhang F, Aravanis A M, Adamantidis A, de Lecea L and Deisseroth K 2015 Circuit-breakers: optical technologies for probing neural signals and systems *Nat. Rev. Neurosci.* **8** 577-581.
[4] Kjonigsen L J, Leergaard T B, Witter M P and Bjaalie J G 2011 Digital atlas of anatomical subdivisions and boundaries of the rat hippocampal region *Frontiers in Neuroinformatics 5* 2.
[5] Cheung K, Djupsund K, Dan Y and Lee L 2003 Implantable multichannel electrode array based on SOI technology *J. Microelectromech. Syst.* **12** 179-184.
[6] Norlin P, Kindlundh M, Mouroux A, Yoshida K, Hofmann U G 2002 A 32-site neural recording probe fabricated by DRIE of SOI substrates *J. Microelctromech. Microeng.* **12** 414-419.
[7] Ruther P, Herwik S, Kisban S, Seidl K and Oliver P 2010 Recent Progress in Neural Probes Using Silicon MEMS Technology *IEEJ Trans. Electr. Electron. Eng.* **5** 505-515.
[8] Bai Q, Wise K D and Anderson D J 2000 A high-yield microassembly structure for three-dimensional microelectrode arrays *IEEE Trans. Biomed. Eng.* **47** 281289.
[9] Grand L, Pongrácz A, Vázsonyi E, Márton G, Gubán D, Fiáth R, Kerekesb B P, Karmosa G, Ulberta I and Battistigc G 2011 A novel multisite silicon probe for high quality laminar neural recordings *Sensors and Actuators A* **166** 1421.
[10] Hajj-Hassan M, Chodavarapu V P and Musallam S 2009 Microfabrication of ultra-long reinforced silicon neural electrodes *Micro & Nano Letters* **4** 5358.
[11] Oh S J, Song J K, Kim J W and Kim S J 2006 A high-yield fabrication process for silicon neural probes *IEEE Trans. Biomed. Eng* **53** 351354.
[12] Yoon T H, Hwang E J, Shin D Y, Park S I and Oh S J 2000 A micromachined silicon depth probe for multichannel neural recording *IEEE Trans. Biomed. Eng.* **47** 10821087.
[13] Seidl K, Herwik S, Torfs T, Neves H P, Paul O and Ruther P 2001 CMOS-Based High-Density Silicon Microprobe Arrays for Electronic Depth Control in Intracortical Neural Recording *J. Microelectromech. Syst.* **20** 1439-1448.
[14] Kindlundh M, Norlin P and Hofmann U G 2004 A neural probe process enabling variable electrode configurations *Sens. Actuators B: Chem.* **102** 5158.
[15] Rakwal D, Heamawatanachai S, Tathireddy P, Solzbacher F and Bamberg E 2009 Fabrication of compliant high aspect ratio silicon microelectrode arrays using micro-wire electrical discharge machining *Microsyst. Technol.* **15** 789797.
[16] Wang M, Maleki T and Ziaie B 2010 A self-assembled 3D microelectrode array *J. Micromech. Microeng.* **20** 035013.
[17] Arif M, Rahman M and San W 2012 A state-of-the-art review of ductile cutting of silicon wafers for semiconductor and microelectronics industries *Int J Adv Manuf Technol* **63**481504.
[18] Jiun H, Ahmad I, Jalar A and Omar G 2006 Effect of laminated wafer toward dicing process and alternative double pass sawing method to reduce chipping *IEEE Trans Electronics Packaging Manufacturing* **29** 1724.
[19] Chen S, Tsai C, Wu E, Shih I and Chen Y 2006 Study on the effects of wafer thinning and dicing on chip strength *IEEE Trans Advanced Packaging* **29** 149157.
[20] Kim S, Lee E, Kim N and Jeong H 2007 Machining characteristics on the ultra-precision dicing of silicon wafer *Int J Adv Manuf Technol* **33** 662667.
[21] Goncalves S B, Oliveira M J, Peixoto A C, Silva A F and Correia J H 2015 Out-of-plane neural microelectrode arrays fabrication using conventional blade dicing *The International Journal of Advanced Manufacturing Technology***85** 431-442.

Journal of Physics: Conference Series **757** (2016) 012012 doi:10.1088/1742-6596/757/1/012012

Circuit Modeling of a MEMS Varactor Including Dielectric Charging Dynamics

P Giounanlis[1], D Andrade-Miceli[1], S Gorreta[2], J Pons-Nin[2], M Dominguez-Pumar[2] and E Blokhina[1]

[1] School of Electrical, Electronic and Communications Engineering,
University College Dublin,
Dublin, Ireland

[2] Micro and Nano Technologies Group Electronic Engineering Department,
Technical University of Catalonia,
Barcelona, Spain

E-mail: `panagiotis.giounanlis@ucdconnect.ie`

Abstract. Electrical models for MEMS varactors including the effect of dielectric charging dynamics are not available in commercial circuit simulators. In this paper a circuit model using lumped ideal elements available in the Cadence libraries and a basic Verilog-A model, has been implemented. The model has been used to simulate the dielectric charging in function of time and its effects over the MEMS capacitance value.

1. Introduction

Capacitive Micro-electromechanical Systems (MEMS) offer excellent performance by combining mechanical and electrical components. They are used in a wide variety of applications such as sensors, resonators, switches, ultrasonic transducers and varactors [1, 2]. However, their micro-scale size brings a number of reliability issues which prohibit their mass commercialization [3].

In particular, MEMS variable capacitors (varactors) suffer from deviations of their capacitance-voltage (C-V) characteristic. Their performance is limited by the dielectric charging effect which is an inherent reliability problem for these devices. The build-up of trapping charges in the dielectric layer can be treated as an offset voltage connected in series with the MEMS, provoking a voltage shift in the characteristic curve. To remove this trapped charge, techniques which apply bipolar width-modulated voltage pulse sequences have been recently introduced and tested with good results [4, 5, 6].

In modern circuit design, the integration of MEMS with CMOS technology is a challenging step in micro-scale evolution. Computer aided design (CAD) tools play a significant role in the design and analysis process of chips [7, 8]. However, the complexity of MEMS, including failure mechanisms and reliability behavior, makes the modelling and simulation of chips which include these devices a task of high complexity.

In this paper, we present an adaptive system to implement a control method for the restriction of the dielectric charge of a MEMS varactor. The system's reliability relies highly on the MEMS capacitance measurement accuracy and resolution [9, 10]. Therefore, an accurate real-time measuring of the capacitance value is essential.

Electrical models for MEMS varactors including the effect of dielectric charging dynamics are not available in commercial circuit simulators. That is, here, for first time, we will present the implementation of a complete circuit model of a MEMS varactor using lumped ideal elements available in the Cadence libraries and a basic Verilog-A model. The model has been used to simulate the dielectric charging as a function of time and its effects over the MEMS capacitance value.

2. Statement of the Problem

The complete description of the bipolar control method to deal with the dielectric charging issue is out of the scope of this study (for a full description see [4]). Here, we will present briefly the whole system, mainly the circuit modelling of the MEMS varactor, which is a crucial component for the successful design and simulation of the chip.

The system consists of a closed control loop, as shown in Fig. 1, and it aims to restrict the build-up charge in the insulator of the MEMS varactor. Briefly, it works as follows: The value of the MEMS capacitance, C_m, is compared with the value of the reference capacitor, C_r. Then the error signal, e, is fed to an analogue processing stage which amplifies the capacitance difference. Finally, the output is converted into a digital word. This digital output is then used by the FPGA to set the duty cycle and the polarity of the voltage to be applied to the MEMS using the DC voltage source, V_B. The control loop is grouped into different sets: the MEMS varactor,

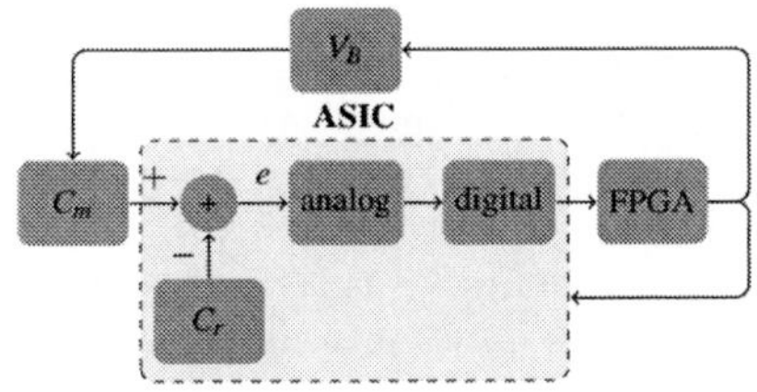

Figure 1: Control loop basic description at block level

the ASIC, the FPGA and the bias voltage. The ASIC consists of the input stage (including the reference capacitor), the analogue processing and the digital conversion. The FPGA reads and saves the digital output from the ASIC, and actuates the DC voltage, V_B. The capacitance difference measurement, the analogue processing and the digital conversion are performed inside the ASIC to reduce the second order effects and deviations introduced by parasitics. Since the accumulation of the dielectric charge is time dependent, in order to have reliable simulations for the ASIC design, it is essential to include an electrical model for the MEMS that not only reflects the voltage dependence, but also the charge dynamics.

3. MEMS varactor DC model

The MEMS varactor consists of a moveable electrode which deflects due to the presence of an electrostatic force, and a fixed bottom electrode which includes a thin dielectric layer on the top (for a detailed description see [6, 11]). The device operates in two regimes, the contact-less operation (below pull-in) and the contacting (ON) operation (beyond pull-in). Pull-in denotes the value of the voltage where the varactor changes mode of operation.

The behavior of the MEMS C-V characteristic in the contact-less regime can be approximated by a parabola, eq. (1), whilst eq. (2) can be used to simulate the operation of the device in the regime beyond pull-in.

$$\text{Below pull-in: } C_m(V) = C_0 + aV^2 \tag{1}$$

$$\text{Beyond pull-in: } C_m(V) = b_1 - \frac{b_2}{\sqrt{V}} \tag{2}$$

Here, the values C_0, a, b_1 and b_2 can be calculated by the fitting of experimental data. These equations were integrated into a Verilog-A script to simulate the MEMS in a simulator for electric circuits.

Fig. 2 visualizes the curve fitting along with the simulation results obtained with the use of a Verilog-A model. The C-V curve for the Verilog-A model was simulated with the use of basic circuit consisting of the MEMS model itself, connected to a DC power supply and performing a DC sweep in Cadence ADE-L.

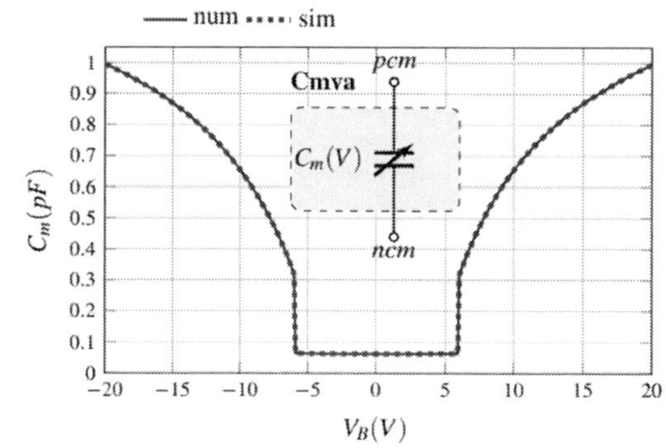

Figure 2: MEMS $C(V)$: numerical simulation in solid blue, cadence simulation in dashed red. $V_B = V_{pcm} - V_{ncm}$.

4. Equivalent electrical model for the dielectric charge dynamics

In principle, the charge dynamics of the dielectric is a complex dynamical procedure which can include more than one mechanisms [12] In this study, we will use a semi-empirical multi-exponential model (for a complete description see [13, 14]). The total charge consists of positive and negative multi-exponential components, in the most general case. Those components evolve in time, in principle, with different time constants, depending also on the polarity and the value of the applied voltage:

$$Q^p(t) = \begin{cases} Q^p_{max} \sum_i \zeta^p_i e^{-t/\tau^p_{Di}} & V > 0 \\ Q^p_{max}(1 - \sum_i \zeta^p_i e^{-t/\tau^p_{Ci}}) & V < 0 \end{cases}$$
$$Q^n(t) = \begin{cases} Q^n_{max}(1 - \sum_i \zeta^n_i e^{-t/\tau^n_{Ci}}) & V > 0 \\ Q^n_{max} \sum_i \zeta^n_i e^{-t/\tau^n_{Di}} & V < 0 \end{cases} \quad (3)$$

where Q^p_{max} is the maximum value of the positive charge component, Q^n_{max} the maximum value of the negative charge component, τ_{Ci} and τ_{Di} the charging and discharging time constants and ζ_i coefficients which express the contribution of each exponential to the total charge. By definition $\sum_i \zeta_i = 1$ for each component. It is seen from this model that the time varying expression of the total charge $Q_d(t)$ in the dielectric is given as: $Q_d(t) = Q^p(t) + Q^n(t) = \sum q^p_n + \sum q^n_n$, and it is the sum of multiple *independent* charge components. For the simulations of this paper, a two exponential model have been used. Based on the form of this equation set, an equivalent

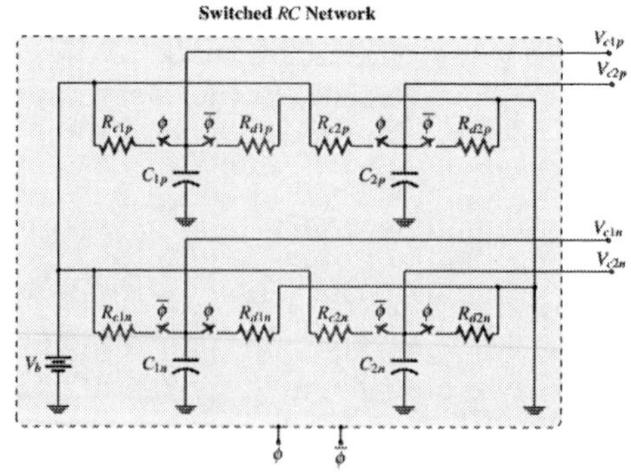

Figure 3: Equivalent switched RC network modelling the RF MEMS dielectric charging dynamics.

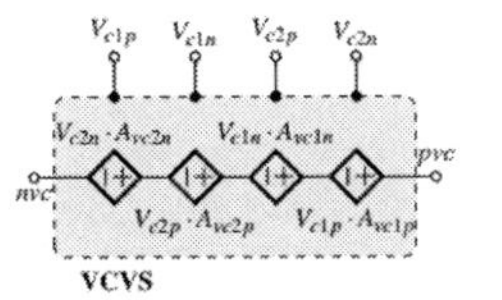

Figure 4: VCVS to generate equivalent V_{sh}.

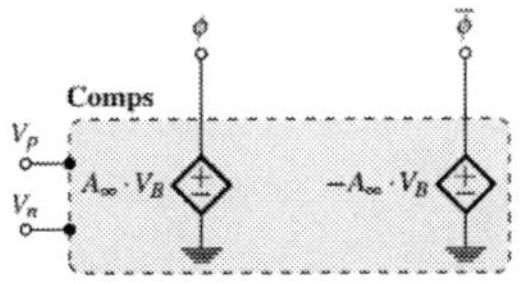

Figure 5: Comparators to generate $\phi - \overline{\phi}$.

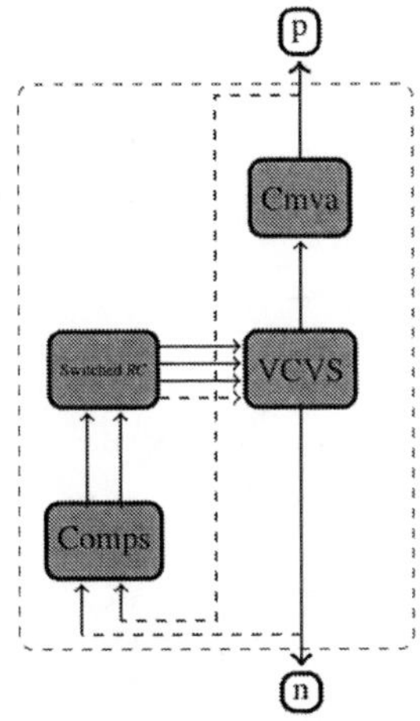

Figure 6: MEMS integrated model

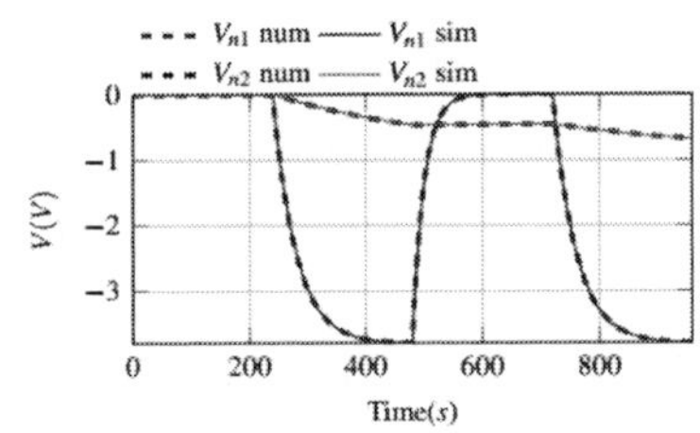

Figure 7: Negative charge contribution to V_{sh}

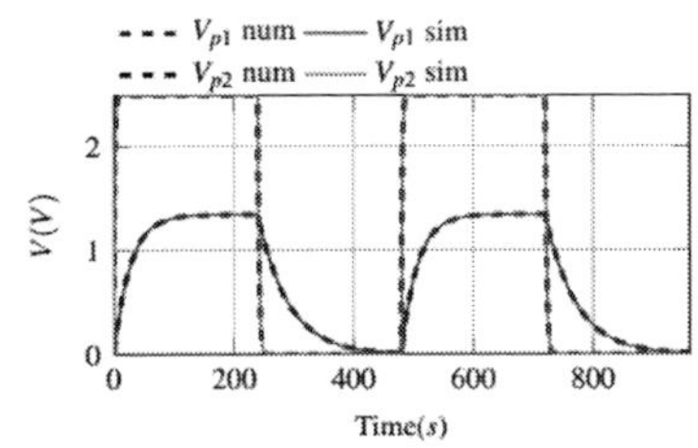

Figure 8: Positive charge contribution to V_{sh}

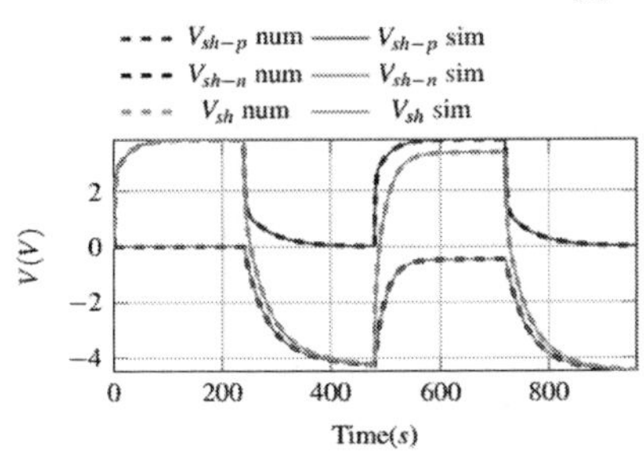

Figure 9: Positive and negative voltage shift and total V_{sh}

electrical network consisting of resistors and capacitors can be found, see eqs.(4) and (5) (where for simplicity we use only two components, but these expressions can be easily generalized):

$$Q_{cp} = C_{c1p}V(1 - e^{-\frac{t}{C_{c1p}R_{c1p}}}) + C_{c2p}V(1 - e^{-\frac{t}{C_{c1p}R_{c1p}}}) \quad (4)$$

$$Q_{dp} = C_{d1p}Ve^{-\frac{t}{C_{d1p}R_{d1p}}} + C_{d2p}Ve^{-\frac{t}{C_{d1p}R_{d1p}}} \quad (5)$$

where the same sub-indexing rules for the positive and negative components apply. Note that the value of voltage V can be set arbitrarily because is not directly related with the bias voltage applied to the MEMS. The values for R and C can be calculated in order the values between the numerical model and the corresponding electrical equivalent model to be matched. It must be taken into account that:

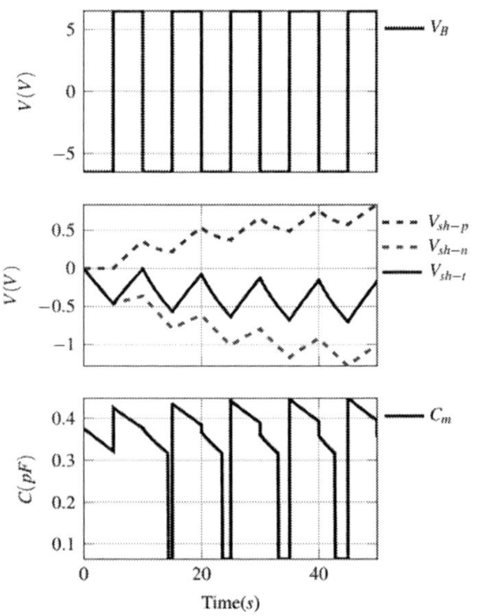

Figure 10: MEMS capacitance for a pulse voltage.

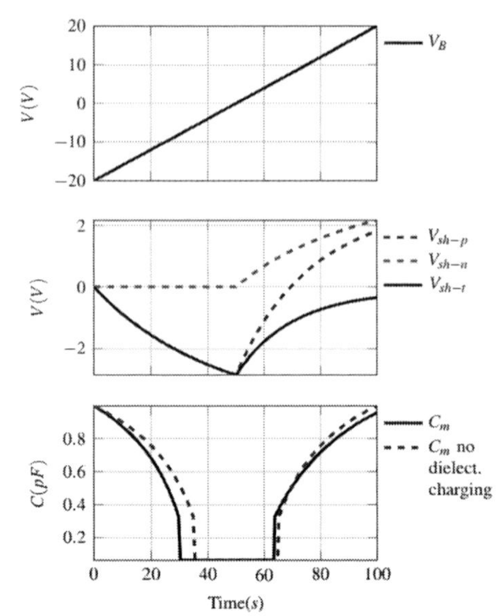

Figure 11: MEMS capacitance for a ramp voltage.

- **for positive bias**, the capacitors C_{1p} & C_{2p} are *charged* to a positive voltage through the resistors R_{c1p} & R_{c2p}, respectively; and simultaneously, the capacitors C_{1n} & C_{2n} are *discharged* from the previous stored voltage (if exists) through the resistors R_{c1n} & R_{c2n}
- **for negative bias**, the capacitors C_{1p} & C_{2p} are *discharged* through the resistors R_{d1p} & R_{d2p}, respectively; and simultaneously, the capacitors C_{1n} & C_{2n} are *charged* through the resistors R_{c1n} & R_{c2n}

That is, the equivalent circuit must be switched according with these rules, and the control signals must be derived from the voltage applied to the MEMS. The equivalent switched RC network is shown in Fig. 3. The analog switches are modeled using Verilog-A with very high off-resistance, zero on-resistance and threshold voltage of $0.1mV$.

The resulting switched RC network models the dielectric charging and discharging times according to the applied voltage polarity. Moreover, in order to convert the voltage stored in every capacitor into an equivalent votlage shift, it must be amplified by a gain factor determined by the ratio between the corresponding capacitor in the network and the MEMS dielectric capacitance C_d. Once amplified, the resulting voltages must be added to calculate the total voltage shift. This must be connected to C_m as a voltage offset source.

To perform these tasks, a series-connected Voltage Controlled Voltage Sources (VCVS), every of them sensing the voltage on every capacitor and with the corresponding gain factor, is implemented, as depicted in Fig. 5. This source array is connected to the negative terminal of $Cmva$. In order to generate the control signals $\phi - \overline{\phi}$, two ideal comparators made of VCVS has been included. Comparators have extremely high voltage gain, no hysteresis, threshold voltage of $0.1mV$, minimum voltage of $0V$, and maximum voltage output of $1V$. The voltage applied to the MEMS integrated electrical model, between the node *pcm* in $Cmva$ and node nvc in $VCVS$ block, is used as control voltage for the comparators. When the polarity is positive and larger than $0.1mV$, ϕ is high and $\overline{\phi}$ is low. When the polarity changes, the signals switch their values. Finally, the whole MEMS electrical model can be integrated as shown in Fig. 6.

5. Simulation results

The electrical model was simulated using the same circuit described in Section 3 for DC sweep, but now for transient analysis. The input signal (not shown in the plots) is a square pulse with maximum voltage of $10V$ and minimum of $-10V$ and a period of $240s$. All the results are compared with the corresponding numerical simulations. In Fig. 7 the contribution of the components of the negative charge to the voltage shift are shown. Similarly, in Fig. 8 for the positive charge. Additionally, in Fig. 9 the results for the total positive and negative charge are visualized along with the voltage shift V_{sh} considering all the contributions. The error is minimum, so the circuit effectively models the dielectric charging effects.

In Figs. 10 and 11 the resulting capacitance as a function of time is shown. In Fig. 10, a pulse bias signal from $\pm 6.5V$ has been applied (top plot) provoking the voltage shift changes (middle plot) altering the value of the capacitance (bottom plot). When the bias voltage is near to the pull-in value, it can be observed that the charge accumulation makes the V_{sh} large enough to provoke the MEMS capacitance to switch from a high to a low value. In Fig. 11, a ramp is applied with voltage value of $\pm 20V$ (top plot). One can see that the evolution of the V_{sh} (middle plot) alters the C-V characteristic (bottom plot), in solid black, which deviates from the ideal behaviour (dashed blue) when no charge injection in the dielectric is considered.

6. Future work

In this paper an electrical model for a MEMS varactor, including the effects of dynamic dielectric charging has been presented. The model consists of ideal lumped elements available in the most common EDA Tools and Verilog-A code. A switched RC network that can be parametrized for different devices characteristics has been implemented. To improve the model, effects like temperature could be included either via Verilog-A code or by introducing some temperature factors for the lumped elements.

References

[1] X. Jin, I. Ladabaum, and B. Khuri-Yakub, "Surface micromachined capacitive ultrasonic immersion transducers," in *Micro Electro Mechanical Systems, 1998. MEMS 98. Proceedings., The Eleventh Annual International Workshop on*, Jan 1998, pp. 649–654.

[2] J. Wu, G. Fedder, and L. Carley, "A low-noise low-offset capacitive sensing amplifier for a 50- mu;g/ radic;hz monolithic cmos mems accelerometer," *Solid-State Circuits, IEEE Journal of*, vol. 39, no. 5, pp. 722–730, May 2004.

[3] W. de Groot, J. Webster, D. Felnhofer, and E. Gusev, "Review of device and reliability physics of dielectrics in electrostatically driven mems devices," *Device and Materials Reliability, IEEE Transactions on*, vol. 9, no. 2, pp. 190–202, June 2009.

[4] M. Dominguez-Pumar, S. Gorreta, J. Pons-Nin, E. Blokhina, P. G., and O. Feely, "Real-time characterization of dielectric charging in contactless capacitive mems," *Analog Integrated Circuit Signal Processing*, 2014.

[5] Z. Peng, X. Yuan, J. Hwang, D. Forehand, and C. Goldsmith, "Dielectric charging of rf mems capacitive switches under bipolar control-voltage waveforms," in *Microwave Symposium, 2007. IEEE/MTT-S International*, June 2007, pp. 1817–1820.

[6] S. Gorreta, J. Pons-Nin, E. Blokhina, O. Feely, and M. DomÃŋnguez-Pumar, "Delta-sigma control of dielectric charge for contactless capacitive mems," vol. 24, no. 4, August 2014.

[7] E. Blokhina, D. O'Connell, D. Andrade-Miceli, S. Gorreta-Marine, J. Pons-Nin, M. Dominguez-Pumar, O. Feely, and D. Galayko, "Understanding complexity in multiphysics systems-on-a-chip: Modern approaches for design," in *Circuits and Systems (ISCAS), 2015 IEEE International Symposium on*, May 2015, pp. 1546–1549.

[8] M. Mita and H. Toshiyoshi, "An equivalent-circuit model for mems electrostatic actuator using open-source software qucs," *IEICE Electronics Express*, vol. 6, no. 5, pp. 256–263, 2009.

[9] G. Ding, D. Molinero, W. Wang, C. Palego, S. Halder, and J. H. C. Goldsmith, "Intelligent bipolar control of mems capacitive switches," *Microwave Theory and Techniques, IEEE Transactions on*, vol. 61, no. 1, pp. 464–471, Jan 2013.

[10] A. Heidary and G. C. M. Meijer, "An integrated interface circuit with a capacitance-to-voltage converter as front-end for grounded capacitive sensors," *Measurement Science and Technology*, vol. 20, no. 1, 2009.

[11] P. Giounanlis, E. Blokhina, O. Feely, M. Dominguez, J. Pons Nin, and S. Gorreta, "Modelling of a charge control method for capacitive mems," in *Circuit Theory and Design (ECCTD), 2013 European Conference on*, Sept 2013, pp. 1–4.

[12] G. J. Papaioannou and J. Papapolymerou, "Dielectric charging mechanisms in rf-mems capacitive switches," in *Microwave Integrated Circuit Conference, 2007. EuMIC 2007. European*, Oct 2007, pp. 359–362.

[13] S. Gorreta, J. Pons-Nin, E. Blokhina, O. Feely, and M. DomÃŋnguez-Pumar, "Delta-sigma control of dielectric charge for contactless capacitive mems," vol. 23, no. 4, August 2014.

[14] M. Dominguez-Pumar, S. Gorreta, J. Pons-Nin, E. Blokhina, P. Giounanlis, and O. Feely, "Real-time characterization of dielectric charging in contactless capacitive mems," 2014.

Automation of Silica Bead-based Nucleic Acid Extraction on a Centrifugal Lab-on-a-Disc Platform

David J Kinahan, Faith Mangwanya, Robert Garvey, Danielle WY Chung, Artur Lipinski, Lourdes AN Julius, Damien King, Mehdi Mohammadi, Rohit Mishra, May Al-Ofi, Celina Miyazaki and Jens Ducrée

School of Physical Sciences, National Centre for Sensor Research, Dublin City University, Glasnevin, Dublin, Ireland

david.kinahan@dcu.ie, jens.ducree@dcu.ie

Abstract. We describe a centrifugal microfluidic 'Lab-on-a-Disc' (LoaD) technology for DNA purification towards eventual integration into a Sample-to-Answer platform for detection of the pathogen *Escherichia coli* O157:H7 from food samples. For this application, we use a novel microfluidic architecture which combines 'event-triggered' dissolvable film (DF) valves with a reaction chamber gated by a centrifugo-pneumatic siphon valve (CPSV). This architecture permits comprehensive flow control by simple changes in the speed of the platform innate spindle motor. Even before method optimisation, characterisation by DNA fluorescence reveals an extraction efficiency of 58%, which is close to commercial spin columns.

1. Introduction

The centrifugal microfluidic platform [1-7] has successfully demonstrated sample-to-answer automation of bioassays for a wide range of applications including biomedical point-of-care diagnostics [8], cell analysis [9, 10], environmental monitoring [11-13] and quality control of industrial processes such as the production of biologics [14]. The cartridge holding all liquids is typically of a similar geometry as optical data storage media (e.g. CD or DVD) and spun on a miniaturised system similar to a commercial Discman™.

The LoaD platform offers a number of benefits. The system-innate centrifugal force field can be harnessed for pumping, liquid handling operations such as metering, aliquoting and mixing, as well as for sample preparation such as high-purity plasma extraction from whole blood [15, 16]. The disc can also be loaded and operate while open to atmosphere; a simple pipette can be used to introduce sample as opposed to a cumbersome 'world-to-chip' interfacing based on liquid connectors. Similarly, low-cost spindle motors are rather ubiquitous and generally significantly cheaper than specialised pumps.

Furthermore, as it scales with the square of the spin rate, the centrifugal field can be easily varied over several orders of magnitude, thus providing a large range of forces on fluids and suspended particles. These features contribute to provide a compact, highly autonomous, rugged, portable, user friendly and affordable bioanalytical sample-to-answer system which is particularly suitable for decentralised point-of-use testing.

For nucleic acid purification and concentration according to the established Boom method [17], a sequence of sample, wash and elution buffers are run through a chip-integrated silica-based solid phase. Many recent advances using the LoaD for nucleic acid analysis incorporate sample purification, nucleic acid extraction and subsequent nucleic acid based amplification to identify specific markers

for pathogens such as *Salmonella enterica* [18], *Bacillus atrophaeus* [19], *Escherichia coli* O157:H7, *Salmonella typhimurium*, *Vibrio parahaemolyticus* and *Listeria monocytogenes*[20]. However, due to the complexity of the laboratory unit operations (LUOs) [7] involved, typically LoaD systems use support instrumentation, such as external heating, to operate the valves required to automate these processes. These additional modules add to the complexity and cost of the systems, and thus tend to undermine the simple concept inherent to the LoaD platform.
In this work we combine our recently introduced event-triggered valves [21, 22], which are based upon dissolvable film (DF) technology [23], with a sample incubation chamber which is combined with a centrifugo-pneumatic siphon valve (CPSV) [16, 24-26]. The event-triggered valves, where liquid is released through one DF, called the 'Load Film' (LF) when a second DF located at a remote location, called the Control Film (CF), is wetted, offer performance akin to an electrical relay. As they are decoupled from external instrumentation and from the changes in spin rate, the number of serial LUOs which can be placed on the disc is practically only limited by the available disc real estate. Yet, a major drawback of the event-triggered approach is the dependence of valve actuation on the dissolution times of the DFs. Thus, the event-triggered valves are unsuitable to implement the long incubations required for many bio-analytical processes. Optimisation of process timing therefore entails the replacement of the formulation of the DFs to be integrated in the system. A possible solution is to place a CPSV on the outlet of the incubation chamber so the timing is controlled by the programmable rotation of the spindle motor rather than the material-specific dissolution time of the DF.
We demonstrate our technology by processing a dilution series of DNA standard through this disc. We incubate our DNA standard with silica beads, then wash this sample with a high concentration of ethanol (75% EtOH) to remove chaotropic salts, and then eluting the DNA from the silica beads using buffer adapted from a commercial kit.

2. Material and Methods

2.1 Disc Manufacture and Assembly
The microfluidic cartridges (Fig. 1a) used in this study were assembled using multi-lamination methods from four layers of Poly(methyl methacrylate) (PMMA) and four layers of PSA (Pressure Sensitive Adhesive, Adhesives Research, Limerick, Ireland) [21]. Voids in the 86-µm thick PSA layers [27] representing microchannels and other small features, were defined by a commercial knife-cutter (Graphtec, Yokohama, Japan). Larger features such as reservoirs for reagent storage and sample collection were created in 1.5-mm thick PMMA layers using a laser cutter (Epilog Zing, USA). The layers were aligned on a custom assembly jig. Between each alignment step, the attachment of layers was reinforced using a hot-roll laminator (Hot Roll Laminator, Chemsultant Int., US). Each DF was mounted on PSA tab using previously described methods [21] and then manually positioned within the disc during manufacture.

2.2 Experimental Test Stand
All discs in this study were imaged using an experimental test platform commonly referred to as a "spin stand" [21, 28]. Discs are mounted on a computer controlled motor (Faulhaber Minimotor SA, Switzerland) and spun according to programmed rates. An encoder embedded on the spindle provides two signals (1 pulse / rotation and 1000 pulses / rotation); custom hardware is used to define the angular position where a digital trigger activates a stroboscopic light source (Drelloscop 3244, Drello, Germany) and a sensitive, short-exposure time camera (Pixelfly, PCO, Germany). The discs are accelerated and slowed down at 12.5 Hz s^{-1}, except for the pneumatically driven disc which was decelerated at 1 Hz s^{-1} (Fig. 1b).

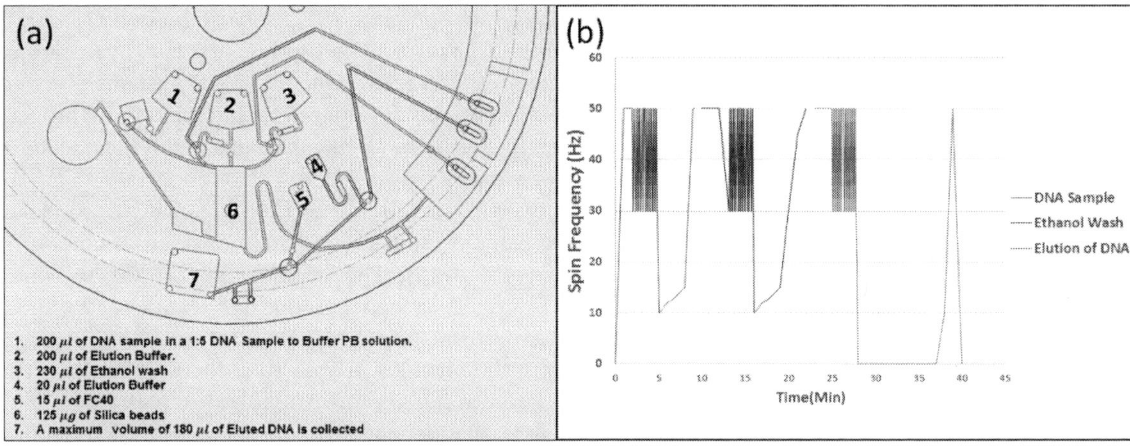

Figure 1. (a) Disc architecture. (b) Spin protocol for the disc-based nucleic acid extraction.

2.3 DNA Purification and Quantification

The DNA purification disc uses reagents adapted from QIAquick PCR Purification kit (Qiagen, Hilden, Germany); the extracted DNA in the eluate and was quantified using a Quant-iT™ PicoGreen® dsDNA (ThermoFisher). First, the Lamda DNA standard provided with the PicoGreen kit was diluted to a concentration of 2 µg mL^{-1} using the method and buffer provided (TE Buffer). A working solution of Quant-iT™ PicoGreen® reagent is also created as per supplier's protocol.

For both, processing in the QIAquick spin column and on disc, the 2 µg mL^{-1} is diluted 5:1 using buffer PB (provided in the QIAquick kit). A sample of 60 µl is loaded. For the QIAquick columns, this sample is then washed using 750 µl of PE Buffer and Ethanol (EtOH) based buffer provided in the QIAquick kit while on disc two 100 µl washes are loaded into the chambers called EtOH #1 and EtOH #2. For both methods, the elution step is performed using 60 µl of EB buffer (again provided in the QIAquick kit). For quantification, 30 µl of the sample recovered from the QIAquick column, from the disc and from the reference DNA (2 µg mL^{-1}) are each added to 30 µl of the working solution of Quant-iT™ PicoGreen®. Fluorescence is measured using a commercial plate reader (Tecan) per the Quant-iT protocol and, for reporting, fluorescent signal is normalised relative to the reference (2 µg mL^{-1} DNA) fluorescent signal.

3. Disc Operation and Performance

3.1 Disc Architecture and Operation

A schematic of the disc architecture and the associate spin protocol are shown in Figure 1, a frame sequence of disc operation is portrayed in Figure 2. The disc is manufactured from eight layers of PMMA and PSA (as described previously) and the DFs are selected to be compatible with the reagents which are loaded in each chamber. The disc is loaded per the volumes described in Figure 1. Initially, the disc is accelerated to 50 Hz. At this point, the dense FC-40 liquid creates a plug in front of the route to the DNA collection chamber (Chamber 7) which ensures that the DNA sample and EtOH are both routed to the waste chamber. The sample flows into the incubation chamber, where it is transiently retained by a centrifugo-pneumatic siphon valve (CPSV). Due to the high spin-rate, centrifugal pressure forces some liquid into the dead end pneumatic chamber (which is a constituent part of the CPSV). By rapidly cycling the spin rate from 50 Hz to 30 Hz, part of this pneumatic chamber can be alternatingly emptied and refilled; this pneumatically enhanced 'shake mode' [29] mixing ensures that the DNA sample and the silica beads interact efficiently. After an arbitrary incubation time, defined by programming the spindle motor, the frequency of rotation is decreased to 10 Hz. This decrease in centrifugally induced pressure expels liquid from the dead end pneumatic chamber, increasing the liquid 'height' in the incubation chamber above the crest of the siphon and

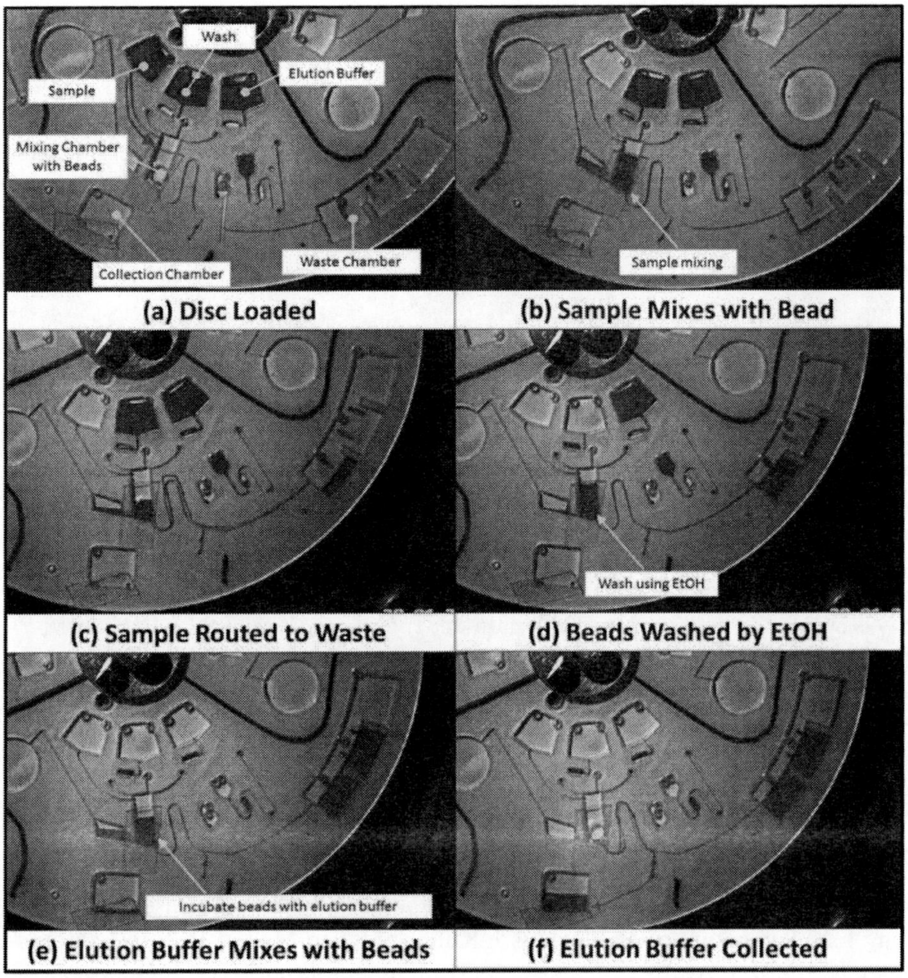

Figure 2. Frame sequence of the disc operation. Dyed water is used for visualisation. (a)The disc is loaded, (b) the spin rate is increased to 50 Hz to deliver the sample into the incubation chamber where it mixes with acid-washed silica beads. The hydrodynamics in this chamber can be agitated by 'shake mode' mixing through rapidly alternating the rotational frequency between 30 Hz and 50 Hz. (c) The disc is decelerated to 10 Hz to prime the siphon. The disc is slowly accelerated to 15 Hz so the liquid empties to the waste (d). This protocol is repeated to wash the beads with Ethanol (75% EtOH). (e) The elution buffer is incubated with the beads. Simultaneously, the release of the |FC-40plug opens a route to the sample collection chamber. (f) Upon lowering the spin rate, the siphon is primed and now the elution buffer is routed to a collection chamber.

thus priming the siphon valve (CPSV). Next, the spin rate is increased so the content of the incubation chamber is emptied into the waste chamber. Here, the first CF is wetted, triggering the release of the 75% EtOH wash. In a manner similar to the DNA sample, the wash can again be incubated with the sample and be transferred to the waste chamber.

The 75% EtOH wash then wets and opens the second and third CFs located into the waste chamber. This releases the elution buffer, which again can be incubated with the bead substrate. Simultaneously, a liquid element is released which removes the FC-40 plug blocking the route to the sample collection chamber. Thus, upon deceleration of the disc, the elution buffer is preferentially routed away from the waste chamber to the sample collection chamber.

3.2 Disc Characterisation and Performance

The disc design underwent iterative optimisation to ensure proper function, manufacturability and correct process control. Changes were made to ensure correct volumes were used, proper siphon priming and efficient routing to the sample collection chamber. Once the reliable liquid handling was accomplished, the disc was tested using a DNA standard using the protocol described above. Normalised relative to the DNA standard (100% ±7.2, n = 3), the commercial spin columns demonstrated an extraction efficiency of 84.7 ± 4.2% (n = 3) while we measured an extraction efficiency of the disc of 58.4 ± 7.5% (n = 3).

4. Conclusion

The disc architecture presented in this work presents several advantages compared to other methods: The combination of event-triggered valves and CPSVs permits good process control to be combined with a reliable and expandable reagent storage mechanism. The extraction efficiency is comparable to the state-of-the-art for LoaD. Performance might be improved, relative to the commercial platforms, by optimisation of reagents, bead concentration, surface treatments, and incubation times.
Combined with appropriate nucleic acid amplification methods, such as LAMP, this platform will be used for multiplex serotyping of common *E. Coli* pathogens.

Acknowledgements
This work was supported by the European Union under Grant number FP7-KBBE-2013-7-613908-DECATHLON and the Science Foundation Ireland under Grant No 10/CE/B1821.

References

[1] Madou M, Zoval J, Jia G, Kido H, Kim J and Kim N 2006 Lab on a CD *Annu. Rev. Biomed. Eng.* **8** 601-28
[2] Ducrée J, Haeberle S, Lutz S, Pausch S, Von Stetten F and Zengerle R 2007 The centrifugal microfluidic Bio-Disk platform *Journal of Micromechanics and Microengineering* **17** S103
[3] Gorkin R, Park J, Siegrist J, Amasia M, Lee B S, Park J-M, Kim J, Kim H, Madou M and Cho Y-K 2010 Centrifugal microfluidics for biomedical applications *Lab on a Chip* **10** 1758-73
[4] Kong L X, Perebikovsky A, Moebius J, Kulinsky L and Madou M 2015 Lab-on-a-CD A Fully Integrated Molecular Diagnostic System *Journal of laboratory automation* 2211068215588456
[5] Smith S, Mager D, Perebikovsky A, Shamloo E, Kinahan D, Mishra R, Torres Delgado S M, Kido H, Saha S and Ducrée J 2016 CD-Based Microfluidics for Primary Care in Extreme Point-of-Care Settings *Micromachines* **7** 22
[6] Tang M, Wang G, Kong S-K and Ho H-P 2016 A Review of Biomedical Centrifugal Microfluidic Platforms *Micromachines* **7** 26
[7] Strohmeier O, Keller M, Schwemmer F, Zehnle S, Mark D, von Stetten F, Zengerle R and Paust N 2015 Centrifugal microfluidic platforms: advanced unit operations and applications *Chemical Society Reviews* **44** 6187-229
[8] Glynn M, Kirby D, Chung D, Kinahan D J, Kijanka G and Ducree J 2013 Centrifugo-Magnetophoretic Purification of CD4+ Cells from Whole Blood Toward Future HIV/AIDS Point-of-Care Applications *J Lab Autom* **19** 285-96
[9] Burger R, Kirby D, Glynn M, Nwankire C, O'Sullivan M, Siegrist J, Kinahan D, Aguirre G, Kijanka G, Gorkin R a and Ducrée J 2012 Centrifugal microfluidics for cell analysis *Current Opinion in Chemical Biology* **16** 409-14
[10] Glynn M, Nwankire C, Lemass K, Kinahan D J and Ducrée J 2015 Cluster size distribution of cancer cells in blood using stopped-flow centrifugation along scale-matched gaps of a radially inclined rail *Microsystems & Nanoengineering* **1**
[11] Kong M C and Salin E D 2012 Spectrophotometric determination of aqueous sulfide on a pneumatically enhanced centrifugal microfluidic platform *Analytical chemistry* **84** 10038-43

[12] Hwang H, Kim Y, Cho J, Lee J-y, Choi M-S and Cho Y-K 2013 Lab-on-a-Disc for Simultaneous Determination of Nutrients in Water *Analytical chemistry* **85** 2954-60

[13] Czugala M, Maher D, Collins F, Burger R, Hopfgartner F, Yang Y, Zhaou J, Ducrée J, Smeaton A and Fraser K J 2013 CMAS: fully integrated portable centrifugal microfluidic analysis system for on-site colorimetric analysis *RSC Advances* **3** 15928-38

[14] Nwankire C E, Donohoe G G, Zhang X, Siegrist J, Somers M, Kurzbuch D, Monaghan R, Kitsara M, Burger R and Hearty S 2013 At-line bioprocess monitoring by immunoassay with rotationally controlled serial siphoning and integrated supercritical angle fluorescence optics *Analytica chimica acta* **781** 54-62

[15] Kinahan D J, Kearney S M, Glynn M T and Ducrée J 2014 Spira mirabilis enhanced whole blood processing in a lab-on-a-disk *Sensors and Actuators A: Physical* **215** 71-6

[16] Kinahan D J, Kearney S M, Kilcawley N A, Early P L, Glynn M T and Ducrée J 2016 Density-Gradient Mediated Band Extraction of Leukocytes from Whole Blood Using Centrifugo-Pneumatic Siphon Valving on Centrifugal Microfluidic Discs *PloS one* **11** e0155545

[17] Boom R, Sol C, Salimans M, Jansen C, Wertheim-van Dillen P and Van der Noordaa J 1990 Rapid and simple method for purification of nucleic acids *Journal of clinical microbiology* **28** 495-503

[18] Kim T-H, Park J, Kim C-J and Cho Y-K 2014 Fully integrated lab-on-a-disc for nucleic acid analysis of food-borne pathogens *Analytical chemistry* **86** 3841-8

[19] Roy E, Stewart G, Mounier M, Malic L, Peytavi R, Clime L, Madou M, Bossinot M, Bergeron M G and Veres T 2015 From cellular lysis to microarray detection, an integrated thermoplastic elastomer (TPE) point of care Lab on a Disc *Lab on a Chip* **15** 406-16

[20] Oh S J, Park B H, Choi G, Seo J H, Jung J H, Choi J S and Seo T S 2016 Fully automated and colorimetric foodborne pathogen detection on an integrated centrifugal microfluidic device *Lab on a Chip*

[21] Kinahan D J, Kearney S M, Dimov N, Glynn M T and Ducrée J 2014 Event-triggered logical flow control for comprehensive process integration of multi-step assays on centrifugal microfluidic platforms *Lab on a Chip* **14** 2249-58

[22] Kinahan D J, Kearney S M, Faneuil O P, Glynn M T, Dimov N and Ducrée J 2015 Paper imbibition for timing of multi-step liquid handling protocols on event-triggered centrifugal microfluidic lab-on-a-disc platforms *RSC Advances* **5** 1818-26

[23] Gorkin III R, Nwankire C E, Gaughran J, Zhang X, Donohoe G G, Rook M, O'Kennedy R and Ducrée J 2012 Centrifugo-pneumatic valving utilizing dissolvable films *Lab on a Chip* **12** 2894-902

[24] Godino N, Gorkin III R, Linares A V, Burger R and Ducrée J 2013 Comprehensive integration of homogeneous bioassays via centrifugo-pneumatic cascading *Lab on a Chip* **13** 685-94

[25] Zehnle S, Roth G, Stetten F v, Zengerle R and Paust N 2012 Centrifugo-dynamic inward pumping of liquids on a centrifugal microfluidic platform *Lab on a Chip* **12** 5142-5

[26] Gorkin III R, Clime L, Madou M and Kido H 2010 Pneumatic pumping in centrifugal microfluidic platforms *Microfluidics and nanofluidics* **9** 541-9

[27] Bartholomeusz D A, Boutté R W and Andrade J D 2005 Xurography: rapid prototyping of microstructures using a cutting plotter *Microelectromechanical Systems, Journal of* **14** 1364-74

[28] Grumann M, Brenner T, Beer C, Zengerle R and Ducrée J 2005 Visualization of flow patterning in high-speed centrifugal microfluidics *Review of Scientific Instruments* **76** 1-6

[29] Grumann M, Geipel a, Riegger L, Zengerle R and Ducrée J 2005 Batch-mode mixing on centrifugal microfluidic platforms *Lab on a chip* **5** 560-5

Journal of Physics: Conference Series **757** (2016) 012014 doi:10.1088/1742-6596/757/1/012014

Micro system comprising 96 micro valves on a titer plate

S. Krabbe[1], D. Flitsch[2], J. Büchs[2], W.K. Schomburg[1]

RWTH Aachen University, 52074 Aachen
[1]Konstruktion und Entwicklung von Mikrosystemen, KEmikro, Steinbachstraße 53 B
[2]Aachener Verfahrenstechnik, AVT, Biochemical Engineering, Worringer Weg 1

Krabbe@KEmikro.RWTH-Aachen.de

Abstract. A system of 96 micro valves has been developed and mounted on top of a 48-well micro titer plate providing two valves for each well controlling its air inlet and outlet. Testing of the valve system showed that all valves are working and are opened and closed reliably. A pneumatic system is switching inlet and outlet valves independently of each other. The geometry of the feed channels ensures an equal air flow through all wells, when the valves are open. Between the micro valves, one optical fibre was inserted through the lid of each well allowing measuring the oxygen partial pressure in the enclosed air volume by fluorescence sensor spots. *Escherichia coli* bacteria were grown inside the wells and their metabolism was observed by the oxygen partial pressure change due to respiration. In all 48 wells, the same oxygen transfer rate was observed within an averaged standard deviation of 1 mmol/L/h. The oxygen transfer rate differences compared to a macroscopic standard shake flask system were overall compatible within their uncertainties.

1. Introduction

The metabolism of cells in bio reactors is often observed by measuring their oxygen consumption. A macroscopic commercially available device providing such measurements is the Respiratory Activity Monitoring System (RAMOS) presented in [1] and [2]. It determines the respiratory activity of aerobic microorganisms by measuring their oxygen transfer rate in eight shaking flasks, each with an inner volume of 0.125 to 1 L. The capacity of RAMOS now has been enlarged from the parallel observation of 8 vessels to the 48 wells of a micro titer plate [3]. This new device is called μRAMOS.

To achieve this, each well was equipped with an inlet and an outlet valve (cf. Fig. 1). The oxygen consumption in the wells was determined within measurement circles. Each circle started with the closing of both valves. Then the change of the oxygen partial pressure was measured for 3 minutes. After that time first the outlet and then the inlet vales of the wells were opened. A time difference of 1 s between switching of the two valves made sure that no overpressure in the wells could occur. After venting the wells for 17 minutes, a new

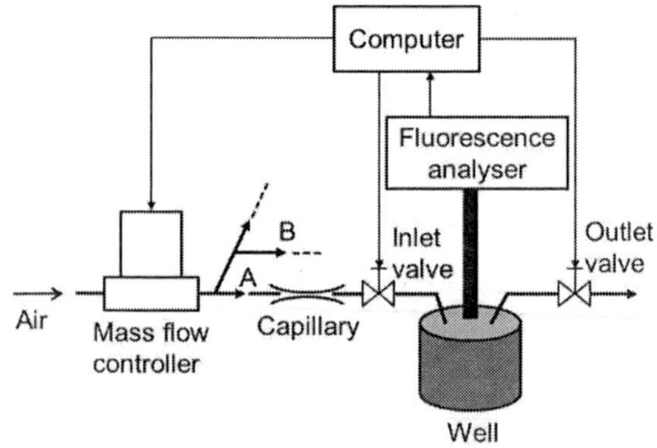

Figure 1. Schematic drawing of the setup realized for each of the 48 wells of the μRAMOS system.

Journal of Physics: Conference Series **757** (2016) 012014 doi:10.1088/1742-6596/757/1/012014

measurement cycle was started.

To provide such a measurement cycle, it was necessary to equip every well with two valves, which can be switched independently of each other and with an oxygen sensor. The oxygen sensor was provided by Pyro Science GmbH in Aachen, Germany. It consists of a polymer whose fluorescence is a function of the oxygen partial pressure in the well. The oxygen sensor was glued on the lid of the wells and was excited to fluorescence and read out by an optical fibre, 2 mm in diameter. All these components need to fit into a circle, 12.4 mm in diameter, available at the top of each well.
In this paper, there are described the design and testing of the valve system.

2. Design of the micro valve system

The micro valve system consists of three plates milled from poly methyl methacrylate (PMMA) and three casted silicone membranes with micro structures (Fig. 2). This stack of plates and membranes is pressed onto a micro titer plate by screws and an aluminium holder. The upper and the bottom plates comprise separate fluidic systems. The upper plate again consists of two pneumatic microfluidic systems separately switching outlet and inlet valves. The bottom part provides an equal air flow to each of the 48 wells.

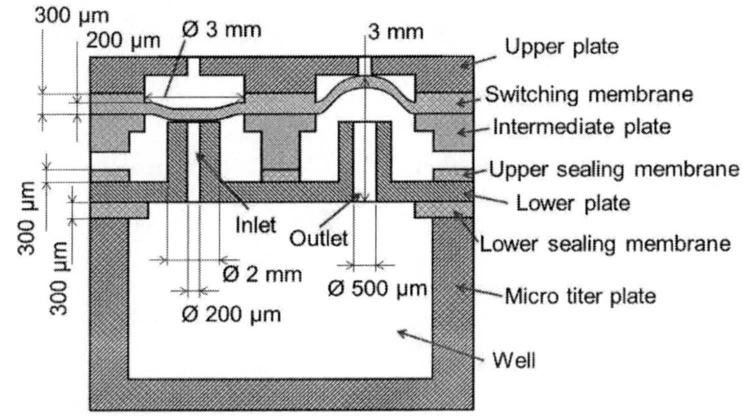

Figure 2. Schematic drawing of the two valves on top of each well of the 48-well micro titer plate. Dimensions are not to scale.

To close and open the valves, a switching membrane (cf. Fig. 2) is pressed onto or sucked off a valve seat integrated into the lower plate of the valve system. When the entire stack of plates and membranes is pressed onto the titer plate, the soft silicone is partly squeezed out of the gap between upper plate and intermediate plate. As a consequence, the switching membranes are compressed in lateral direction and bulge down onto the valve seat when a valve is closed. Therefore, it is necessary to pull the switching membrane up with 35 kPa underpressure when the valve shall be opened. On the other hand, when the valves shall be closed, an overpressure of 65 kPa is applied making sure that all micro valves are tightly closed.

2.1. Pneumatic switching system

The upper PMMA plate contains four micro fluidic channel systems. Two of them are employed for switching all inlet valves and the other two for switching all outlet valves. The respective two fluidic systems are merged by tubes and tube connectors above the valve system. The connector for the air flow supply into the wells is located in the centre. All channels are 1.5 and 2 mm in width and depth, respectively. They widen up to circles, 3 mm in diameter, in the valve areas.

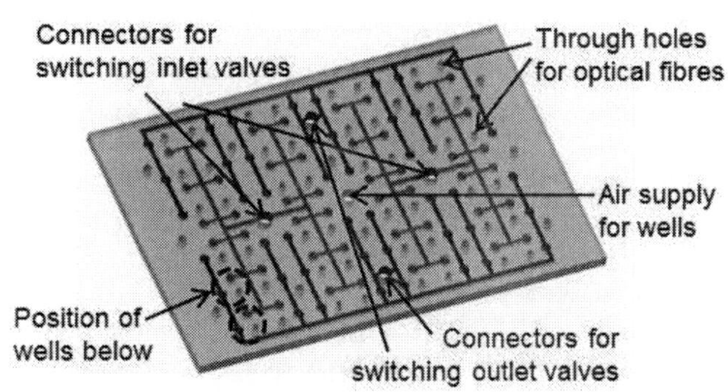

Figure 3. Drawing of the upper PMMA plate comprising the valve systems for switching the inlet (red) and outlet (blue) valves.

The thickness of the switching membrane is 200 µm where it needs being deflected. To focus the pressure provided by the screws onto the sealing areas of the membrane, this area was designed with a width of 700 µm around all channels and valve chambers. In the sealing areas the thickness of the switching membrane is 300 µm and where it is neither employed as a valve membrane nor for sealing, the membrane thickness is 500 µm.

2.2. Air supply system

The micro fluidic system providing the air flow into and out of the wells is formed by intermediate and lower PMMA plate and the upper sealing membrane. All channels are of the same size as the channels of the switching system (1.5 x 2 mm^2 and widening to circles in the valve areas).

To ensure an equal flow through all wells, it was defined by the dimensions of the orifices in the inlet valves. Diameter and length of these orifices are 200 µm and 3 mm, respectively. The length of the orifices in the outlet valves is the same as for the inlet valves but their diameter is 500 µm. This way, it is made sure that the flow resistance of the entire system is defined by the size of the inlet orifices alone. The flow resistances of the orifices of inlet and outlet valves and the feed channels at 37 °C are 94, 2.4, and 5 Pa s/mL, respectively, as calculated with the Hagen Poiseuille equation (see, e.g., [4]).

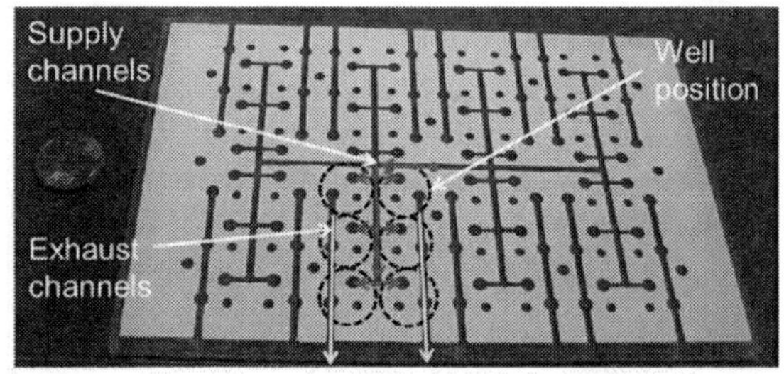

Figure 4. Photo of the lower PMMA plate covered by the upper sealing membrane from silicone.

The exhaust channels lead to the rim of the intermediate plate disposing the used air to the environment.

The upper sealing membrane is 300 µm thick. Its thickness also defines the distance of the valve seats to the non-deflected switching membrane which is 100 µm. In an open valve, the switching membrane is deflected 1.2 mm off the valve seat.

3. Fabrication

The PMMA plates as well as the moulds for the silicone membranes and the aluminium holder were milled by the milling machine M7HP from Datron AG, Mühltal, Germany. The milling tools were purchased also from Datron AG. The drilling heads, 200 µm and 500 µm in diameter, required for drilling the orifices of the valves were purchased from vhf camfacture AG, Ammerbuch, Germany.

The drilling parameters for the 200 µm holes were a feed rate of 400 mm/s and 10,000 rotations per minute while cooling with air at 300 kPa. The parameters for the 500 µm wide holes differed only in the feed rate of 500 mm/s.

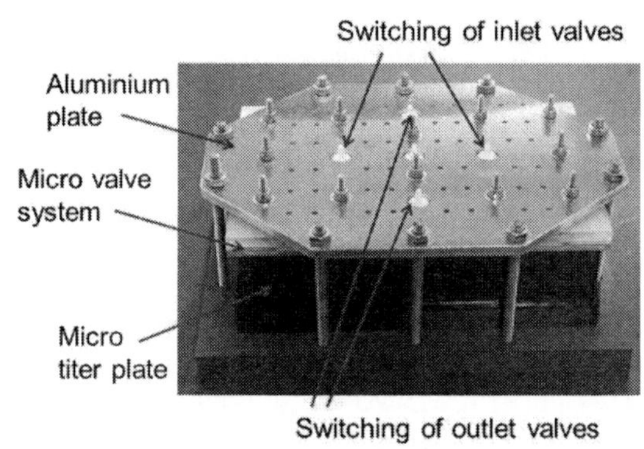

Figure 5. Photo of the micro valve system mounted on top of a micro titer plate.

Milling and drilling was performed in extruded PMMA plates, 4 mm in thickness. Optimum fabrication parameters are a strong function of the material used and its processing (casting or extrusion).

The membranes were cast from the silicone Elastosil RT 625 A mixed at a 9 tol ratio with its hardening component RT 625 B from Wacker Chemie AG, Munich, Germany, in an open air process at room temperature. As moulds PMMA plates were employed which had been milled as described above. After casting the silicone into the moulds they were closed with a PMMA plate and pressed together by tongs squeezing out surplus silicone. After hardening the membranes were removed from the moulds and cleaned.

Four micro systems each with 96 valves have been fabricated up to now and all of their micro valves are opening and closing tightly. One micro titer plate with the micro valve system mounted on top is shown in Fig. 5.

4. Testing

For testing, the gas flow through the wells was changed from air to N_2 and back while the valves were switched and the oxygen partial pressure was measured in the wells. In Fig. 6, the converted oxygen concentrations in four wells are drawn. In phase I, the wells were filled with air and then all valves were closed. Then (phase II), the outlet and 1 s later the inlet valves opened and N_2 was flowing in. In phase III, first the inlet and 1 s later the outlet valves closed. In phase IV, first the outlet and then the inlet valves opened again and air was flooding the wells. During phases I and III no leakage was observed although the N_2 and the air pressure, respectively, already had been applied to the inlet channels. The average difference of oxygen concentration in the wells was 0.15 %.

Mean values of the measured oxygen transfer rates of 48 wells and 4 shake flasks cultivations are shown in Fig. 7a [3]. No significant deviations of the mean oxygen transfer rates are found between RAMOS and μRAMOS (Fig. 7b).

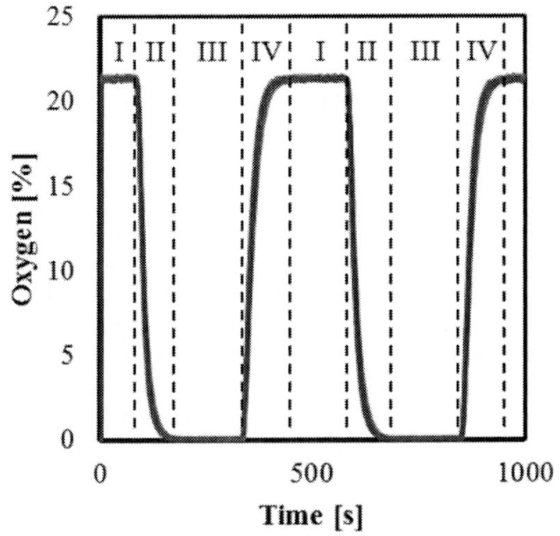

Figure 6. Oxygen concentration in four wells measured during two switching cycles with air and N_2.

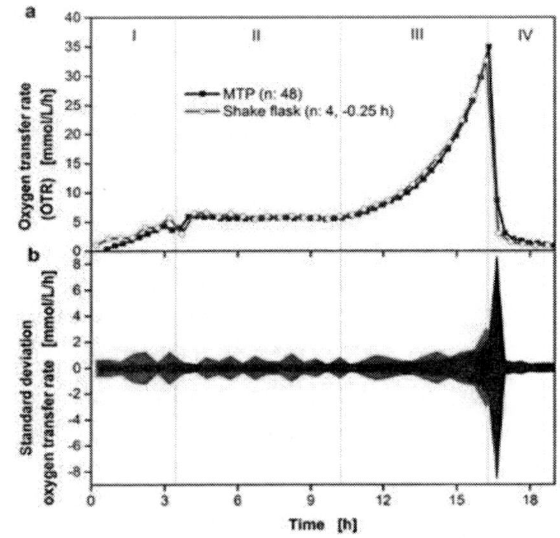

Figure 7. Comparison of *E. coli* BL21 (DE3) pRotHi-YFP cultivations using RAMOS shake flask system and μRAMOS MTP system.

5. Conclusions

Micro valve systems each with 96 micro valves have been manufactured and successfully tested. The measured oxygen transfer rates in all 48 wells deviate by less than 1 mmol/L/h and, therefore, show that reliable investigations of cell cultures can be performed in such devices. These results were achieved by conventional milling of PMMA plates and casting of silicone membranes.

References

[1] T. Anderlei, J. Büchs, "Device for sterile online measurement of the oxygen transfer rate in shaking flasks", Biochem Eng J **7** (2001) 157-162, doi: 10.1016/S1369-703X(00)00116-9.

[2] T. Anderlei, W. Zang, M. Papaspyrou, J. Büchs "Online respiration activity measurement (OTR, CTR, RQ) in shake flasks", Biochem Eng J **17** (2004) 187-194, doi: 10.1016/S1369-703X(03)00181-5.

[3] D. Flitsch, S. Krabbe, W. Schomburg, J. Büchs, "Respiration activity monitoring system for each well of a 48 well microtiter plate", Dechema Himmelfahrtstagung, 2.-4. May 2016 Koblenz, Germany.

[4] W.K. Schomburg, "Introduction to Microsystem Design", (2015), 2nd issue, Springer, Heidelberg, Berlin, doi: 10.1007/978-3-662-47023-7.

Design, numerical simulation and experimental investigation of an SU-8 microgripper based on the cascaded V-shaped electrothermal actuators

Rodica-Cristina Voicu

National Institute for Research and Development in Microtechnologies-IMT Bucharest, Erou Iancu Nicolae str., 126A, 077190, Romania

rodica.voicu@imt.ro

Abstract. This paper presents the numerical simulation results and the experimental investigations of a polymeric microgripper designed using the cascaded V-shaped electrothermal actuators. The microgripper was simulated using electro-thermo-mechanical finite element method (FEM) based on Coventorware 2014 software in order to check the performance of the gripper. As structural material of the microgripper, the SU-8 biocompatible polymer was used during the fabrication process. The metallic micro-heaters were encapsulated in the polymeric actuation structures of the microgrippers to reduce the undesirable out-of-plane displacement of the gripper tips, to reduce the mechanical stress and to improve the thermal efficiency. Experimental testing has been performed to determine the openings of the microgripper tips as function of electrical current. A displacement of the tips of more than 50 μm can be obtained at an electrical current of around 25-26 mA. Over 27-28 mA the heaters are still working but a softening and a damaging status in the polymer were observed.

1. Introduction

Microgrippers used as end-effectors are important tools for handling, pick and place fragile objects. The multiplicity of the microgripper's applications includes biological micro-manipulation such as cells, blood vessels and tissues, micro-assembly of Microelectromechanical Systems (MEMS) and MOEMS components (lenses, fibers) and micro-robotics.

The actuators play a significant part of MEMS that are used for energy conversion, motion generation and force production [1-3]. The V-shaped actuators are widely used for grippers, micro-valves, micro-pumps and other devices. V-shaped electrothermal actuators have the advantages of generating a large force (up to several hundred mN), the simple structure design, a lower dive voltage and a large deformation. Que et al. [2] developed single and cascaded V-shaped electrothermal actuators and present the experimental results. Shen et al. [3] present an analytical model for cascaded V-shaped actuators bringing a complete description of the mechanical performance. Usually the structural material is silicon, polysilicon or aluminium.

A variety of microgrippers have been studied using the SU-8 based electrothermal actuators designed on different configurations such as, U-shape or V-shape. This is proving the interest in the bio-micro-manipulation domain [1, 4-19]. SU-8 is a highly crosslinked epoxy-type photo-patternable polymer which has been used extensively as the preferred polymer material for fabrication of biocompatible devices. The SU-8 polymer has a relatively large coefficient of thermal expansion (CTE) of 52 ppm, good mechanical strength with a modulus of elasticity of around 4.02 GPa and a good thermal stability with a glass transition temperature of 210°C [15] which make it a good polymer material for fabrication of electrothermal actuators. The polymer V-shaped actuators are preferred for the better performance in aqueous medium [4].

In this work, we report the design, numerical simulation results and the experimental investigations of an SU-8 polymeric microgripper designed using the cascaded V-shaped electrothermal actuators. The SU-8 microgripper can be used for bio-manipulation and assembly applications. The microgripper was numerically investigated using the coupled electro-thermo-mechanical simulations based on finite element method (FEM) and using the Coventorware 2014 software in order to confirm the performance of the microgripper. In this paper an improved technological process was realized in order to fabricate the microgripper using a sandwich structure actuator. As structural material of the microgripper, the SU-8 biocompatible polymer was used during the fabrication process. The metallic micro-heaters were encapsulated in the polymeric actuation structures of the microgrippers to reduce the undesirable out-of-plane displacement of the gripper tips, the mechanical stress and to improve the thermal efficiency. Experimental testing and characterisation have been performed to determine the openings of the microgripper tips as function of electrical current. A displacement of the tips of more than 50 μm can be obtained at an electrical current of around 25-26 mA. Over 27-28 mA the heaters are still working but a softening and a damaging status in the polymer were observed.

2. Design

The polymer microgripper was designed using the principle of the cascaded V-shaped electrothermal actuators. The gripper is designed with an initial opening of 50 μm (figure 1). When the gripper is electro-thermally actuated, the tips will close and can handle a micro-object. The total length of the gripper arms used to grasp an object is of 920 μm. The arms were designed with a width of 20 μm. A metallic micro-heater is implanted between two SU-8 layers. The heater wires have a width of 10 μm and were designed to be fabricated using Cr/Au/Cr materials.

The optimized design consists of symmetrically disposed of three material layers. A metallic layer for the heater is implanted between two SU-8 based structure layers having the same thickness, as described previously [17-20]. The thicknesses of the Cr/Au/Cr films were 30 nm/300 nm/30 nm. For each SU-8 layer we obtained a thickness of 9 μm. The details of the fabrication work have been reported also previously in [20].

The microgripper was designed symmetrically with encapsulated metallic micro-heaters in the structural material of the grippers, the SU-8 polymer, in order to reduce the undesirable out-of-plane displacement of the gripper, to obtain the electrical isolation of the heaters and to reduce the mechanical stress that can occur in the structure.

3. Simulation

Finite element modeling using Coventorware was performed to optimize the microgripper design and to evaluate the performance in air. Coupled electro-thermo-mechanical simulations were completed using the MemMech simulator. A simplified 3D microgripper model (figure 2) was meshed using hexahedral elements (Extruded bricks). The number of volume elements was optimized choosing the proper size of the mesh elements using the Split and Merge algorithm.

The materials properties and the surface boundary conditions were set for the simulations. The initial temperature of the whole structure and the temperature of the environment were considered to be T_0=27 °C, with respect to the Coventorware settings requirements for such kind of analyses. The radiation losses from the device are negligible in comparison with the heat loss by convection to the surrounding media [4], since the maximum temperature reached in the microgripper, in order to operate, is lower than 800°C. The air convection coefficient was set to 250 W/m^{2}K [4]. The Young's modulus of the SU-8 were measured with the nanoindentation technique and set for a value of 4.6 GPa. The thermal coefficient of expansion was set at 52 (ppm/°C) and the thermal conductivity at 2 X 10^5 pW/μmK. For the gold layer we used a Young's modulus of 77 GPa reported for thin films. For the thin Cr/Au/Cr films, a temperature coefficient of resistance (TCR) of 0.0014/°C was used [20].

Journal of Physics: Conference Series **757** (2016) 012015 doi:10.1088/1742-6596/757/1/012015

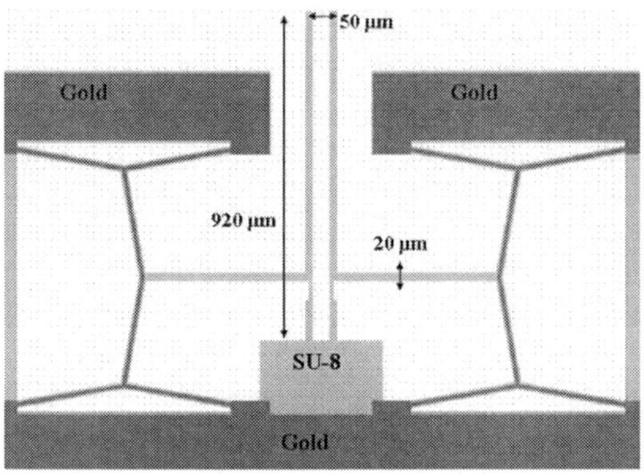

Figure 1. Schematic design of the SU-8 microgripper.

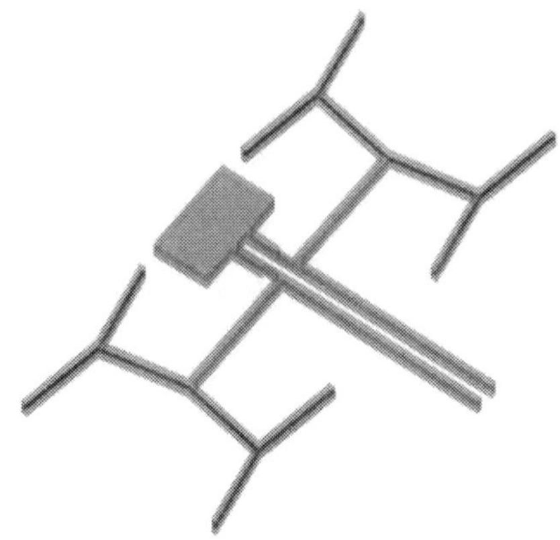

Figure 2. The simplified 3D model of the microgripper with encapsulated heaters in SU-8 polymer: the layers sandwich structure used in FEM simulations (Coventorware2014).

The simulated values of the temperatures reached on the arms and the in-plane deflections of the tips as function of electrical current were presented (figure 3 and 4). The simulation results show that the gripper can work up to a maximal temperature of 160 °C reached in the actuation area, when a complete closing tips is occurred. The gripper can continue to work up to 205 °C in order to obtain a higher displacement or a higher pressure on the griped micro-object. The simulation results indicate that the polymer microgripper can work at low operation temperatures of the tips and with high in plane displacement. A displacement of 25 µm for each microgripper polymeric arm was obtained for a current value of 25 mA. At 25 mA a complete closing tips was observed. The tips temperatures remain at 30-35 °C nearby the initial temperature settings of the media. The microgripper can be optimized by increasing the initial opening.

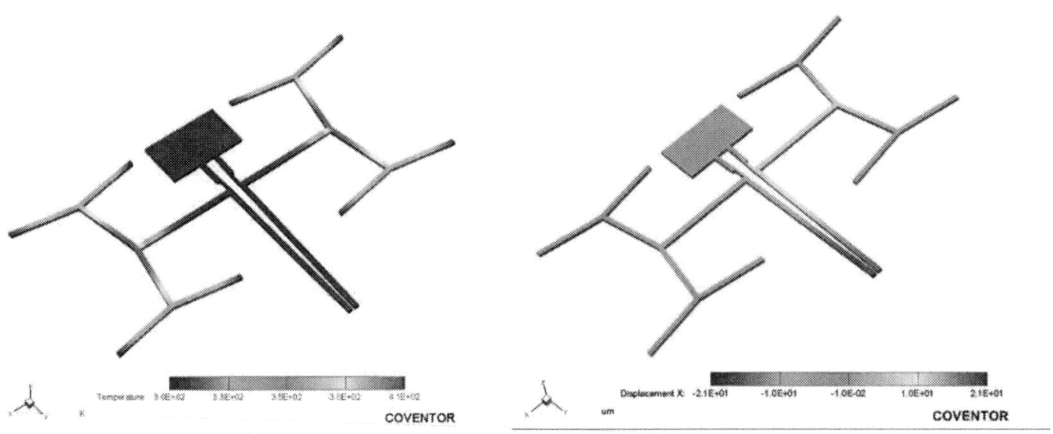

a) b)

Figure 3. FEM coupled electro-thermo-mechanical simulations results: a) The temperatures distribution in the microgripper at 22 mA; b) The in-plane deflections at 22 mA (Coventorware 2014 simulation).

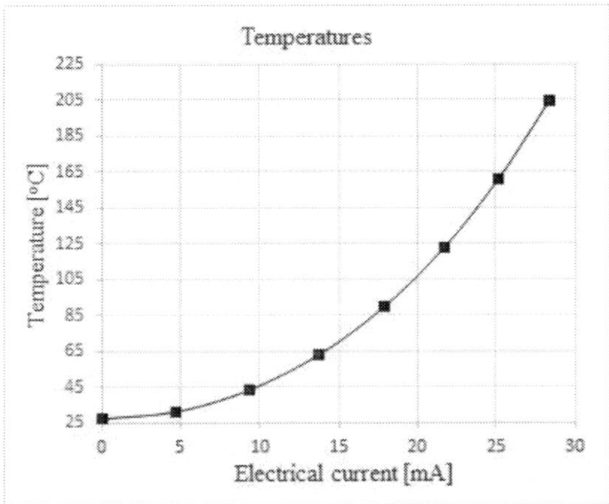

a)

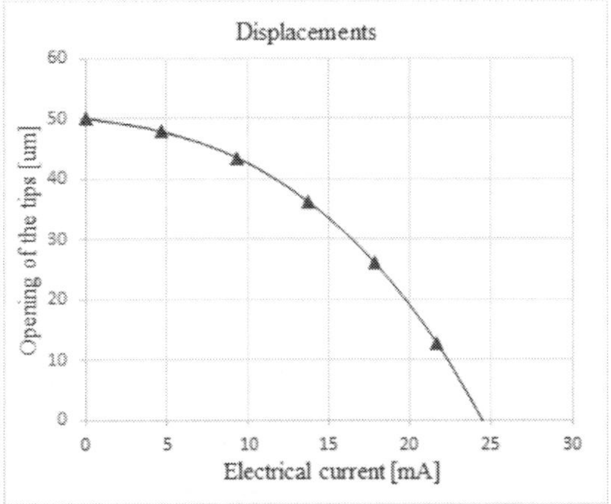

b)

Figure 4. Simulation results: a) The temperature values versus electrical current; b) The jaw displacement versus electrical current;

4. Fabrication and experimental investigations

4.1. Fabrication

For fabrication of the microgrippers we used a 3 mask process and the OmniCoat stripper (MicroChem) in order to completely release the final structures [20].

A thin layer of Omnicoat was deposited on a silicon wafer by spin-coating and baked at 200°C on a hotplate. Then the SU-8 (MicroChem) polymer was deposited by spinning on the wafer in order to obtain a thickness of 9 μm. The wafer was soft-baked at 65°C and at 95°C. The SU-8 layer was then exposed using the first mask. After the exposure, the wafer was post-baked at 65°C and at 95°C and then developed. The polymer structure was hard-baked at 185°C in order to complete cross-linking of the SU-8 polymer. The metal layer consists of a sandwich of Cr/Au/Cr films of 30nm/300nm/30nm thicknesses. The metals were evaporated and the heater and pads were obtained using a lift-off process based on AZ photoresist. The second SU-8 layer was obtained using the same conditions as for the first layer. In this step the access to the metallic pads was created using the third mask for SU-8. The final thermal process of the polymer in this step was the hard-baking for cross-linking of the SU-8 polymer. To release the microgripper structures the Omnicoat layer was developed. Figure 5 shows an optical picture of the fabricated microgripper.

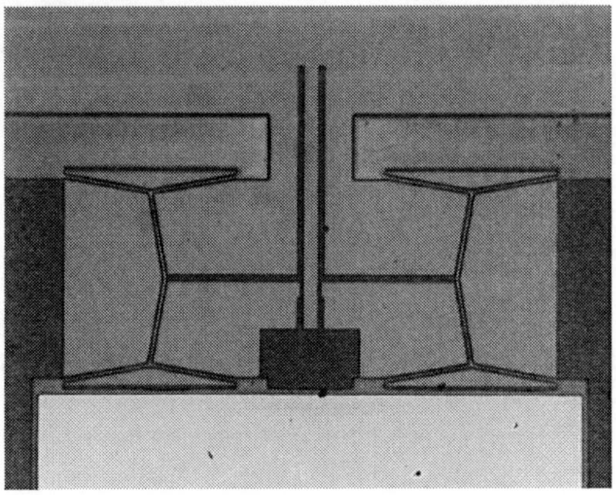

Figure 5. Optical microscope picture of the fabricated electrothermal SU-8 microgripper

4.2. Experimental Testing

In order to validate the model, the experiments were performed in air. Each structure was fixed manually one a silicon substrate. The in-plane deflection change with drive current was observed with an optical microscope and a camera with the associated viewing software. For each actuation step, the displacements of the gripper tips were measured using the optical images. Figure 6 show the first and the last stage of the opening-closing tips of the microgripper. The out of plane displacement is not observed during experiments and due to the simulation results is less than 54 nm. Over 27-28 mA the polymer SU-8 become softer and the device is damaged due to the increased temperature over Tg of 210 °C (figure 7).

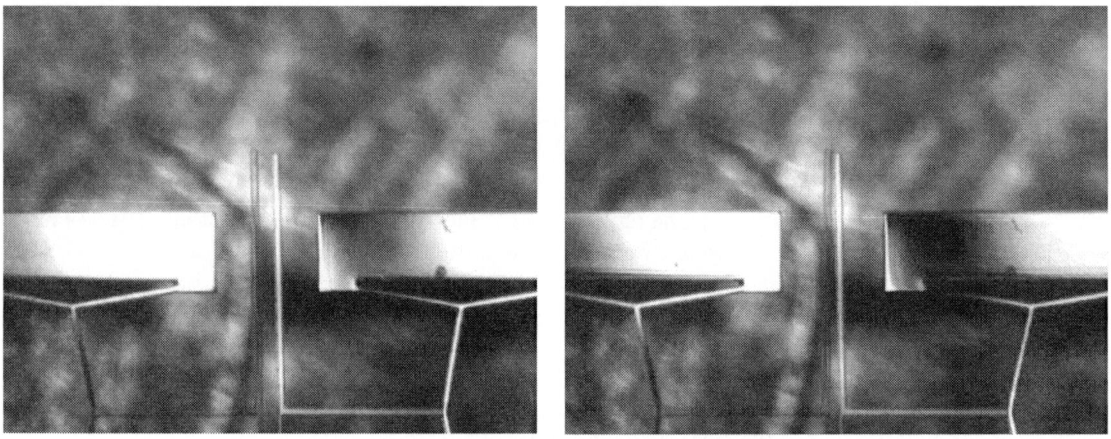

a) b)

Figure 6. Optical images of the actuated microgripper and the tips in the open and close stage: a) the initial stage of the microgripper tips with the initial opening of 50 μm; b) closing tips stage at 24 mA.

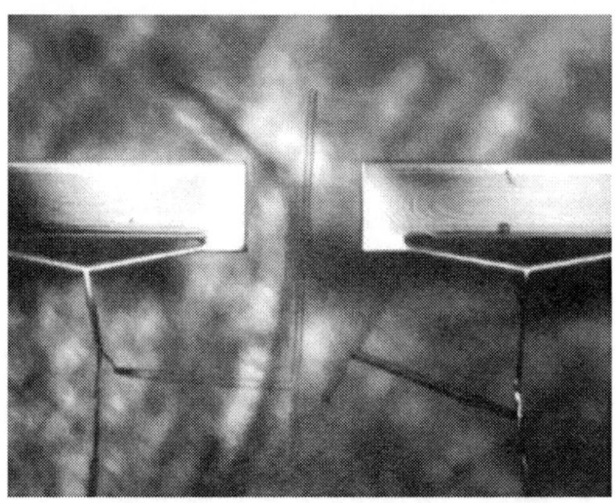

Figure 7. Optical image of the damaged SU-8 microgripper at 28 mA.

5. Conclusions

An SU-8 electrothermally actuated microgripper based a cascaded V-shaped configuration were designed, numerically and experimentally investigated. The results show that the microgripper can work in air in his maximal stage for an electrical current of 25-26 mA and for a highest temperature of the tips of 35 °C. Over 27-28 mA the device is damaged due to the SU-8 transformations over the glass transition temperature reached in the structure. Further investigations will be the comparison between the simulation results and the measurements.

Acknowledgments
This work was supported by a grant of the Romanian National Authority for Scientific Research and Innovation, CCCDI – UEFISCDI, project number 22/2016, within PNCDI III, ERA-MANUNET-II-Robogrip project "Microgrippers as end-effectors with integrated sensors for microrobotic applications (ROBOGRIP)" (2016-2017).

References

[1] Jia Y and Xu Q 2013 MEMS microgripper actuators and sensors: The state-of-the-art survey *Recent Patents on Mechanical Engineering* **6** 2
[2] Que L et al 2001 Bent-beam electrothermal actuators. Part I: Single beam and cascaded devices *J. Microelectromech. Syst.* **10** pp. 247–254
[3] Shen X and Chen X 2013 Mechanical Performance of a Cascaded V-shaped Electrothermal Actuator *Int. J. Adv. Robot. Syst.* **10** p. 379
[4] Colinjivadi K S, Lee J-B, Draper R 2008 Viable cell handling with high aspect ratio polymer chopstick gripper mounted on a precision manipulator *Microsyst. Technol.* **14** pp. 1627-1633
[5] Nguyen N, Ho S and Low C 2004 A polymeric microgripper with integrated thermal actuators *J. Micromech. Microeng.* **14** 7 pp. 969-974
[6] Kim K et al 2004 Metallic microgripper with SU-8 adaptor as end-effectors for heterogeneous micro/nano assembly applications *Microsyst. Technol.* **10** 10 pp. 689-693
[7] Hamedi M et al 2013 Design, analysis and fabrication of silicon microfixture with electrothermal microclamp cell *Microelectron. Eng.* **111** pp. 160-165
[8] Wang Z, Shen X and Chen X 2015 Design, modeling, and characterization of a MEMS electrothermal microgripper *Microsyst. Technol.* **21** 11 pp. 2307-2314
[9] Voskerician G et al 2003 Biocompatibility and biofouling of MEMS drug delivery devices *Biomaterials* **24** pp. 1959–1967
[10] Somà A et al 2014 Design and building up of an electro thermally actuated cell microgripper *Proc. Recent Advances in Mechanical Engineering and Mechanics* pp. 125-130
[11] Chu Duc T et al 2008 Electrothermal Microgripper with Large Jaw Displacement and Integrated Force Sensors *J. of Microelectromech. S.* **17** 6 pp. 1546-1555
[12] Solano B and Wood D 2007 Design and testing of a polymeric microgripper for cell manipulation *Microelectron. Eng.* **84** 5-8 pp. 1219-1222
[13] Zhang W et al 2008 A polymer V-shaped elctrothermal actuator array for biological applications *J. Micromech. Microeng.* **18** 075020
[14] Zhang R et al 2012 A multipurpose electrothermal microgripper for biological micro-manipulation *Microsyst. Technol.* **19** 1 pp. 89-97
[15] MicroChem 2016 SU-8 2000 Permanent Epoxy Negative Photoresist [Online] http://www.microchem.com/
[16] Lorenz H et al 1997 SU-8: a low-cost negative resist for MEMS *J. Micromech. Microeng.* **7** 3 pp. 121-124
[17] Voicu R et al 2007 Method for overcoming the unwanted displacements of an electro-thermally actuated microgripper *Proc. Int. Conf. on Multi-Material Micro Manufacture (4M)* 2007 pp. 39-42
[18] Voicu R and Műller R 2013 New electro-thermally actuated micromanipulator with optimized design and FEM simulations analyses *Proc. Conf. Design, Test, Integration & Packaging of MEMS/MOEMS (DTIP)* pp. 1-6
[19] Voicu R and Műller R 2013 Design and FEM analysis of a new micromachined electro-thermally actuated micromanipulator *Analog Integr. Circ. Sig. Process* **78** 2 pp. 313-321
[20] Al-Zandi M H, Voicu R-C, Muller R, Wang C 2016 Testing and Characterisation of Electrothermal Microgrippers with Embedded Microheaters, *Proc. Conf. Design, Test, Integration & Packaging of MEMS/MOEMS (DTIP), in press*

Journal of Physics: Conference Series **757** (2016) 012016 doi:10.1088/1742-6596/757/1/012016

Development and fabrication of a very High-g sensor for very high impact applications

Piotr Mackowiak[1], Biswajit Mukhopadhyay[1], Xiaodong Hu[2], Oswin Ehrmann[1, 2], Klaus-Dieter Lang[1, 2], Stefan Linke[3], Anthony Chu[3], Ha-Duong Ngo[1, 4]

[1] Fraunhofer Institute for Reliability and Microintegration, [2] Technische Universität Berlin, Germany, Germany, [3] TE Connectivity [4] University of Applied Sciences Berlin, Germany

piotr.mackowiak@izm.fraunhofer.de

Abstract. In this paper we present the first time our development work of a family of silicon on insulator (SOI)- based piezoresistive MEMS very high G sensors for measurement of accelerations up to 60.000 g. Two sensors have been realized, one for 20.000g and one for 60.000g.

1. Introduction

Piezoresistivity is first discovered by Lord Kelvin in 1856, it is defined strictly to whose electrical resistivity changes with applied mechanical stress or stain. Many material exhibit piezoresistivity when strained, but the effect is most pronounced in semiconductors. Piezoresistive acceleration sensor is one of attempts by using the piezoresistive effect. The sensor in this work does have a silicon beam with 4 integrated piezoresistors. The sensor geometry of the sensor is optimized for high acceleration range up to 60.000g. Figure 1 shows a sketch of the top view of the sensor design.

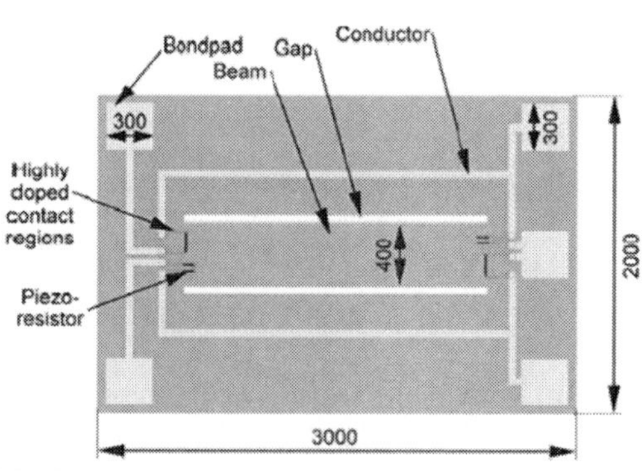

Figure 1: Sketch of top view of sensor design featuring one double-clamped beam carrying one transversal and one longitudinal piezoresistor on each end of the beam. Piezoresistors and conductors are connected to form an open full Wheatstone bridge

2. Simulation

A piezoresistive sensor with a Wheatstone bridge transforms mechanical strain into a change of output voltage when the bridge is voltage feeded. The output voltage changes according to equation (Eq. 1).

$$\Delta U = U0 \cdot \varepsilon\, k \qquad \text{(Eq. 1)}$$

(with ΔU representing the change of output voltage and and U0 the supply voltage of the bridge). To increase the sensitivity the piezoresistive gauge factor k and the mechanical strain ε can be increased. As the change of the gauge factor in silicon depends on the doping concentration but high doping concentration is not desirable because of the increase of thermal coefficient of resistivity (TCR) According to beam theory it is very effective to change geometrical parameters in order to obtain higher mechanical strain.

$$\varepsilon = (\rho \cdot f \cdot g)/(2 \cdot E \cdot h) \cdot l^{\wedge}2 \qquad \text{(Eq. 2)}$$

(With density ρ and Young's modulus E being fixed material parameters for a silicon sensor and load f and acceleration due to gravity on Earth g describing the load case) Only the geometrical parameters of the beam thickness h and beam length l can be changed to increase the strain.

The design of the sensor was evaluated using Finite Element simulation. The simulation was performed using a bottom up approach in ANSYS: Figure 2 shows the model used for the simulation. Because of the symmetry a half model could be used. The element size in the beam was set to 20µm. All degrees of freedom of the model was limited at the ground of the sensor and one at the symmetric areas. Figure 3 shows the strain on top of the beam with 60.000g applied on a 1000µm long beam. Strain strongly increases at the border of the beam. To find the maximum signal output an integral of the strain was calculated using the piezo resistor length as integration length. Figure 4 shows the simulation results of the strain depending on the beam length. The strain clearly increases with increasing beam length.

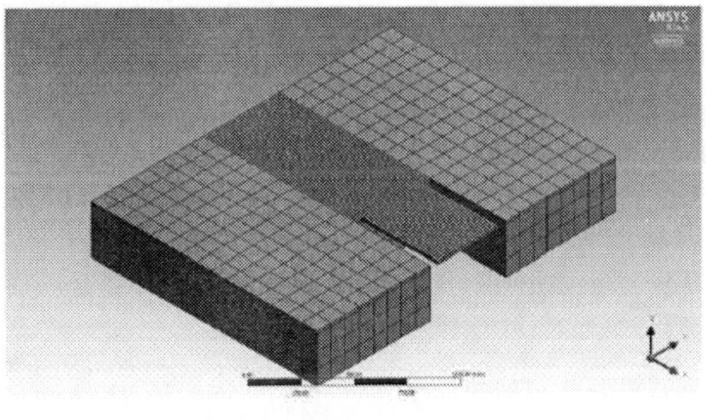

Figure 2 Meshed simulation model for FEA with fine meshing in the beam region

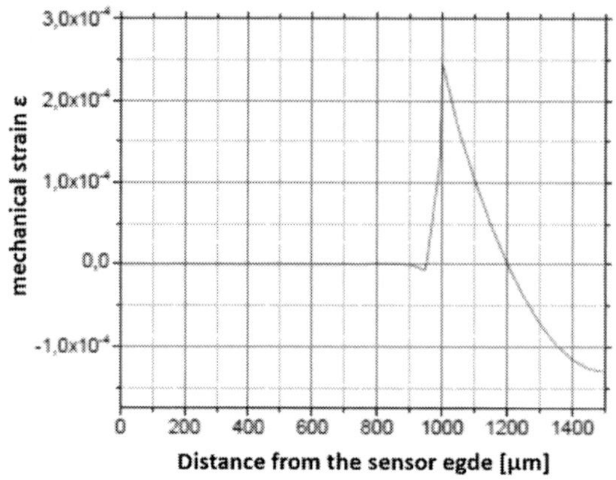

Figure 3 Mechanical Strain along the beam for a 1000 µm long beam accelerated with 60.000g.

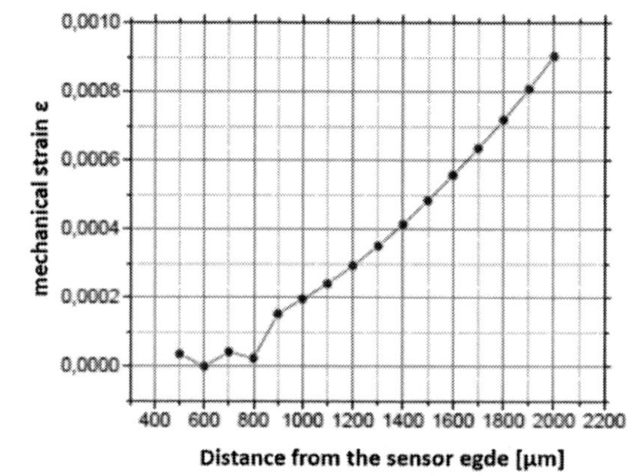

Figure 4 Mechanical strain at 60.000g in longitudinal direction for different beam length

Figure 5 summarizes the simulation results. It shows the calculated output voltage (V_{input}=5V) as a function of the beam length and the acceleration g. The output voltage rises with increasing beam length and with increasing acceleration. Limiting factor of the output voltage is the fracture stress of the silicon beam.

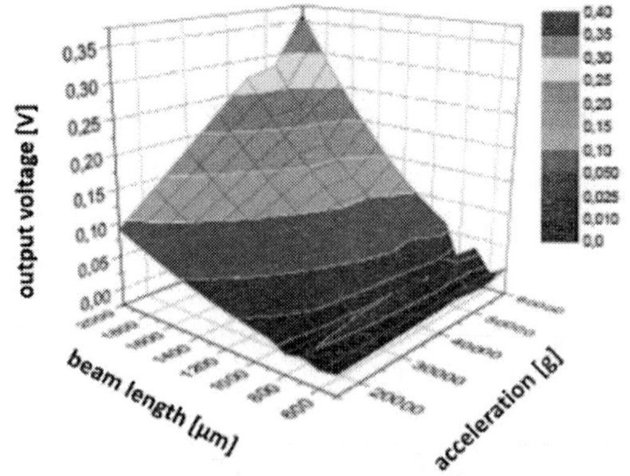

Figure 5 Sensor output voltage as a function of beam length and acceleration. The beam length was varied from 500µm to 2000µm and the acceleration from 15.000g to 60.000g

While the sensor sensitivity increases with increasing the beam length the resonance frequency decreases. The sensor design is always a compromise between the sensor sensitivity and its resonance frequency. In Figure 6 the simulation results of a modal analysis are shown. It can be clearly seen that the modal frequencies decrease with increasing beam length.

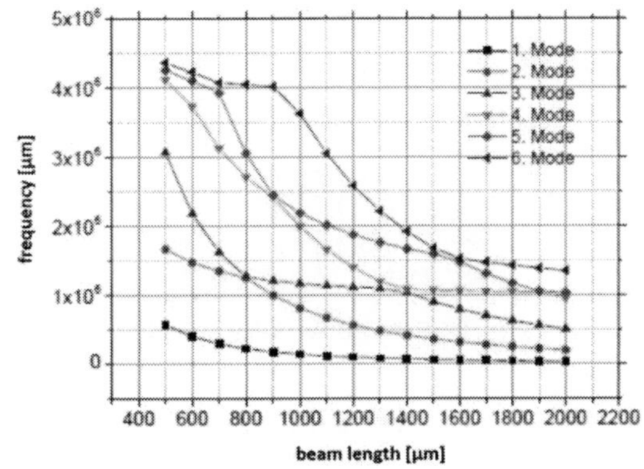

Figure 6 Modal analysis for the acceleration sensors for beam lengths from 500µm to 2000µm for the first six modes.

3. Device Fabrication

The sensors are produced from silicon on insulator wafers with a device layer thickness of 20µm. First implantation of the piezoresistors is performed with the help of a thin oxide layer. After the implantation the Wafer is passivated with a thermal Silicon oxide and a LPCVD Silicon nitride layer. These passivation layers are opened locally using lithography and dry etching before sputtering Aluminum. Afterwards this Aluminum is structured using lithography and chemical wet etching. Next the passivation layers on the backside are structured using dry and wet etching before a membrane will be etched from the backside using DRIE. The buried oxide layer is used a stopping layer. For processing the aluminum on the backside of the wafer is removed and the wafer is mounted on a carrier wafer. Using a further lithography step the passivation layers on top and the device layer is etched to generate slots which makes a double side clamped beam out of the sides clamped membrane. The carrier wafer is removed and anodic bonding is performed.

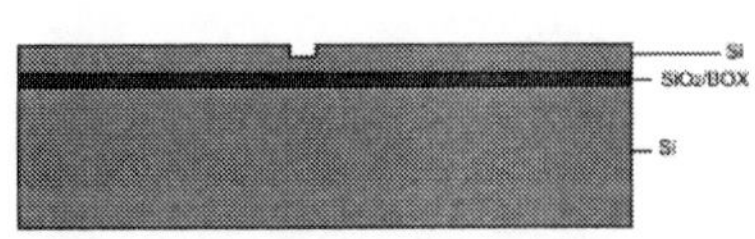

a) SOI Wafer with alignment marks

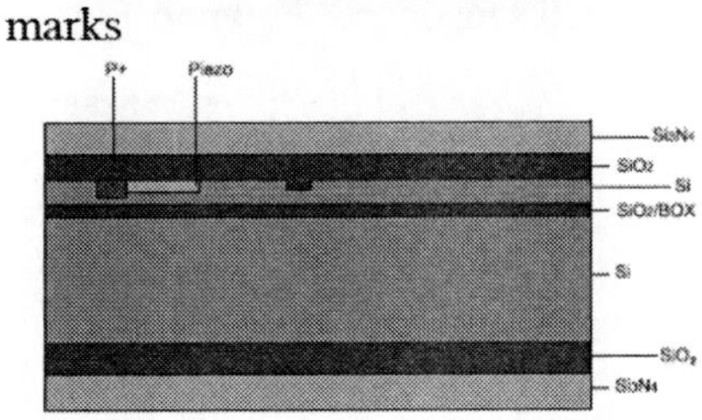

b) After implantation of piezoresistors silicon oxdie and silicon nitride passivation layer was deposited

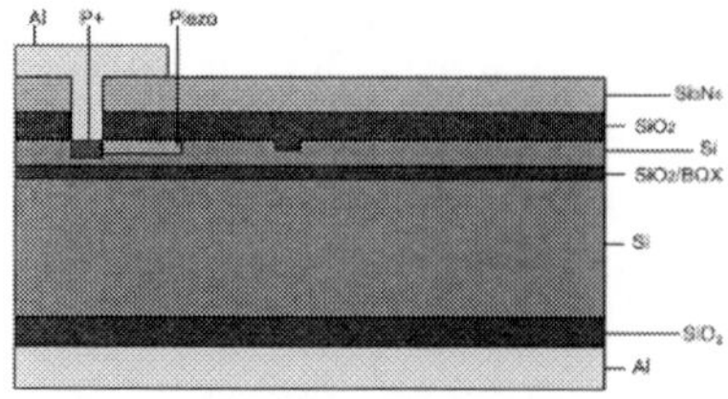

c) Passivation layer was opened using dry and wet etching to contact the piezoresistors using aluminium.

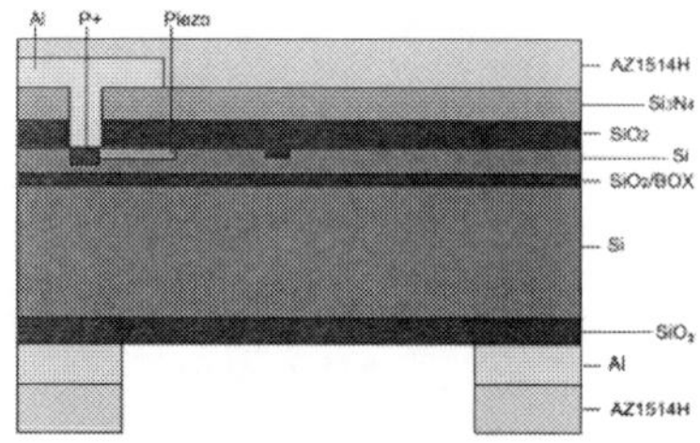

d) Frontside protection using photoresist and structuring aluminium on the backside using lithography.

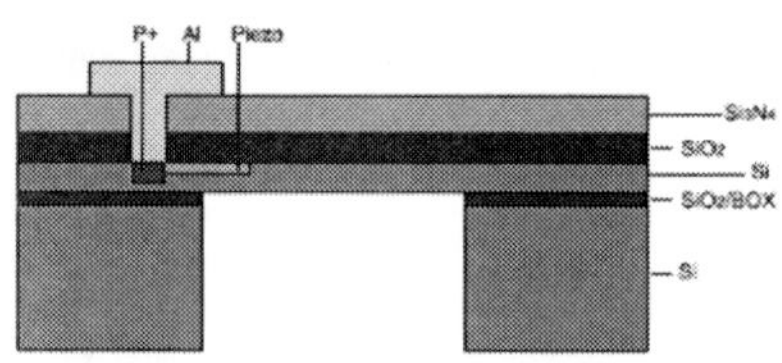

e) Structuring the oxide on the backside and etching the membrane using DRIE

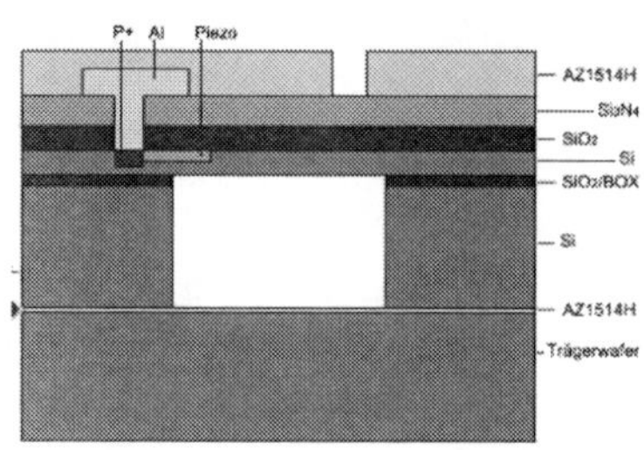

f) Bonding temporarily to a carrierwafer and using lithography to generate the slots on the frontside generating the beam.

Figure 7 Main steps of device fabrication process.

The realized sensor element and packaged sensor are shown in Figure 8.

a) b)

Figure 8 a) A realized MEMS high g sensor element (3x2x1mm3) b) Packaged high-g sensor.

4. Device Characterization

In order to generate the necessary accelerations, a Shock Wave Bare and a Polytech Vibrometer were used as an acceleration reference. The MEMS sensor element was attached to a stainless steel test fixture using standard die attach method. The die was wire bonded to the PCB where the output wires were soldered. A cover was added for protection. The test fixture was mounted to the end of the Shock Wave Bar. It also serves as the reference for the laser. Figure 9 shows the time domain by excitation at about 60.000 g. The sensor wave form tracks the laser signal very well.

A second approach for sensor characterization is to calculate the beam deflection caused by the acceleration of the sensor and use a micro actuator to deflect the beam and measure the resistance change. This allows linearly increase the deflection and observe the resistive change of the Wheatstone bridge.

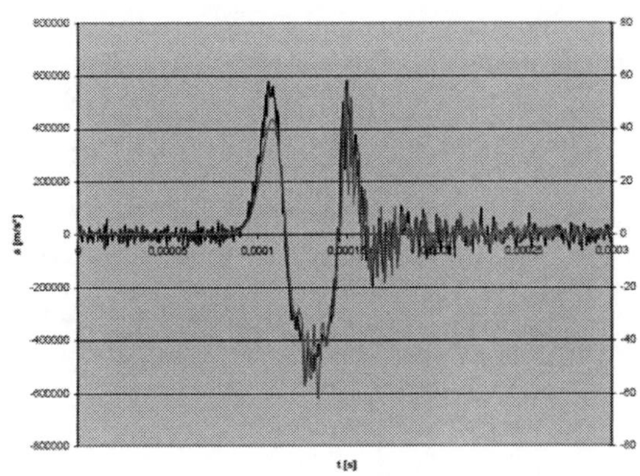

Figure 9: Test results of the 60.000 g sensor. Purple: MEMS sensor signal. Blue: reference sensor.A subsection

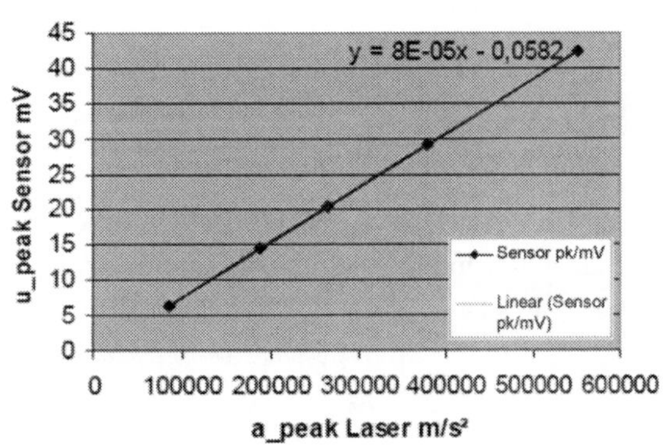

Figure 10: Calculated linearity of the sensor

Figure 11 shows the micro actuator mounted on a three axis table. With the three axis stage the actuator needle was positioned at the center of the beam of the sensor. The actuator is able to deflect the beam up to 30µm. Figure 12 shows the schematic of the measurement setup to deflect and measure the resistive change.

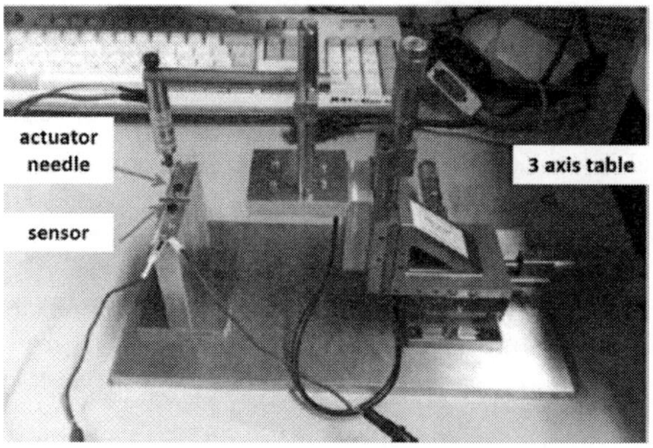

Figure 11 Photograph of the micro actuator used to displace the beam of the sensor. On the left side the sensor I located and the needle deflects the sensor beam. The position of the micro actuator needle can be positioned using the 3 axis table.

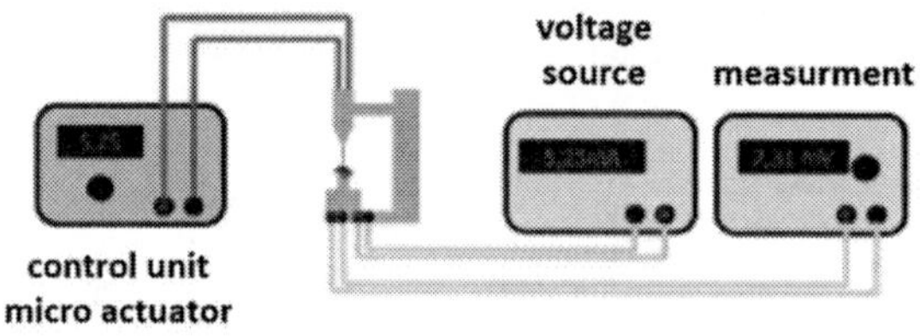

Figure 12 Schematic of the measurement setup to deflect and measure the resistive change of the sensor.

5. Conclusion

We present first time very high impact sensors for impact up to 60.000g. Before the sensor realization intense simulation work has been performed to find the best sensor design. The simulation results show a relationship between beam length and acceleration. The calculated output voltage increases with increasing acceleration and beam length.

The characterization of the realized sensor show a good match with the simulation results. The deviation of the measured output signal and the calculated output voltage is lower than 10mV for a beam length of 900µm and 60.000g. The sensors can be used in aerospace applications or in control and detection of impact levels.

Acknowledgement

Authors would like to thank TE Connectivity for financial support of the work.

Journal of Physics: Conference Series **757** (2016) 012017 doi:10.1088/1742-6596/757/1/012017

Characterizing coupled MEMS resonators with an electrical resonator

Guowei Tao[1,2], **Bhaskar Choubey**[1]

[1] Department of Engineering Science, University of Oxford, Oxford OX1 3PJ, UK
[2] Vlatacom d.o.o., Bul. Milutina Milankovica 5, 11000 Belgrade, Serbia

E-mail: guowei.tao@eng.ox.ac.uk

Abstract. Rapid development in micro/nano fabrication has enabled the shrinking of MEMS devices and the ability to fabricate them in large arrays. However, process variations and device mismatch have also raised testability issues in the MEMS industry. MEMS resonators have been coupled to simplify the characterization of the fabrication process and device performance using their collective behaviour. Perturbation analysis using eigenvalues can therefore be applied to extract the system matrix of coupled resonators. We propose a new way of perturbation analysis by coupling an electrical resonator to an array of MEMS resonators. The electrical resonator is simple in structure and easy to readout. It can also precisely control the amount of perturbation based on two available techniques. Coupling between MEMS resonators and electrical resonator opens a new window for process characterization, device testing, material characterization, as well as large sensors array actuation.

1. Introduction

Process induced variabilities have been widely seen in MEMS fabrication [1, 2], which are caused by environment factor, temperature difference, as well as uncertainty during lithography or etching. Variabilities would degrade sensor performance, increase device mismatch, reduce yield and add cost to characterization. Device characterization has become a growing concern for the MEMS industry as it can consume up to 70% of the manufacturing cost. Hence there is a need for an accurate and inexpensive way to monitor fabrication processes and characterize device properties. One way of enhancing testability is to optimise MEMS design at the beginning considering the presence of variability [1]. This method, however, is prone to device mismatch, even if they are on the same wafer with identical shape and geometry design. Therefore, it is important to regularly monitor these fabricated devices or test structure. Another solution is to test MEMS devices at the wafer level and predict their performance at the package level [2].

MEMS resonators have been used for extensive applications spanning from mass sensing [3] to acceleration sensing [4]. With the continuing reduction in feature size and advancement in micro/micro fabrication techniques, two or even more resonators can be fabricated on the same substrate thereby providing multi degree-of-freedom or multi function sensing. Individual sensors hence need to be fully characterized before packaging. However, for a large array of sensors, testing each element individually would significantly increase connection wires, bond pads, and the bandwidth for signal processing. Coupling these individual resonators emerges as a simple and economical way to record the collective behaviour of all sensors and hence

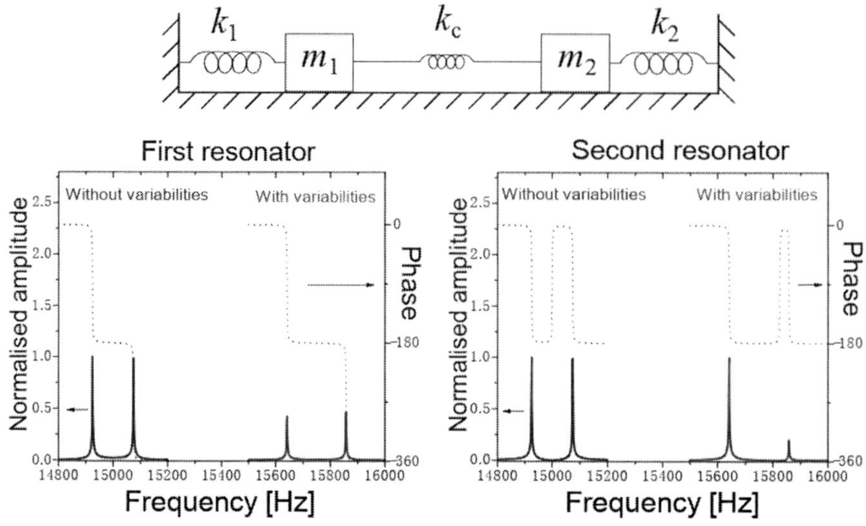

Figure 1. Simulated responses of 2 coupled resonators with/without process variabilities.

characterize them [5]. Resonators can be coupled intentionally through mechanical string or electrostatic force. They can also be coupled by simply using the inherent parasitic effect [6].

2. System matrix reconstruction using inverse eigenvalue analysis

2.1. System matrix, eigenvalue and eigenvector

An array of n nearest-neighbour coupled resonators can be modelled by differential equations

$$\ddot{\mathbf{x}} + \mathbf{B_n}\dot{\mathbf{x}} + \mathbf{S_n}\mathbf{x} = \mathbf{F}, \tag{1}$$

with $\mathbf{x}$ being the displacement vector, $\mathbf{F}$ the driving force and $\mathbf{B_n}$ the damping matrix. $\mathbf{S_n}$ denotes the system matrix of the array. In the ideal situation without variabilities, all resonators have identical spring constant k_s, mass m and coupling strength k_c in between, hence

$$\mathbf{S_n} = \begin{bmatrix} (k_s+k_c)/m & -k_c/m & \cdots & 0 \\ -k_c/m & (k_s+2k_c)/m & \cdots & 0 \\ \vdots & \vdots & \ddots & \vdots \\ 0 & 0 & -k_c/m & (k_s+k_c)/m \end{bmatrix}, \tag{2}$$

whose eigenvalues μ_i represent the eigenfrequencies (i.e. resonance frequencies) f_i as $\mu_i = (2\pi f_i)^2$. The eigenvectors are the amplitude ratio of all resonators at different modes.

Figure 1 simulates 2 coupled resonators, with each having a nominal natural resonance frequency of 15 KHz ($Q = 10000$, 1% weak coupling). In practice, the inevitable process variations and device mismatch would modify $\mathbf{S_n}$ slightly, hence the actual responses are always different, as can be seen from the responses in figure 1 when considering variabilities.

Eigenvector change has been observed in figure 1 due to device mismatch. However, since the coupling strength can hardly be measured, it is impossible to know the amount of change through eigenvectors. Eigenvectors are also very unstable and do not provide frequency information. Furthermore, they require amplitudes measured from every resonator in order to calculate their ratio, hence increasing the complexity and cost of readout.

In contrast, eigenvalues can be measured from the response of any one single resonator due to the collective behaviour of the coupled system. They can be used to extract full information of the system matrix after a perturbation has been introduced to the array, and hence fully characterize coupled MEMS resonators. In this paper, we propose two inverse eigenvalue techniques to perform perturbation analysis using an electrical resonator.

2.2. Perturbation analysis

With variabilities taken into account, the system matrix in equation 2 then has $2n-1$ unknown parameters, among which n are the diagonal elements containing spring constant and mass information, whereas $n-1$ are off-diagonal elements related to the coupling strength. Measuring one set of n eigenvalues is therefore impossible to extract all $2n-1$ elements. Another set of eigenvalues is hence needed by perturbing the array slightly. There are two techniques to obtain a second set of eigenvalues and reconstruct the system matrix.

The first technique perturbs one diagonal element of the system matrix, by changing the spring constant or mass of the related resonator. This change would give another n eigenfrequencies that could be measured from the response of any one of the resonators, and hence n new eigenvalues μ_i^*. The eigenvector $x_{i,n}$ for the nth eigenvalue is calculated by:

$$x_{i,n}^2 = \frac{\mu_i^* - \mu_i}{s^*} \prod_{j=1, i\neq j}^{n} \frac{\mu_i - \mu_j^*}{\mu_i - \mu_j} \qquad i = 1, \cdots, n. \tag{3}$$

where s^* is the difference between the traces of the initial and the changed system matrix. The initial $\mathbf{S_n}$ can be extracted from the eigenvalues and eigenvectors using Lanczos' algorithm [5].

The second technique couples an additional resonator to the array and hence adds a new element to the existing array of resonators. Another set of eigenvalues λ can be measured from such new array. This technique first computes the weights ω_i from λ and μ [7]:

$$\omega_i = \prod_{j=1}^{n} \frac{\lambda_i - \mu_j}{\lambda_i - \mu_{j+1}} \qquad i = 1, \cdots, n+1. \tag{4}$$

A set of orthogonal polynomials are then calculated, the first two of which are

$$p_0(\lambda_i) = 1, \qquad p_1(\lambda_i) = \lambda_i - \frac{\sum_{j=1}^{n+1} w_j \lambda_j}{\sum_{j=1}^{n+1} w_j} \qquad i = 1, \cdots, n+1. \tag{5}$$

Three-term recurrence relationships are then used to derive the diagonal elements d_i and off-diagonal elements o_i of $\mathbf{S_n}$, as well as the subsequent orthogonal polynomials $p_k(\lambda_i)$:

$$d_{n+2-k} = \frac{\sum_{i=1}^{n+1} \lambda_i w_i p_{k-1}^2(\lambda_i)}{\sum_{i=1}^{n+1} w_i p_{k-1}^2(\lambda_i)}, \qquad o_{n+1-k}^2 = \frac{\sum_{i=1}^{n+1} w_i p_k^2(\lambda_i)}{\sum_{i=1}^{n+1} w_i p_{k-1}^2(\lambda_i)}, \tag{6}$$

$$p_k(\lambda_i) = (\lambda_i - d_{n+2-k}) p_k(\lambda_i) - o_{n+2-k}^2 p_{k-2}(\lambda_i) \qquad k = 2, \cdots, n-1. \tag{7}$$

3. Characterization method

3.1. Electrical resonator

Inside the dotted box of figure 2 shows a simple electrical resonator that can be represented by

$$\ddot{x}_e + \frac{1}{R_2 C}\dot{x}_e + \frac{1}{R_1 R_3 C^2} x_e = \frac{1}{R_3 R_f C^2} F_e, \tag{8}$$

where F_e is the driving voltage and x_e is the output voltage. The centre frequency and quality factor can be easily adjusted. The above equation shares the similar form as equation 1, making it

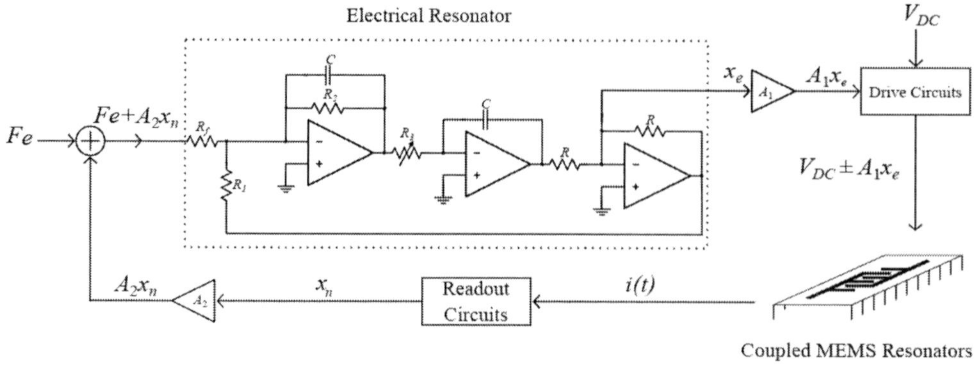

Figure 2. Coupling an electrical resonator to an array of MEMS resonators.

possible to couple the electrical resonator to MEMS resonators. To electrically couple them, the driving mechanism of the MEMS resonators needs to be voltage controlled, e.g. via electrostatic force or converse piezoelectric effect. Conversely, the sensor readout needs to be converted to voltage signal to feed back to the circuit.

Figure 2 shows an example of coupling an electrical resonator to an array of electrostatically driven MEMS resonators. These resonators are capacitive type with opposing comb structures. The electrical circuit is coupled to the array of MEMS resonators by adding its output voltage signal to electrostatic force to drive one of the coupled MEMS resonators. Conversely, the mechanical output of the same MEMS resonator, in this case a current signal, is converted to a voltage signal to feed back to the circuit. Once coupled to an array of n MEMS resonators, readouts from both the electrical circuit and the mechanical resonators contain $n+1$ resonant peaks. These can be measured from the response of the mechanical resonator or simply from the electrical resonator. The amount of the feed in and feed back signals can be any different values, as long as the resonance peaks are detectable. It can be proven mathematically that unequal feed in/back amount would not affect the diagonal elements of the system matrix [7]. Only the diagonal elements contain useful information.

Perturbation can be carried out using both techniques mentioned in section 2.2. For the first technique, perturbation can be performed by coupling the electrical resonator and changing its centre frequency. For the second technique, the array can be perturbed by connecting/disconnecting the electrical resonator.

3.2. Characterizing an example of ten coupled MEMS resonators

We simulated ten nearest-neighbour coupled MEMS resonators to demonstrate both techniques using an electrical resonator. Each resonator has a nominal natural resonance frequency of 15 KHz and 4% coupling ratio in between. We intentionally modified $\mathbf{S_n}$ by adding a Gaussian distributed variability ($\sigma = 5\%$) to the spring constants. The additional electrical resonator to be coupled to such array has a centre frequency of 15 KHz. The feed in and feed back amount for the electrical resonator are intentionally set to be different to simulate real situation.

We designed a series of experiments to characterize such MEMS array using both techniques. First, before coupling the electrical resonator, ten eigenfrequencies were recorded from any one of the ten coupled MEMS resonators. Second, the electrical resonator was coupled to the last element of the ten MEMS resonators, forming a new eleven-element array. Eleven new eigenfrequencies were hence recorded. Third, the centre frequency of the electrical resonator was then changed from 15 to 16 KHz, thereby providing another set of eleven eigenfrequencies.

The acquired data for the above three steps are shown in table 1. From the eigenfrequencies

Table 1. Eigenfrequencies recorded for different modes.

Mode	1	2	3	4	5	6	7	8	9	10	11
Step 1	13231	13839	14626	15139	15320	15351	15844	16492	16914	17871	–
Step 2	13231	13839	14626	14959	15139	15320	15351	15844	16492	16946	17875
Step 3	13231	13839	14626	15139	15320	15351	15844	15928	16492	16975	17877

Table 2. Nominal, actual and calculated natural resonance frequencies.

Resonator	1	2	3	4	5	6	7	8	9	10
Nominal	15000	15000	15000	15000	15000	15000	15000	15000	15000	15000
Actual	15403	16375	13306	15647	15239	14019	14675	15257	17684	17077
1st technique	15403	16365	13313	15652	15239	14019	14675	15257	17684	17077
2nd technique	15403	16373	13307	15648	15239	14019	14675	15257	17684	17077

obtained through step 1 and step 2, the second technique can be applied, giving a set of natural resonance frequencies of the ten MEMS devices. From the eigenfrequencies obtained through step 2 and step 3, the first technique can be applied, giving a similar set of natural resonance frequencies. These two sets of frequencies are compared with the actual natural resonance frequencies, as shown in table 2, with the standard deviations of frequency errors as low as 0.03% and 0.005% for the first and second technique respectively. These values match well with the actual natural resonance frequencies. The slight difference is caused by numerical errors.

4. Conclusions

A new inverse eigenvalue analysis approach based on an electrical resonator has been revealed to characterize coupled MEMS resonators. With the help of the electrical resonator, perturbation analysis can be easily achieved. We have proposed two different techniques using such electrical resonator, both of which can provide full information about the system matrix. Coupling an additional electrical resonator to the MEMS array can be used for a wide range of applications, including process monitoring, device testing, material characterization, and single-input-single-output sensors actuation. Future work will be dedicated to the fabrication of a large array of coupled resonators to experimentally verify both techniques. Comparison of both techniques in terms of accuracy and sensitivity will also be performed.

Acknowledgments

This work was supported by the European Commission, grant no: 611362, FP7-PEOPLE.

References

[1] Xu J and Tsai J M 2012 *J. Micromech. Microeng.* **22** 105029
[2] Osterberg P M and Senturia S D 1997 *J. Microelectromech. Syst.* **6** 107–18
[3] Kumar M and Bhaskaran H 2015 *Nano Lett.* **15** 2562–67
[4] Zhang H, Li B, Yuan W, Kraft M and Chang H 2016 *J. Microelectromech. Syst.* **25** 286–96
[5] Choubey B, Anthony C, Saad N H, Ward M, Turnbull R and Collins S 2010 *Appl. Phys. Lett.* **97** 133114
[6] Choubey B, Boyd E, Armstrong I and Uttamchandani D 2012 *J. Microelectromech. Syst.* **21** 1252–60
[7] Golub G H and Van Loan C F 2012 *Matrix computations* (JHU Press)

Journal of Physics: Conference Series **757** (2016) 012018 doi:10.1088/1742-6596/757/1/012018

Design and fabrication of 45° inclined mirrors for wafer-level optical absorption spectroscopy

N P Ayerden, M Ghaderi and R F Wolffenbuttel

Faculty of EEMCS, Delft University of Technology, Mekelweg 4, 2628 CD, Delft, Netherlands

E-mail: n.p.ayerden@tudelft.nl

Abstract. The increasing demand for small, robust and low-cost gas sensors triggers the batch fabrication of highly selective and sensitive miniaturized devices. A linear variable optical filter (LVOF) based microspectrometer enables selectivity in a wide wavelength range, while maintaining the robustness and low cost. To achieve sensitivity in an LVOF based absorption spectrometer, a long gas cell is required. In this paper, we propose an on-chip absorption path that also serves as a gas cell, where the light beam is steered using 45° inclined mirrors. The fabrication of 45° inclined mirrors is demonstrated and optical efficiency of the system is analyzed using ray tracing.

1. Introduction

The composition measurement of gases is of great importance in a diverse range of applications, such as the assessment of indoor air quality [1, 2], atmospheric studies [3, 4] and breath analysis [5, 6, 7, 8]. The increasing demand for field use provides an incentive for the development of small, robust and low-cost gas sensors.

Optical absorption spectroscopy offers a promising compromise between nonselective and low-cost methods such as pellistors, and high-resolution and high-cost methods such as gas chromatography [9]. An optical absorption based spectrometer is composed of a light source, a wavelength-selective device, a sample cell and a detector. The light is passed through a sample and the ratio of absorbed to incident radiation is recorded. The sample is identified by comparing the acquired spectrum with a database. Thanks to its self-referencing, non-destructive, fast response and high-selectivity properties, optical absorption spectroscopy stands out among other gas sensing methods.

The Beer-Lambert Law states that the attenuation of light (i.e. absorbance) passing through a medium is a linear function of the absorption coefficient and the concentration of the attenuating species in the medium as well as the optical path length [10]. The absorption coefficient is a wavelength-dependent intrinsic property of the sample gas. However, the resolving power of the measurement instrument determines its actual value. The remaining two parameters; the concentration and the optical path length determine the absorbance for a particular instrument resolution. Therefore, at a given detector sensitivity, the low concentration of the sample can be compensated either by a wavelength-selective device with high resolution or a long optical absorption path.

The integration of a miniaturized light source with the LVOF and the detector array enables low-cost microspectroscopy. The sensitivity of such a microspectrometer can be enhanced by improving the resolution of the filter. However, given the operating principle of the LVOF, a pixel in the detector array must not be wider than the full-width-half-max (FWHM) of the filter. In addition to the constraints on the pixel size imposed by the fabrication tolerances, the inversely proportional relation between the pixel width and the detector sensitivity allows for only limited improvement in the overall sensor sensitivity.

The second parameter that can be used for improving the sensor sensitivity is the length of the optical absorption path. External multipass absorption cells such as White cell [11] and Herriott cell [12] composed of a set of spherical mirrors have been widely used for path length elongation through multiple reflections. An improved configuration known as the circular multipass cell allows a star-shaped trajectory of the light beam [13, 14]. Moreover, the multiple reflections in high-finesse optical cavities can be exploited to elongate the effective optical absorption path [15, 16]. However, the demanding requirements on the alignment and surface profile of the mirrors complicates the actual implementation in miniaturized systems.

An on-chip gas cell, where the light beam is propagated along the surface of the wafer allows for efficient use of space, while improving the sensitivity in wafer-level spectroscopy. Lab-on-a-chip devices with 45° inclined mirrors to steer the beam along the wafer surface have been presented in the literature to analyze fluids with high sensitivity using small sample volumes [17, 18, 19]. However, these devices usually employ a low-cost LED as the light source without a wavelength-selective device, thereby limiting the selectivity of the sensor. Instead, we propose a system that is realized by implementing the absorption path and the optics on one wafer and the light source, the LVOF and the detector array on the other. By bonding these two wafers, a sample cell to trap the gas inside is created. Therefore, the sensitivity of the sensor is improved by the on-chip absorption path, while the selectivity is ensured by the wideband operation of the LVOF at the wafer level. In this paper, the fabrication and the optical efficiency of an on-chip gas cell with 45° inclined mirrors are investigated.

2. Fabrication

The increasing use of waveguides in integrated optical devices necessitated the fabrication of 45° inclined mirrors for efficient coupling. Several methods can be used to fabricate 45° inclined mirrors, such as laser ablation [20], soft molding [21], reactive ion etching [22] and microdicing [23, 24]. Among all these, the method based on photolithography and subsequent wet etching is the most compatible with MEMS fabrication [25].

The fabrication of 45° inclined mirrors can be achieved by aligning the etch mask with the $< 100 >$ direction in a (100) wafer and subsequent anisotropic etching with a surfactant [26, 27, 28]. The mask alignment allows for revealing the {110} planes according to the crystal orientation, while the surfactant improves surface quality of the mirror [29].

The process flow for the fabrication of 45° inclined mirrors is given in Fig. 1(a)-(d). Initially, a 300 nm thick SiN is deposited using LPCVD. To be able to use this layer as a mask during wet etching, a lithography step with 10 mm by 10 mm checker pattern mask is applied. Subsequently, the SiN on the windows is removed by reactive ion etching, using 65 sccm C_2F_6 at an etch rate of approximately 5 nm/s. After the removal of the photoresist in oxygen plasma, the wafer is subjected to wet etching using KOH and TMAH with the surfactant Triton X-100.

The patterned wafer before the wet etching step is shown in Fig. 1(e). The checker pattern is aligned with the $< 100 >$ direction, which is equivalent to rotating the mask for 45° with respect to the primary flat. The crystallographic alignment could be further improved compared to the angular alignment with respect to the primary flat of the wafer by observing pre-etched test structures [30, 31, 32]. However, this was not required in our experiments.

First, wet etching experiments were performed with TMAH, where a solution of 25% TMAH + 200 ppm Triton X-100 at 75 °C was used. 45° inclined mirrors were achieved on the {110} plane at an etch rate of 0.21 µm/min. However, randomly located pyramids were formed on the

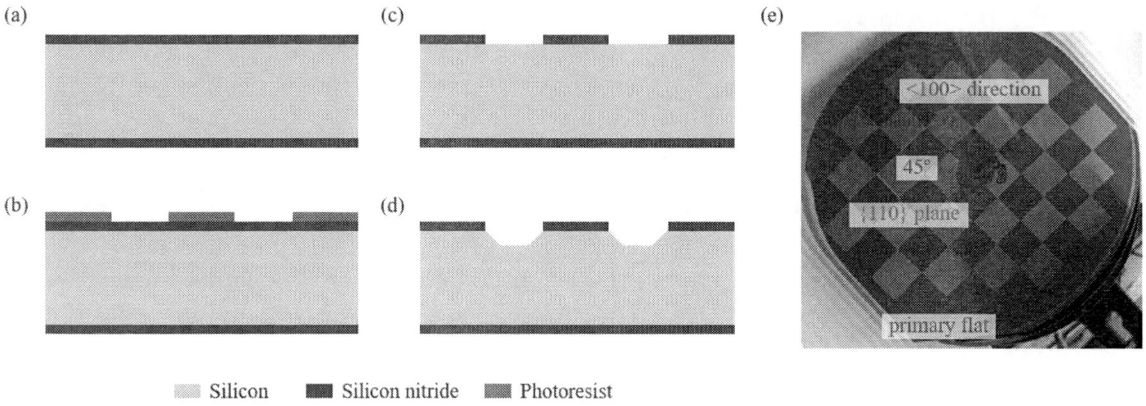

Figure 1. The process flow for the fabrication of 45° inclined mirrors: (a) LPCVD SiN deposition, (b) patterning the SiN mask, (c) plasma etching of the SiN mask and (d) anisotropic etching of silicon. (e) The wafer after lithography and development steps with a 10 mm by 10 mm checker pattern.

{100} plane as shown in Fig. 2. Since these pyramids are on the light path, they could block the light and decrease the throughput. Therefore, a second set of experiments were performed with KOH as the etching agent.

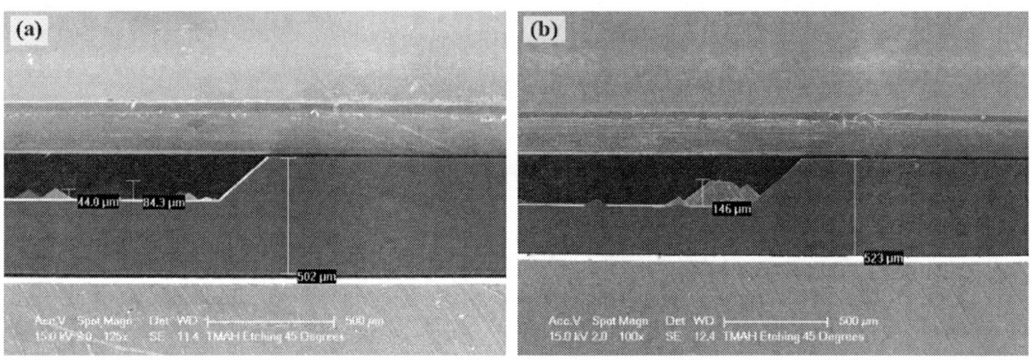

Figure 2. SEM images of 45° inclined mirrors etched (a) 198 µm and (b) 262 µm deep using 25% TMAH + 200 ppm Triton X-100 at 75 C.

For KOH etching, a solution with 2 M KOH + 60 ppm Triton X-100 at 90 °C is prepared. KOH etching with the surfactant Triton X-100 resulted in smooth inclined surfaces as shown in Fig. 3. The etch rate is calculated as 0.10 µm/min. In contrast to the partial blockage in TMAH etching, the KOH etching resulted in roughness-free {100} planes, thereby allowing a light path without randomly formed structures.

The surface roughness of the etched mirrors are measured using atomic-force microscopy. The mirrors are diced along the edges and placed on a 45° inclined wedge to obtain a flat surface for roughness analysis. The measured root-mean-square (RMS) roughness values are below 15 nm, safely within the optical flatness limit of $\lambda/10$, even in the UV band of the spectrum.

3. Optical Efficiency Analysis

The power efficiency of a system with 45° inclined mirrors is analyzed using the ray tracing software, Zemax. The simplified system is composed of a light source, two 45° inclined mirrors

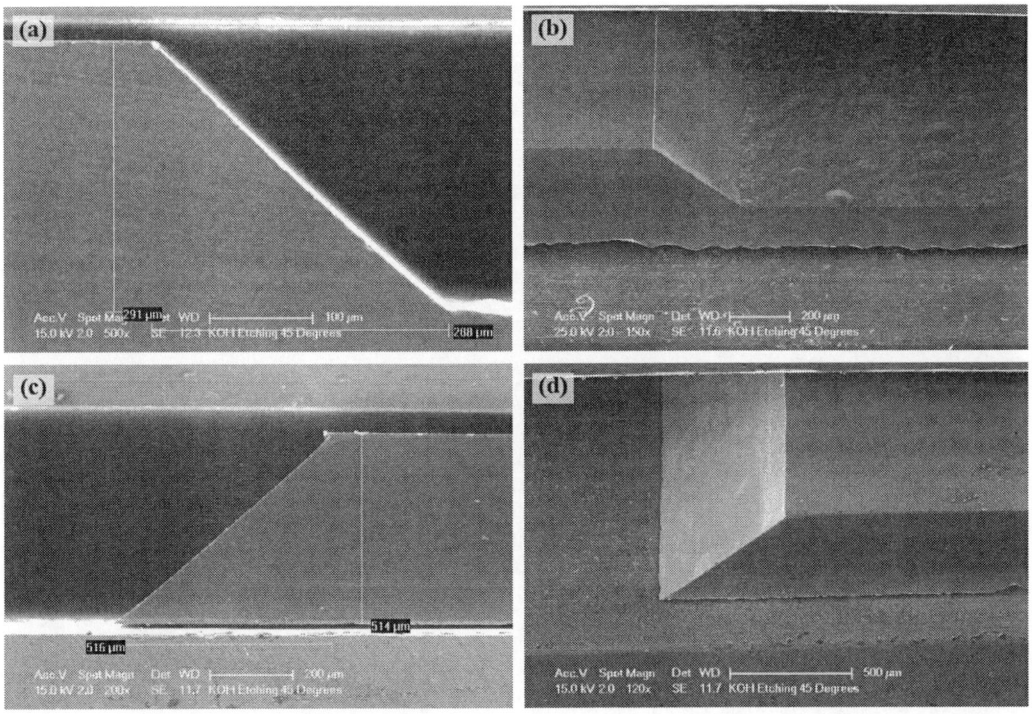

Figure 3. SEM images of 45° inclined mirrors etched (a, b) 290 µm deep and (c, d) through the Si wafer using 2 M KOH + 60 ppm Triton X-100 at 90 °C.

with a 10 mm long optical light path in-between and an image plane as shown in Fig. 4. The light source is implemented as a cylindrical lens to realize a filament-like emitter. The cylindrical lens is designed as a toroidal surface with infinite radius of rotation. The diameter of the cylinder is 0.06 mm, while the length is 3 mm. The size of the entrance pupil diameter is selected as 3 mm so that the entire cylindrical lens surface is illuminated. The size of the 45° inclined mirrors is maximized to demonstrate the best-case efficiency. The largest mirror can be achieved by through-wafer etching. Therefore, the width of the mirror is selected as 0.74 mm assuming a standard 4" wafer thickness of 524 µm. The cylindrical lens and the image plane are placed 0.5 mm away from the mirrors.

Geometric image analysis in the sequential mode of Zemax is used for power simulations. This type of analysis assumes a unit power of 1 W at the initial surface and calculates the power distribution at every surface based purely on geometrical optics. Power analysis results for the system with 45° inclined mirrors are given in Fig. 5. Since the cylindrical lens is the light source in the actual implementation, the power at the image plane is normalized to the power at the cylindrical lens rather than the entrance pupil. This results in the best-case power efficiency of 1.1%. However, through-wafer etching, despite maximizing power efficiency, impairs the robustness of the entire system. Moreover, the stray light that does not follow the predefined optical path makes the quantitative material identification, which highly depends on the length of the optical absorption path, unreliable. In addition, the fabrication tolerances could easily decrease the efficiency below 1.1%, which is undesirable in systems with low-power wideband emitters. Therefore, an on-chip gas cell with 45° inclined mirrors must be combined with collimating optics.

Figure 4. Zemax layout of the design with 45° inclined mirrors.

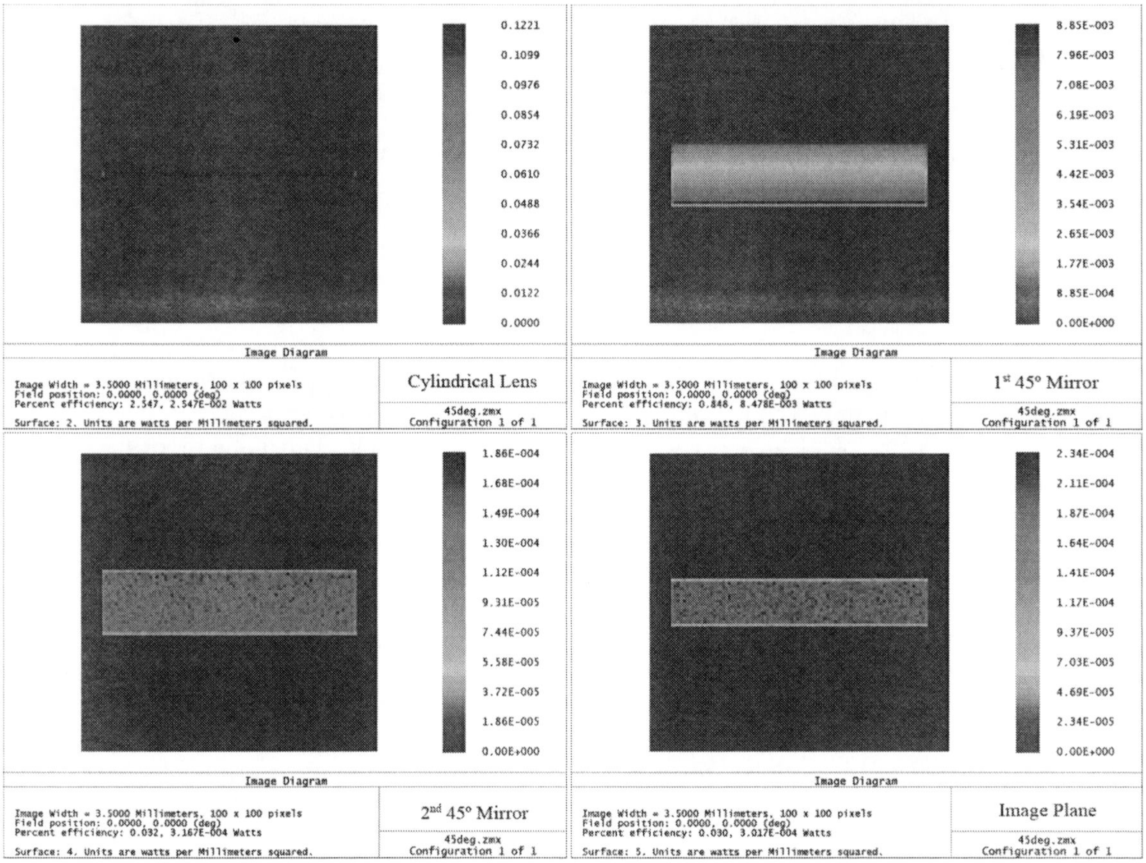

Figure 5. Power analysis results of the design with 45° inclined mirrors.

4. Conclusions

The feasibility of an on-chip absorption path with 45° inclined mirrors is investigated in terms of fabrication and optical efficiency. 45° inclined mirrors are realized by aligning the mask with the $< 100 >$ direction in a (100) wafer and subsequent anisotropic etching. The surfactant Triton X-100 is added to the wet etching solution to improve the surface quality of the mirrors.

Experiments with KOH and TMAH solutions resulted in an RMS surface roughness of less than 15 nm, while randomly located pyramids were formed on the {100} plane in TMAH etching. Optical efficiency of the on-chip absorption path with 45° inclined mirrors is analyzed using ray tracing in Zemax. The results imply that without collimating and focusing optics, the optical throughput is 1.1% in the best-case scenario. The fabrication of lenses on the light source and the LVOF to collimate and focus the light respectively, further complicates the process and puts an imbalance on the fabrication complexity between the top and the bottom wafers. Therefore, instead of steering the light beam with 45° inclined mirrors, off-axis parabolic mirrors can be employed to both steer and collimate or focus the light. Fabrication of such aspherical optical components have been previously presented in the literature [33, 34]. As a result, by integrating the wafer with the mirrors and the optical absorption path to the wafer with the light source, the LVOF and the detector array, a highly sensitive and selective microspectrometer can be built at the wafer-level.

References

[1] Kostiainen R 1995 *Atmos. Environ.* **29** 693–702
[2] Pushkarsky M B, Webber M E and Patel C K N 2003 *Appl. Phys. B-Lasers O.* **77** 381–385
[3] Akimoto H 2003 *Science* **302** 1716–1719
[4] Martin R V 2008 *Atmos. Environ.* **42** 7823–7843
[5] Lovett A M, Reid N M, Buckley J A, French J B and Cameron D M 1979 *Biol. Mass Spectrom.* **6** 91–97
[6] Thorpe M J, Balslev-Clausen D, Kirchner M S and Ye J 2008 *Opt. Express* **16** 2387–2397
[7] Wang C and Sahay P 2009 *Sensors* **9** 8230–8262
[8] Arslanov D D, Swinkels K, Cristescu S M and Harren F J M 2011 *Opt. Express* **19** 24078–24089
[9] Hodgkinson J and Tatam R P 2013 *Meas. Sci. Technol.* **24** 012004
[10] Swinehart D F 1962 *J. Chem. Educ.* **39** 333
[11] White J U 1942 *J. Opt. Soc. Am.* **32** 285–288
[12] Herriott D R and Schulte H J 1965 *Appl. Optics* **4** 883–889
[13] Thoma M L, Kaschow R and Hindelang F J 1994 *Shock Waves* **4** 51–53
[14] Mangold M, Tuzson B, Hundt M, Jgersk J, Looser H and Emmenegger L 2016 *J. Opt. Soc. Am. A* **33** 913–919
[15] Ayerden N P, de Graaf G and Wolffenbuttel R F 2016 *Opt. Express* **24** 2981–3002
[16] Ayerden N P, Ghaderi M, Enoksson P, de Graaf G and Wolffenbuttel R F 2016 *Sensor. Actuat. B-Chem.*
[17] Rushworth C M, Davies J, Cabral J T, Dolan P R, Smith J M and Vallance C 2012 *Chem. Phys. Lett.* **554** 1–14
[18] Noda T, Takao H, Yoshioka K, Oku N, Ashiki M, Sawada K, Matsumoto K and Ishida M 2006 *Sensor. Actuat. B-Chem.* **119** 245–250
[19] Tiggelaar R M, Veenstra T T, Sanders R G P, Gardeniers J G E, Elwenspoek M C and van den Berg A *Talanta* **56** 331–339
[20] Hendrickx N, Erps J V, Steenberge G V, Thienpont H and Daele P V 2007 *IEEE Photonic. Tech. L.* **19** 822–824
[21] Wang L, Wang X, Jiang W, Choi J, Bi H and Chen R 2005 *Appl. Phys. Lett.* **87** 141110
[22] Kagami M, Kawasaki A and Ito H 2001 *J. Lightwave Technol.* **19** 1949–1955
[23] Glebov A L, Roman J, Lee M G and Yokouchi K 2005 *IEEE Photonic. Tech. L.* **17** 1540–1542
[24] Maciel M J, Costa C G, Silva M F, Peixoto A C, Wolffenbuttel R F and Correia J H *Sensor. Actuat. A-Phys.* **242** 210–216
[25] Wang F, Liu F and Adibi A 2009 *Opt. Express* **17** 10514–10521
[26] Strandman C, Rosengren L, Elderstig H G A and Backlund Y 1995 *J. Microelectromech. S.* **4** 213–219
[27] Zubel I and Barycka I 1998 *Sensor. Actuat. A-Phys.* **70** 250–259
[28] Drago R, Danilo V, Uros A, Matej M and Slavko A 2005 *J. Micromech. Microeng.* **15** 1174
[29] Rola K, Ptasinski K, Zakrzewski A and Zubel I 2012 *Procedia Eng.* **47** 510–513
[30] Mattias V and Ylva B 1996 *J. Micromech. Microeng.* **6** 279
[31] Jun Hyun H, Shankar R and Chung-Hoon L 2013 *J. Micromech. Microeng.* **23** 055017
[32] Singh S S, Veerla S, Sharma V, Pandey A K and Pal P 2016 *J. Micromech. Microeng.* **26** 025012
[33] Kendall D L, Digges Iii T G, Eaton W P and Manginell R P 1994 *Opt. Eng.* **33** 3578–3588
[34] de Lima Monteiro D W, Akhzar-Mehr O, Sarro P M and Vdovin G 2003 *Opt. Express* **11** 2244–2252

Journal of Physics: Conference Series **757** (2016) 012019 doi:10.1088/1742-6596/757/1/012019

Split-Ring Resonator-Based Strain Sensor on Flexible Substrates for Glaucoma Detection

Gizem Ekinci[1], Arda Deniz Yalcinkaya[1,2], Gunhan Dundar[1], Hamdi Torun[1,2]

[1]*Department of Electrical and Electronics Engineering, Bogazici University, Bebek 34342 Istanbul, Turkey*
[2]*Center for Life Sciences and Technologies, Bogazici University, Kandilli 34684 Istanbul, Turkey*

Corresponding author: hamdi.torun@boun.edu.tr

Abstract. This paper presents split-ring resonator-based strain sensors designed and characterized for glaucoma detection application. The geometry of the sensor is optimized such that it can be embedded in a contact lens. Silver conductive paint is to form the sensors realized on flexible substrates made of cellulose acetate and latex rubber. The devices are excited and interrogated using a pair of monopole antennas and the characteristics of devices with different curvature profiles are obtained. The sensitivity of the device, i.e. the change in resonant frequency for a unit change in radius of curvature, on acetate film is calculated as -4.73 MHz/mm and the sensitivity of the device on latex is 33.2 MHz/mm. The results indicate that the demonstrated device is suitable for glaucoma diagnosis.

1. Introduction

Metamaterials exhibit electromagnetic properties, which are not present in ordinary materials, such as negative values of electric permittivity and magnetic permeability, simultaneously [1]. The unique properties bring metamaterials into consideration for sensing applications. One of the most popular structures of metamaterials is split-ring resonators (SRRs). A conventional SRR is simply a ring with a split, which is made of a preferably high-conductivity metal, and is fabricated on a dielectric substrate. In addition to their favorable electromagnetic characteristics, the SRR devices can be realized in a simple and cost-effective manner.

Wireless strain measurement attracts attention of many researchers and finds application in numerous fields when it comes to material characterization. In civil engineering, the sensors of the sort are used to ensure maintenance in infrastructure, by measuring the strain on the structure and take precaution in case of any abnormal strain [2]. In biomedical engineering, these sensors are utilized by observation of the healing process of a fractured bone [3]. Moreover, in aerospace industry, wireless strain measurement comes into prominence in crack or abnormal strain detection on metal surfaces [4]. SRR-based sensors have been demonstrated for wireless strain measurement previously [3,4]. The resonant frequency of the device strongly depends on the geometry of the ring and any deviation in geometry due to external strain applied to the device results in change in resonant frequency of the resonator. This enables strain measurement through the observation of the shift in resonant frequency.

Glaucoma is an eye disease, which may damage the optic nerves and leads to vision loss. Even though it might be caused by different factors, in most of the patients, it is caused by the increase of intraocular pressure and might eventually cause irreversible blindness. Today, there are several different methods to detect the symptoms of glaucoma, such as measuring central corneal thickness, measuring peripheral vision, examining the optic nerve and measuring the eyeball pressure [5]. It was demonstrated that the radius of curvature of the sclera is well correlated with the intraocular pressure, while the radius of the curvature of the cornea is insensitive to the changes in intraocular pressure [6]. Noninvasive monitoring of the intraocular pressure has been demonstrated using piezoresistive [7] and capacitive [8] sensors embedded on soft contact lenses. It is possible to monitor the progress of the disease in noninvasive and continuous manner using this approach. However these sensors are electrically active and require application of electrical signals in the contact lens during operation. In addition, the system on the contact lens also includes a transmission circuitry to send the signals to an external unit.

Here, we describe an electrically passive, wireless and low cost sensor to measure intraocular pressure by placing an SRR on top of contact lenses. The sensor on the lens does not require any electrical signal to operate and are interrogated by external antennas that can be located away from the contact lens. We use silver conductive paint to define the rings on flexible substrates. We realized two different SRR-based strain sensors and characterized them experimentally to evaluate their potential for wireless strain sensing. This study suggests that SRR-based sensors are promising for the diagnosis of glaucoma.

2. Design and Fabrication of the SRR-based Sensors

SRRs behave like RLC circuits but in most cases the material is chosen so that the resistance value is negligible. Hence the structure might be reduced to LC resonator and the resonant frequency might be calculated by Eq.1.

$$f_o = \frac{1}{2\Pi\sqrt{LC}} \tag{1}$$

The inductance of the ring is expressed in equation 2 as a function of geometrical parameters [9].

$$L = \mu_0 R_m \left(\log \frac{8R_m}{h+w} - \frac{1}{2} \right) \tag{2}$$

where μ_0 is the free-space permeability, R_m is the effective radius of the ring, w and h are the width and the height of the ring, respectively.

The capacitance of the ring has two components, the gap capacitance and surface capacitance. The gap capacitance is calculated as follows [9]:

$$C_{gap} = \varepsilon_0 \frac{hw}{g} + C_0 \tag{3}$$

where ε_0 is the free-space permittivity. C_0 is the capacitance caused by the fringing fields and can be calculated as $C_0 = \varepsilon_0 (h + w + g)$ [9]. Equation 3 assumes that the ring is in free-space. If the ring is placed on or embedded into a dielectric substrate, then the effective permittivity of the medium should be considered in this equation.

The geometry of the SRR structure is determined so that it can be placed around the boundary between cornea and sclera where the change in intraocular pressure can be measured effectively to diagnose glaucoma [6]. In addition, the sensor is optimized for S-band (2-4 GHz) of the electromagnetic band. Fig. 1(a) shows a photograph of a sensor placed on a flexible substrate made of cellulose acetate. We extended the gap region to increase the gap capacitance and to keep the resonant frequency of the device in S-band. We applied silver conductive paint (SCP, Electrolube, United Kingdom) to define the ring using a hard mask on the substrate. The thickness of the ring is 50 μm and the other relevant geometrical parameters are shown in Fig. 1(a). We prepared another type of sensor on a spherically deformed substrate made of latex rubber with a thickness of 50 μm. Using a hard mask on a spherical substrate is not possible, so we painted the ring on the second substrate.

We simulated the device using commercially available electromagnetic simulation software (CST Microwave Studio, Darmstadt, Germany) to obtain its scattering parameters. We performed the simulations in time domain. Fig. 1(b) shows the surface current density along the conductor path at the resonant frequency observed at 2.54 GHz. The magnetic field is perpendicular to the surface of the device for the simulation that supports the circulating current at resonance. Fig. 1(c) shows the s21 spectrum of the device with these settings.

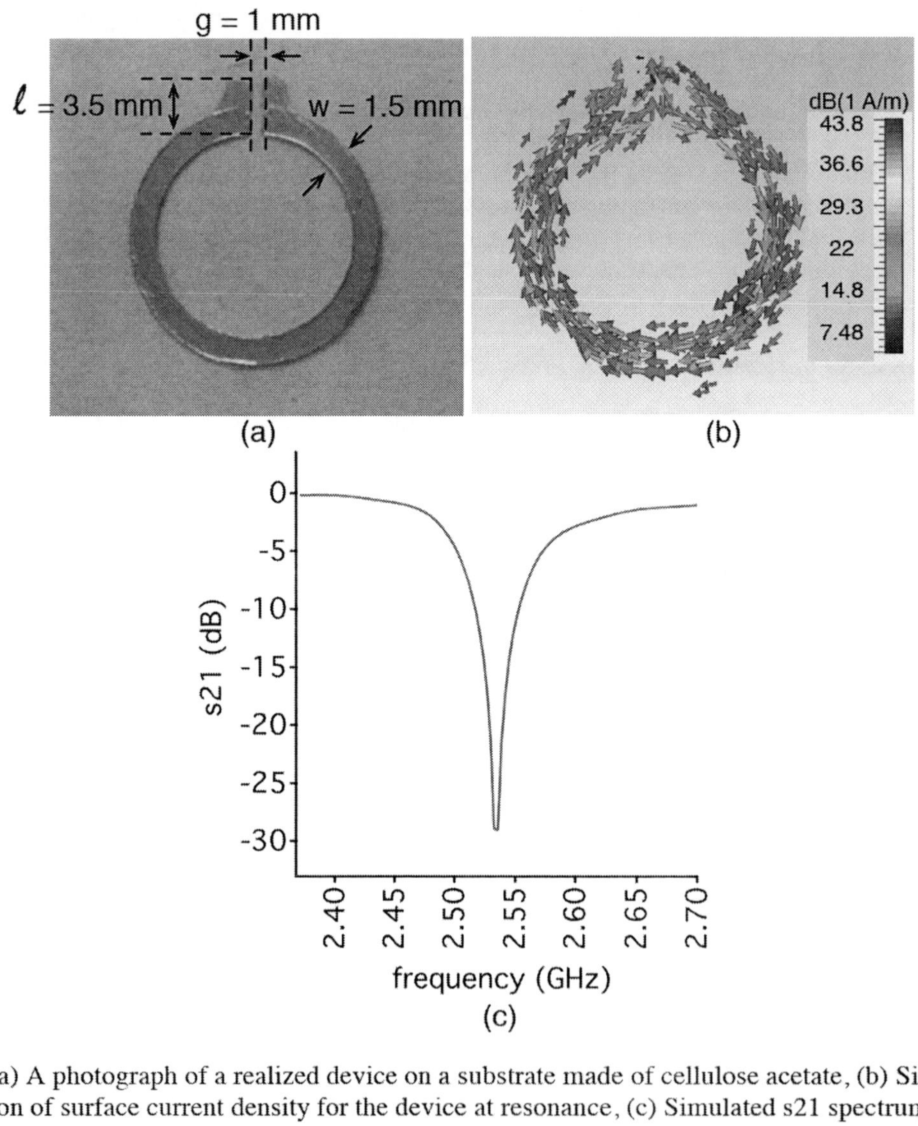

Figure 1.(a) A photograph of a realized device on a substrate made of cellulose acetate, (b) Simulated distribution of surface current density for the device at resonance, (c) Simulated s21 spectrum of the device.

3. Experimental Characterization

We characterized the SRR-based strain sensors using a vector network analyzer (ZVB4, Rohde & Schwarz, Munich, Germany) by measuring their scattering parameters. We conducted the experiments using two different setups. We placed the SRR sensor realized on the substrate of cellulose acetate in the first setup shown in Fig. 2(a). The device was placed on an adjustable frame to introduce controllable bending between two monopole antennas. We adjusted the radius of curvature while measuring the s21 spectrum of the resonator. The nominal resonant frequency of the device was measured as 2.49 GHz, which is in good agreement with our simulations. Fig. 2(b) shows how the s21 spectrum changes with increasing curvature (1/ρ). As the curvature increases, the resonance frequency increases monotonically as shown in Fig. 2(c). The change in resonant frequency linearly depends on curvature for the selected range of curvature. We repeated the bending experiment six times and report the standard error mean of the measurements with the error bars of Fig. 2(c). The gap size of the device does not change significantly with curvature. However, the projected area of the bended device on the plane of antennas decreases with curvature. This results in a decrease in surface capacitance, so the resonant frequency of the device increases.

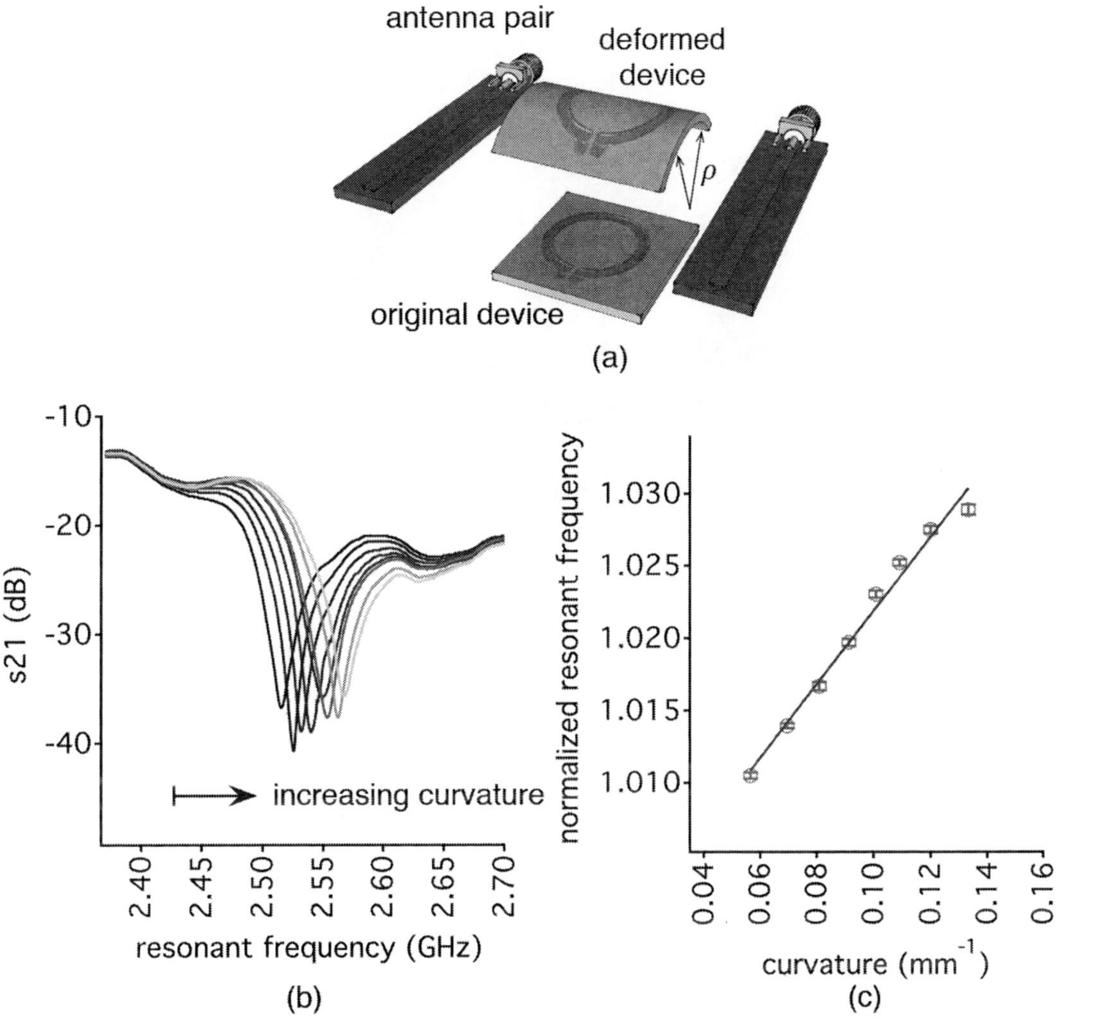

Figure 2.(a) Three-dimensional drawing of the experimental setup for the SRR sensor realized on cellulose acetate, (b) s21 spectra of the device as the curvature (1/ρ) increases, (c) Variation of the resonant frequency as a function of curvature.

We used the second setup to characterize the SRR sensor realized on the substrate of latex rubber. Fig. 3(a) shows the experimental setup where the substrate is stretched on top of a cylindrical injection syringe with a diameter of 9 mm. The syringe is used to pump air in and out of the device. First, we pumped air so that the radius of the curvature of the substrate becomes 13 mm. Then, we paint the SRR structure on the surface and obtained its s21 spectrum using the vector network analyzer. We continued the following measurements by evacuating 0.2 ml air at each step and obtaining the s21 spectrum of the device. The inflated device shrinks with decreasing air volume so does the radius of the curvature. Fig.3 (b) shows how the s21 spectrum of the device changes with decreasing radius of curvature. The gap of the SRR structure decreases as the radius of the curvature decreases. This results in an increase in the capacitance of the resonator, so the resonant frequency decreases as shown in Fig. 3(c). We repeated this experiment six times with different SRR structures painted on the substrate each time separately after we pumped air such that the radius of the curvature became 13 mm. Fig.3(c) shows the mean values and the standard error mean values of the measured resonant frequencies normalized to the value we measured for the radius of curvature of 13 mm. The range of the change in radius of the curvature is physiologically relevant to the diagnosis of glaucoma. The result of an experiment conducted with 16 fresh porcine eyes, by applying five consecutive incremental 100μl injections, suggests that with each incremental injection of fluid, intraocular pressure and radius of curvature of the sclera increases linearly from 9mm to 13mm [6].

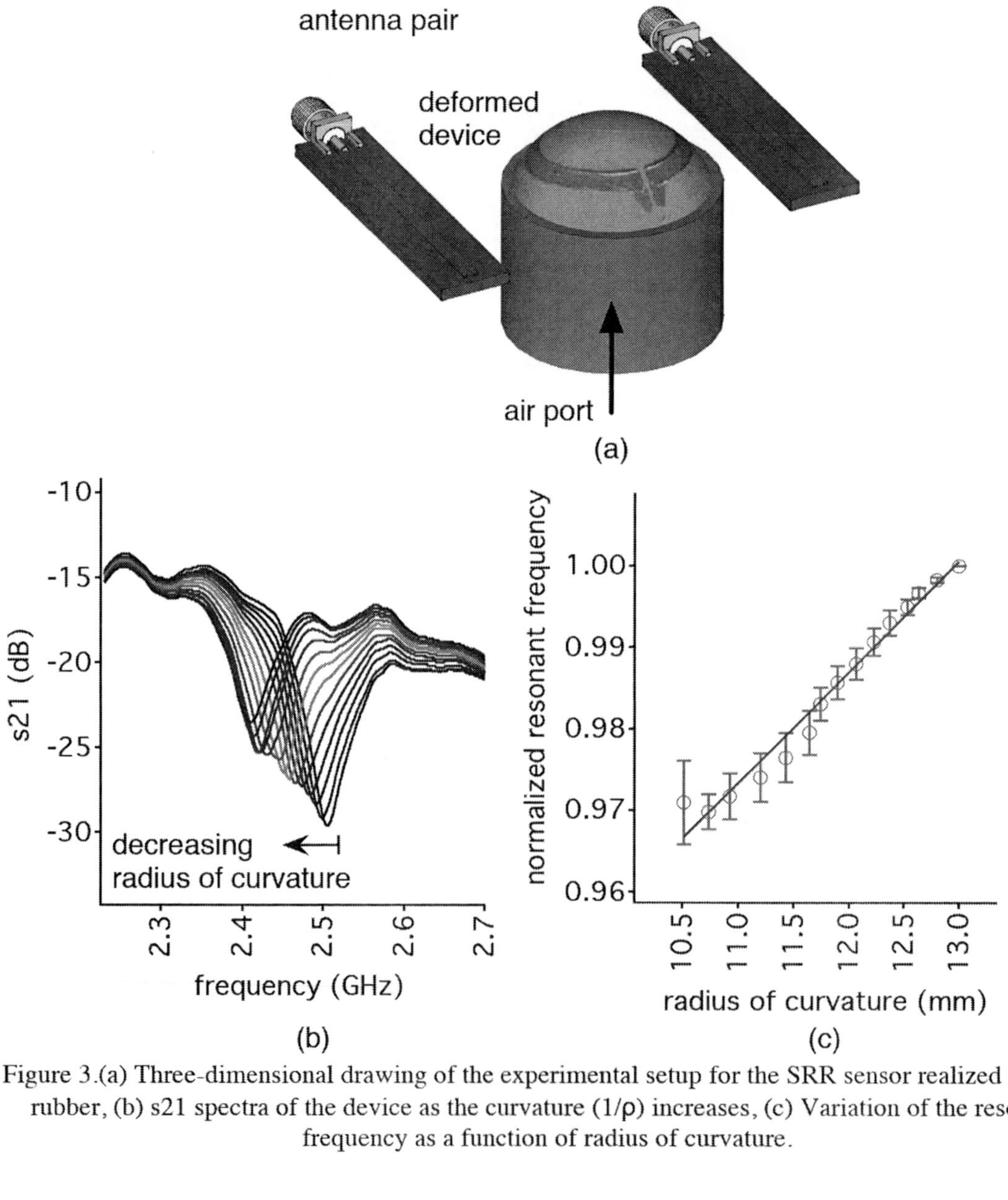

Figure 3.(a) Three-dimensional drawing of the experimental setup for the SRR sensor realized on latex rubber, (b) s21 spectra of the device as the curvature (1/ρ) increases, (c) Variation of the resonant frequency as a function of radius of curvature.

4. Conclusion

Split-ring resonator-based strain sensors enable realization of wireless strain sensors that are electrically passive. In this study, we demonstrate the feasibility of detecting glaucoma using SRR sensors realized on flexible substrates that can be integrated to contact lenses. We present two devices with different substrates that are suitable for distinct curvature profiles. The resonant frequencies of the devices depend on the curvature and the devices exhibit monotonic variation with respect to change in radius of curvature.

The first device, depicted in Fig.2, has a sensitivity of -4.73 MHz/mm. We measure the quality factor of the resonators on the order of 500. This corresponds to a frequency resolution of 5 MHz. Given the sensitivity values, the change in radius of curvature of the first SRR sensor can be measured with a resolution of 1.1 mm. The sensitivity of the second device, illustrated in Fig.3, is obtained as 33.2 MHz/mm and the measurement resolution for the radius of curvature corresponds to 150 μm. The demonstrated device is promising towards a novel method for glaucoma diagnosis.

Acknowledgements
This work was supported by the TUBITAK grant no. 112E250.

References

[1] D. Schurig, J. J. Mock, B. J. Justice, S. A. Cummer, J. B. Pendry, A. F. Starr, and D. R. Smith, "Metamaterial Electromagnetic Cloak at Microwave Frequencies",Science 314(5801), 977 (2006).
[2] T. Nagayama, M. Ruiz-Sandoval, B.F. Spencer Jr., K.A. Mechitov, G. Agha, "Wireless Strain Sensor Development for Civil Infrastructure", Trans. of the Society of Instrument and Control Engineers, Vol. E-3, No.1, 104/109 (2004).
[3] R. Melik, N. K. Pekgoz, E. Unal, C. Puttlitz, and H. V. Demir, "Bio Implantable passive on-chip RF-MEMS strain sensing resonators for orthopaedic applications," J. Micromech. Microeng. 18(11), 115017 (2008).
[4] Rajni, A. Kaur, A. Marwaha, "Crack Detection on Metal Surfaces with an Array of Complementary Split Ring Resonators", International Journal of Computer Applications (0975 – 8887) Volume 119 – No.10, June 2015.
[5] "Five Common Glaucoma Tests", glaucoma.org, Glaucoma Research Foundation, April 22, 2013.
[6] B.K. Pierscionek, M. Asejczyk-Widlicka, R.A. Schachar, "The effect of changing incraocular pressure on the corneal and scleral curvatures in the fresh porcine eye", Br J Ophthalmol 2007;91:801–803. doi: 10.1136/bjo.2006.110221.
[7] M. Leonardi, P. Leuenberger, D. Bertrand, A. Bertsch, P. Renaud, "First Steps toward Noninvasive Intraocular Pressure Monitoring with a Sensing Contact Lens", Invest. Ophthalmol. Vis. Sci., 45(9), 3113 (2004).
[8] D. Piso, P. Veiga-Crespo, E. Vecino, "Modern Monitoring Intraocular Pressure Sensing Devices Based on Application Specific Integrated Circuits", Journal of Biomaterials and Nanobiotechnology, 2012, 3, 301-309
[9] O. Sydoruk, E. Tatartschuk, E. Shamonina and L. Solymar, "Analytical formulation fort he resonant frequency of split rings", American Institute of Physics, Journal of Applied Physics 105, 014903 (2009).

Journal of Physics: Conference Series **757** (2016) 012020 doi:10.1088/1742-6596/757/1/012020

Piezoelectric MEMS resonators for monitoring grape must fermentation

J Toledo[1], F Jiménez-Márquez[1], J Úbeda[2], V Ruiz-Díez[1], G Pfusterschmied[3], U Schmid[3] and J L Sánchez-Rojas[1]

[1] Microsystems, Actuators and Sensors Group, Universidad de Castilla-La Mancha, Ciudad Real, Spain.

[2] Tecnología de los Alimentos, Universidad de Castilla-La Mancha, Ciudad Real, Spain.

[3] Institute of Sensor and Actuator Systems, TU Wien, Vienna, Austria.

javier.toledo.serrano@gmail.com

Abstract. The traditional procedure followed by winemakers for monitoring grape must fermentation is not automated, has not enough accuracy or has only been tested in discrete must samples. In order to contribute to the automation and improvement of the wine fermentation process, we have designed an AlN-based piezoelectric microresonator, serving as a density sensor and being excited in the 4th-order roof tile-shaped vibration mode. Furthermore, conditioning circuits were designed to convert the one-port impedance of the resonator into a resonant two-port transfer function. This allowed us to design a Phase Locked Loop-based oscillator circuit, implemented with a commercial lock-in amplifier with an oscillation frequency determined by the vibrating mode. We were capable of measuring the fermentation kinetics by both tracking the resonance frequency and by determining the quality factor measurements of the microresonator. Moreover, the resonator was calibrated with an artificial model solution of grape must and then applied for the monitoring of real grape must fermentation. Our results demonstrate the high potential of MEMS resonators to detect the decrease in sugar and the increase in ethanol concentrations during the grape must fermentation with a resolution of 100 µg/ml and a sensitivity of 0.16 Hz/µg/ml as upper limits.

1. Introduction

The fermentation of grape must involves the interaction between yeasts, bacteria, fungi and viruses. A correct bio-chemical process is a necessary condition but not sufficient by itself to determine the final quality of the wine, whose assessment relies on a comprehensive analysis of its chemical components, as the basic flavour of wine depends on at least 20 compounds [1]. Therefore, winemakers must carefully supervise the wine fermentation process to ensure a wine of the expected quality. One of the key parameters monitored during wine fermentation is the fermentation kinetics. This provides essential information about the steady transformation of grape must into wine due to the decrease of glucose and fructose that leads to the formation of ethanol, glycerol and carbon dioxide along with biomass, as a result of yeast metabolism [2]. This process is traditionally monitored by enologists, who manually extract and analyse discrete samples at least twice a day using an aerometer, spectrophotometer or a colorimeter. They essentially determine the density and its rate of change since these parameters provide information about the evolution of the fermentation as a result of yeast metabolism.

However, there are different ways for monitoring the grape must fermentation: (1) density determination, based either on differential pressure measurements [3] or on the use of flexural oscillators [4]; (2) monitoring of CO_2 released during the process due to the gradual loss of mass [3,5]; (3) determination of yeast cell population evolution by means of impedance techniques [6] and turbidity measurements [7]; (4) ultrasound measurements conducted to determine the propagation velocity in grape musts [8,9]; (5) refractive techniques based on fiber optics [10]; (6) optoelectronic device based on measurements of the refractive index [11,12].

Overall, these different approaches have not enough resolution or have only been tested in discrete must samples. In this context, the simultaneous determination of the density and viscosity of liquids, through measurement of the resonant frequency (f_r) and quality factor (*Q-factor*) of a mechanical resonator, has already been reported [13–15]. This approach presents several advantages with respect to traditional methods, such as real-time analysis, on-line configuration and low liquid volumes. Piezoelectric resonators with in-plane vibration modes [16], flexural and torsional modes [17] have already been tested. However, it is well-established that a different family, such as the roof tile-shape modes, present better quality factors at moderate frequencies [18,19]. The present work evaluates the performance of this vibration mode, within micromachined self-actuated and self-sensing aluminium nitride (AlN)-based cantilever sensor for the monitoring of grape must fermentation.

2. Resonator design and characterization

The resonator was designed for the 4th order roof-tile shaped vibration mode. It resulted in a cantilever resonator with optimized electrode layout featuring five nodal lines in one direction, and 1 nodal line in the perpendicular direction. Considering Leissa's nomenclature, the vibration mode is named as 15-mode [20]. The resonator was designed with a length of L=2524 μm, width of W=1274 μm and thickness T of about 20 μm. It was fabricated from a SOI wafer with a 20 μm-thick device layer covered with a 650 nm-thin AlN piezoelectric film [21]. The top metallization has four striped electrodes that allow a selective excitation of the vibration modes and act as a filter for higher modes [17]. The considered modal shape and the fabricated resonator are shown in figure 1.

(a)

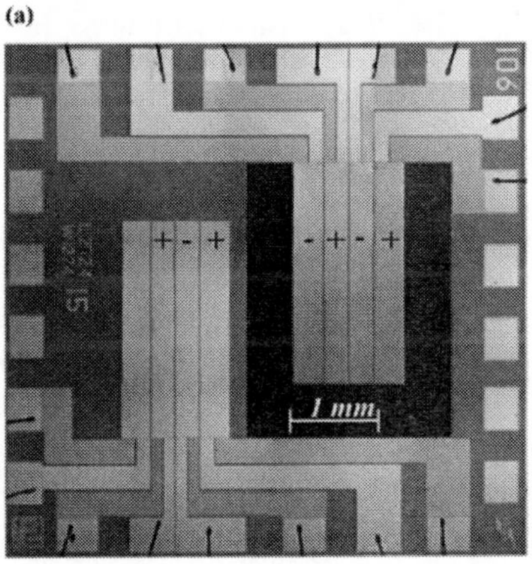

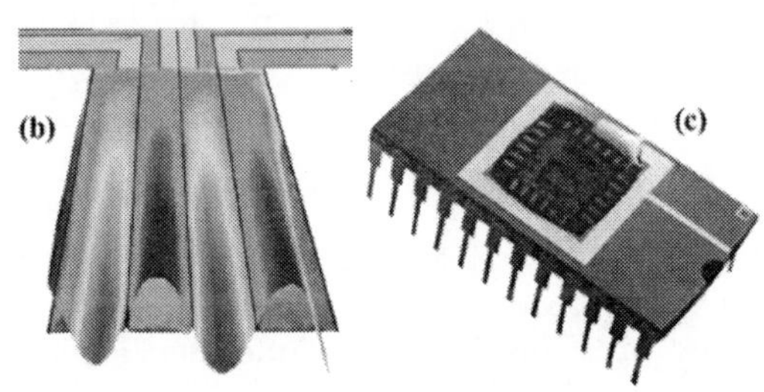

Figure 1. (a) Top-view micrograph of the resonator. (b) Modal shape measured (15-mode) with a laser Doppler vibrometer. (c) The resonator is packaged, wire bonded in a 24-pin DIP (dual inline package) and immersed in grape must.

3. PLL-Based oscillator

The simultaneous determination of density and viscosity of liquids, through measurement of the resonant frequency and the quality factor of a mechanical resonator, is challenging due to the low quality factors and parasitic effects present in liquid media [22]. For this reason, our approach is based on roof tile-shaped resonators in two-port configuration and a parasitic compensating device.

In the two-port scheme, one of the electrodes was used for actuation (+) and the other for sensing (-) (see Figure 1). However, our results show that a capacitive crosstalk across the actuation and sensing ports has a significant contribution to the output current.

To minimize this parasitic component, a compensating device was introduced. This dummy device reproduces the structure of the resonator, but without the backside release, thus preventing any vibration. For this reason, we designed an interface circuit, used in a previous work [15], to subtract the dummy response (v_{dum}) from that of the resonator (v_{res}). Since the materials and dimensions in the dummy and resonator devices are the same, they are expected to show identical electrical behaviour with respect to their parasitic effects. This results in a clear resonance, with low baseline (see Figure 2). However, a phase shift (about -12°) is introduced by the instrumentation amplifier, representing the differences in electrical performance between the dummy, the resonator device and C_{ft}.

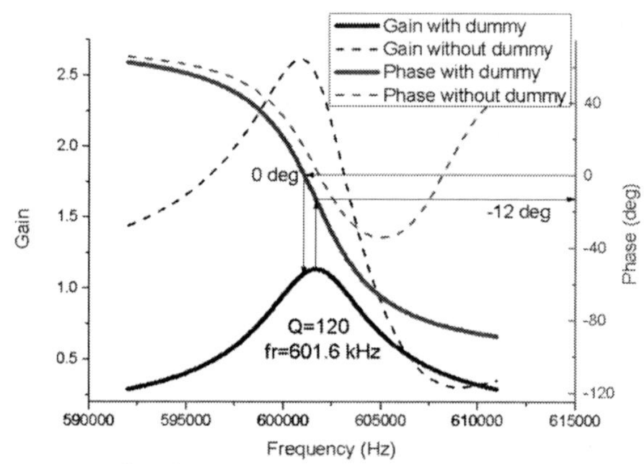

Figure 2. Open loop response for the 15-mode measured in a liquid test (2-Propanol) with and without the dummy compensation.

In order to avoid this issue, the resonators were finally included in an oscillator based on a Phase Locked Loop (*PLL*) instrument, instead of an oscillator circuit based on discrete components [15]. The phase shift was compensated by the *PLL* configuration, so that the oscillation frequency and the natural frequency of the resonator are approximately equal. The *PLL* instrument used (*Zurich HF2LI*) provides software-based voltage controlled oscillator (*VCO*), proportional-integral controller (*PI*) and phase detector (*PD*) blocks that form a control system able to track the resonant frequency.

4. Experimental details

With the goal of estimating the density (ρ) and viscosity (μ), thus being able to monitor the fermentation process, two variables were measured from our interface circuit: the oscillation frequency (f_{osc}) and the gain of the interface circuit module at operation (G_{osc}), given by the ratio V_{out}/V_{in}. Furthermore, for the application of the resonator as a density-viscosity sensor, a calibration procedure was developed with a model solution of artificial grape must [19] and finally applied to the monitoring of one model sluggish fermentation and other real grape must fermentation process. This calibration process was carried out in two steps, each with adjustable parameters [15].

4.1. Preparation of the model solutions of grape must

In order to confirm that piezoelectric resonators are valid for the monitoring of grape must fermentation, two different sets of model solutions were used for pre-investigation purposes. The first set represents a normal fermentation (N_1:N_9) and the second represents a sluggish fermentation (S_1:S_9). A variety of reasons may be attributable to this kind of failed or sluggish fermentations, which occasionally occur: the lack of dissolved oxygen, an unbalanced ratio between sugar and nitrogen, a low fermentation temperature, an inadequate rehydration, a thermal shock of yeast, etc. [23]. These model solutions represent different stages of the corresponding fermentation process according to their particular mixture of glucose, fructose, ethanol and glycerol dissolved in water. The composition of the prepared model solutions for a normal and sluggish fermentation is shown in a previous work [11]. In the normal model solution, a decrease in fructose and glucose concentration from 110&100 g/l (N_1) to 2&1 g/l (N_9) occurs. Similarly, the glycerol and ethanol concentration are increasing from zero to 9 g/l and 14 % v/v. In the sluggish fermentation the decrease in fructose and glucose concentration stops prematurely and settles around 55 and 20 g/l. Along with this development the increase in the ethanol concentration stops and does not exceed the value of 8 % v/v.

4.2. Calibration process with model solutions of grape must

With the aim of monitoring one real grape must fermentation, two different sets of model solutions were prepared in the previous section. The first set represents a normal fermentation (N_1:N_9) and was used as a model for the calibration process. The validation of this process was confirmed with a model sluggish fermentation (S_1:S_9) and finally with a real grape must fermentation (Section 4.4).

In the first step of the calibration process a constant K_{ic}, which is not affected by the loading conditions of the resonator, was used to transform the circuit outputs f_{osc} and G_{osc} into the key parameters of the resonator, namely the resonant frequency (f_r) and the quality factor (*Q-factor*).

Due to the high sensitivity of the density and viscosity to the temperature and electronic noise from surrounding equipment at different frequencies, low, but measurable drift effects are present when determining f_r and *Q-factor*. The random uncertainty of f_r was obtained directly through the Allan deviation of f_{osc} [24]. However, the random uncertainty of G_{osc} was translated into random uncertainty of *Q-factor* by uncertainty propagation [25]. The random uncertainties or resolutions obtained for the model solutions of grape must were below 100 mHz for f_r and 0.7 for *Q-factor*. Figure 4 shows the mean values of f_r and *Q-factor* when the resonator is immersed in the model solutions of grape must: normal and sluggish fermentation process. The measurements of f_r and *Q-factor* were performed with a temperature control unit [15]. Nevertheless, the impact of temperature fluctuations in the resolution of the sensor during the measurement process is minimal compared with the continuous evolution of the real grape must fermentation. As can be seen in figure 3, the significant changes in the composition of the model solutions affects the f_r and the *Q-factor* of the resonator. The combination of unfermented sugars and less than expected ethanol concentration, resulted into a flat curve of f_r for the sluggish fermentation.

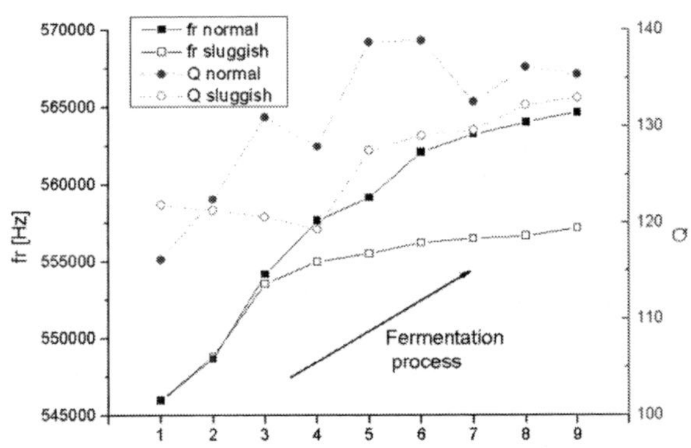

Figure 3. Resonant frequency and quality factor of the resonator immersed in two model solutions of grape must at 20°C.

In the second step of the calibration procedure, the physical properties of the liquid, density and viscosity, were obtained from the key parameters of the resonator, f_r and *Q-factor*. These are related to the mechanical properties of the resonators, i.e. resonator equivalent mass (m) and natural frequency in vacuum ($f_{0,vac}$), and the fluid conditions, i.e. distributed damping associated to the liquid

(g_1) and the distributed equivalent added mass (g_2), both per unit length (L). A model defined in previous works [26–28] allows the determination of these variables separately. Nevertheless, the dependence of g_1 and g_2 on the viscosity and density of the fluid, as a closed-form analytical expression, is only available for simple geometries and particular cases [29]. For this reason, we used a Taylor series expansion of g_1 and g_2 in terms of the viscosity, density and their product [30,31]. The model was solved iteratively to obtain the values of the coefficients (Table 1) that fit best to the experimental data for the normal model solution. The results of the density estimated values (*MEMS*) are compared to those measured from the density-viscosity meter *Anton Paar DMA4100M* (*viscometer*) at 20°C in figure 4.

Table 1. Values of the coefficients obtained in the calibration process for the normal model solution.

	K_{ic}	C_1	C_2	C_3	C_4
Normal model solution	$3.58\cdot10^{-10}$	0.12	-20.99	2.11	-103.22

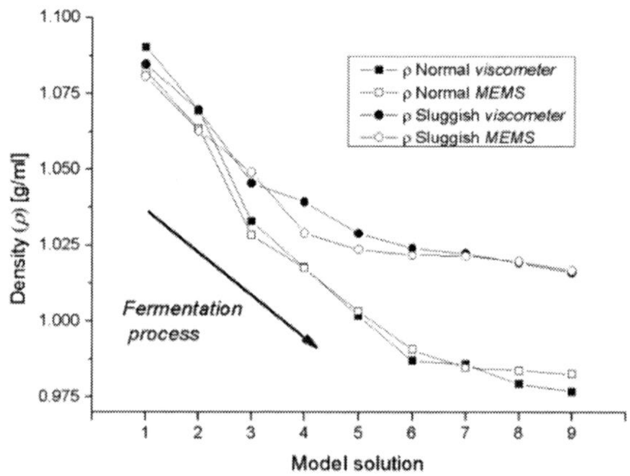

Figure 4. Density values estimated from our resonator (*MEMS*) and measured with a commercial density-viscosity meter (*viscometer*) at 20°C for the two model solutions: normal and sluggish fermentation.

As can be seen in figure 4, an almost linear decrease in the density occurs for the normal model solution and a premature stop in the density decrease for the sluggish solution. These results show the possibility to distinguish between ordinary and sluggish fermentations at an early stage of the fermentation process in artificial grape must.

Once the density and viscosity for each liquid is known, both the error associated with the calibration process and the resolution can be determined. The error terms, are calculated as the deviation between the values of viscosity and density estimated from our calibration (*MEMS*) and the values measured with the commercial instrument (*viscometer*), obtaining a mean error term around 8.3% for the viscosity and 0.38% for the density. The resolutions were obtained from uncertainty propagation from f_r and *Q-factor* to density and viscosity, obtaining a value below 500 µg/ml and 0.017 mPa·s. As the results show, the resolution and the calibration error obtained are better in density than in viscosity. This occurs because the *Q-factor* measurement, which corresponds to the measurement of G_{osc} in the interface circuit, presents a lower resolution.

4.3. Winemaking samples

In the previous section, the theoretical background for using the density and viscosity as valid parameters of the fermentation process was checked. In addition, the performance of the resonator was verified and validated with two model solutions of grape must. Nevertheless, our main goal is the monitoring of a real red grape must fermentation process. In this section, there is a brief description of the winemaking of *Cencibel* grape must variety.

The *Cencibel* grape must was completed using 25% (w/v) of red grape skin. Afterwards, the must was inoculated with 0.2 g/ml of Saccharomyces cerevisiae strain (UCLM S325, Fould-Springer) previously rehydrated following the supplier's guidelines. A cylindrical fermenter (*mini-Bioreactor Applikon*) was filled with 3 L of inoculated must and grape skins, where the temperature was controlled at 28°C.

The fermentation monitoring was carried out by means of the extraction and analysis of 7 mL-samples every 5 hours approximately during 6 days. After extraction the samples were centrifuged for one minute at 1000 rpm (*Universal 32R Hettich*) and kept refrigerated until analysis. Once the measurements were completed, the samples were kept frozen at -20°C, in order to stop the evolution of the fermentation process.

In an industrial fermentation process temperature fluctuations would influence to some extent the fermentation process and thus the density. Nevertheless, it is the variations that help enologists determine the current fermentation status, and whether or not corrective measures are required as it progresses.

4.4. Monitoring grape must fermentation

In order to monitoring one real grape must fermentation, we applied the same procedure that was developed for the model sluggish solution. In this case, through the measurements of f_{osc} and G_{osc} in the interface circuit and using the coefficients obtained for the normal model solution (see Table 1), we were able to obtain the density and viscosity of the grape must.

Figure 5 represents the evolution in terms of density and viscosity of one real red wine fermentation, measured in the same way as in the previous section. Both density and viscosity follow a rather similar trend, as observed previously with the normal model solution. Therefore, it was confirmed that the calibration process performed for the normal model solution is also valid for other liquids in a similar range of density and viscosity.

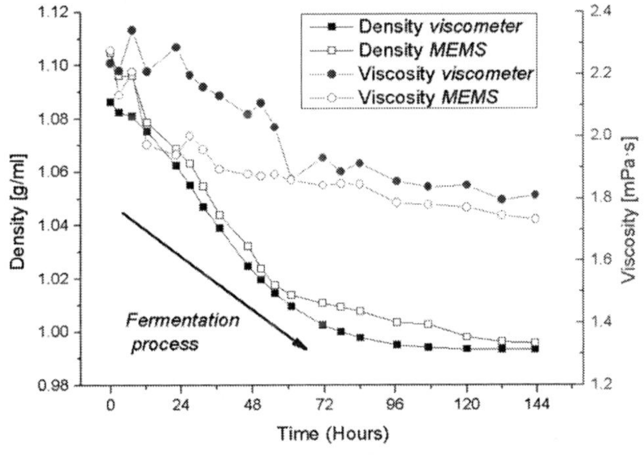

Figure 5. Evolution of density and viscosity values estimated from our resonator (*MEMS*) and measured with the commercial density-viscosity meter (*viscometer*) at 20°C for a real fermentation process.

The error terms and resolutions were calculated as in the previous section. The mean error term obtained was around 6% for the viscosity and 0.7% for the density being the viscosity resolution below 0.006 mPa·s. In the case of the real grape must the obtained density resolution was below 100 μg/ml instead of the 1200 μg/ml value obtained in some previous works using a high-resolution low-cost optoelectronic instrument [11,12].

5. Conclusion

This work presents a novel approach for the monitoring of a grape must fermentation process with a piezoelectric MEMS resonator excited in the 4th order of the roof tile-shaped mode. Two important parameters of the resonator were measured using a PLL-based oscillator circuit: the quality factor and the resonant frequency. Once these two parameters are known, the viscosity and density of the liquid can be determined, requiring only a small amount of test liquid of about 100 μl. In order to measure the viscosity and density of a real grape must, a calibration procedure of the resonator was performed using model solutions of artificial grape must representing an ordinary (for calibration) and a stuck or sluggish (for validation) fermentation process and a commercial density-viscosity meter. Our results demonstrate the high performance of MEMS resonators to detect the decrease in sugar and the increase in ethanol concentrations during the grape must fermentation with a resolution of 100 μg/ml and a sensitivity of 0.16 Hz/μg/ml as upper limits.

Acknowledgments

This work was supported by Spanish Ministerio de Economía y Competitividad project TEC2015-67470-P and the Austrian Research Promotion Agency within the "Austrian COMET-Program" in the frame of K2 XTribology (project no. 849109). Javier Toledo and Víctor Ruiz-Díez acknowledge financial support from FPI-BES-2013-063743 and FPU-AP2010-6069 grants.

References

[1] Ronald S. Jackson. Wine Science, Principles and Applications. Third edit. Elsevier; 2008.
[2] Boulton RB, Singleton VL, Bisson LF, Kunkee RE. Principles and Practices of Winemaking. Springer US; 1999. XIV, 604.
[3] El Haloui N, Picque D, Corrieu G. Alcoholic fermentation in winemaking: On-line measurement of density and carbon dioxide evolution. J Food Eng. 1988;8(1):17–30.
[4] Koukolitschek K, Krieg G, Maier W. Verfahren und Vorrichtung zur präzisen Bestimmung der Alkoholkonzentration in Flüssigkeiten. 4138419, 1991.
[5] Nerantzis ET, Tataridis P, Sianoudis IA, Ziani X, Tegou E. Winemaking process engineering: On line fermentation monitoring - sensors and equipment. e-Journal Sci Technol. 2007;29–36.
[6] Pérez MA, Muñiz R, Torre C, García B, Carleos CE, Crespo R, et al. Impedance spectrometry for monitoring alcoholic fermentation kinetics under wine-making industrial conditions. XIX IMEKO World Congr Fundam Appl Metrol [Internet]. 2009;2574–8. Available from: http://www.imeko2009.it.pt/Papers/FP_228.pdf
[7] Crespo R, Cárcel LM, Pérez MA, Nevares I, del Álamo M. Suitable at-line turbidity sensor for wine fermentation supervision. Int Conf Food Innov. 2010;(May 2016):1–4.
[8] Lamberti N, Ardia L, Albanese D, Di Matteo M. An ultrasound technique for monitoring the alcoholic wine fermentation. Ultrasonics [Internet]. Elsevier B.V.; 2009;49(1):94–7. Available from: http://dx.doi.org/10.1016/j.ultras.2008.06.003
[9] Resa P, Elvira L, De Espinosa FM. Concentration control in alcoholic fermentation processes from ultrasonic velocity measurements. Food Res Int. 2004;37(6):587–94.
[10] Quintáns C, Acevedo JM, Cao y Paz AM, Graña MJ. Experiences in measuring density by fiber optic sensors in the grape juice fermentation process. XIX IMEKO World Congr Fundam Appl Metrol. 2009;(1):2579–82.
[11] Jiménez-Márquez F, Vázquez J, Úbeda J, Sánchez-Rojas JL. High-resolution low-cost optoelectronic instrument for supervising grape must fermentation. Microsyst Technol. 2014;20(4-5):769–82.
[12] Jiménez-Márquez F, Vázquez J, Úbeda J, Sánchez-Rojas JL. Low-cost and portable refractive optoelectronic device for measuring wine fermentation kinetics. Sensors Actuators, B Chem [Internet]. Elsevier B.V.; 2013;178:316–23. Available from: http://dx.doi.org/10.1016/j.snb.2012.12.091
[13] Jakoby B, Vellekoop MJ. Physical sensors for liquid properties. IEEE Sens J. 2011;11(12):3076–85.

[14] Toledo J, Manzaneque T, Hernando-García J, Vázquez J, Ababneh A, Seidel H, et al. Application of quartz tuning forks and extensional microresonators for viscosity and density measurements in oil/fuel mixtures. Microsyst Technol. 2014;20(4-5):945–53.
[15] Toledo J, Manzaneque T, Ruiz-Díez V, Jiménez-Márquez F, Kucera M, Pfusterschmied G, et al. Comparison of in-plane and out-of-plane piezoelectric microresonators for real-time monitoring of engine oil contamination with diesel. Microsyst Technol. Springer Berlin Heidelberg; 2016;22:1781–90.
[16] Manzaneque T, Ruiz V, Hernando-Garcia J, Ababneh A, Seidel H, Sanchez-Rojas JL. Characterization and simulation of the first extensional mode of rectangular micro-plates in liquid media. Appl Phys Lett. 2012;101(15).
[17] Sanchez-Rojas JL, Hernando J, Donoso a, Bellido JC, Manzaneque T, Ababneh a, et al. Modal optimization and filtering in piezoelectric microplate resonators. J Micromechanics Microengineering. 2010;20(5):055027.
[18] Toledo J, Manzaneque T, Ruiz-Díez V, Kucera M, Pfusterschmied G, Wistrela E, et al. Piezoelectric mems resonators for density and viscosity sensing in engine oil with diesel fuel. Transducers. 2015;2591:436–9.
[19] Pfusterschmied G, Kucera M, Toledo J, Steind W, Ruiz-Diez V, Bittner A, et al. Wine fermentation monitoring using piezoelectric MEMS resonators. In: 20th International Conference on Miniaturized Systems for Chemistry and Life Sciences. MicroTAS; 2016.
[20] Leissa AW. Vibration of plates. Scientific and Technical Information Division, National Aeronautics and Space Administration; 1969.
[21] Kucera M, Hofbauer F, Wistrela E, Manzaneque T, Ruiz-Díez V, Sánchez-Rojas JL, et al. Lock-in amplifier powered analogue Q-control circuit for self-actuated self-sensing piezoelectric MEMS resonators. Microsyst Technol. 2014;20(4-5):615–25.
[22] Ferrari V, Marioli D, Taroni A. Improving the accuracy and operating range of quartz microbalance sensors by a purposely designed oscillator circuit. IEEE Trans Instrum Meas. 2001;50(5):1119–22.
[23] Ribéreau-Gayon P, Glories Y, Maujean A, Dubourdieu D. Handbook of Enology. Handbook of Enology. 2006. 1-441 p.
[24] Walls FL, Allan DW. Measurements of Frequency Stability. Proc IEEE. 1986;74(1):162–8.
[25] Manzaneque T, Ruiz-díez V, Hernando-garcía J, Wistrela E, Kucera M, Schmid U, et al. Piezoelectric MEMS resonator-based oscillator for density and viscosity sensing. Sensors Actuators A Phys. Elsevier B.V.; 2014;220:305–15.
[26] Castille C, Dufour I, Lucat C. Longitudinal vibration mode of piezoelectric thick-film cantilever-based sensors in liquid media. Appl Phys Lett. 2010;96(15):1–4.
[27] Maali A, Hurth C, Boisgard R, Jai C, Cohen-Bouhacina T, Aimé JP. Hydrodynamics of oscillating atomic force microscopy cantilevers in viscous fluids. J Appl Phys. 2005;97(7):1–6.
[28] Youssry M, Belmiloud N, Caillard B, Ayela C, Pellet C, Dufour I. A straightforward determination of fluid viscosity and density using microcantilevers: From experimental data to analytical expressions. Sensors Actuators, A Phys [Internet]. Elsevier B.V.; 2011;172(1):40–6. Available from: http://dx.doi.org/10.1016/j.sna.2011.02.025
[29] Landau LD, Lifshitz EM. Fluid Mechanics [Internet]. Image Rochester NY. Elsevier; 1987. 539 p. Available from: http://www.springerlink.com/index/10.1007/b138775
[30] Riesch C, Reichel EK, Keplinger F, Jakoby B. Characterizing Vibrating Cantilevers for Liquid Viscosity and Density Sensing. J Sensors [Internet]. 2008;2008:1–9. Available from: http://www.hindawi.com/journals/js/2008/697062/
[31] Dufour I, Lemaire E, Caillard B, Debéda H, Lucat C, Heinrich SM, et al. Effect of hydrodynamic force on microcantilever vibrations: Applications to liquid-phase chemical sensing. Sensors Actuators, B Chem [Internet]. Elsevier B.V.; 2014;192:664–72. Available from: http://dx.doi.org/10.1016/j.snb.2013.10.106

Direct 3D printed shadow mask on Silicon

S Rahiminejad, E Köhler and P Enoksson

Chalmers University of Technology, Department of Micro and Nanotechnology, Sweden

E-mail: `rahimine@chalmers.se`

Abstract. A 3D printed shadow mask method is presented. The 3D printer prints ABS plastic directly on the wafer, thus avoiding gaps between the wafer and the shadow mask, and deformation during the process. The wafer together with the 3D printed shadow mask was sputtered with Ti and Au. The shadow mask was released by immersion in acetone. The sputtered patches through the shadow mask were compared to the opening of the 3D printed shadow mask and the design dimensions. The patterned Au patches were larger than the printed apertures, however they were smaller than the design widths. The mask was printed in 4 *min*, the cost is less than one euro cent, and the process is a low temperature process suitable for temperature sensitive components.

1. Introduction

One of the most important steps during microfabrication is masking of the wafer. There exists different masking techniques, the most common one is photolithography.

Photolithography can be done by writing the pattern into the photoresist with e.g. an e-beam or a laser writer. Photolithography can also be done with a chrome mask placed between the photoresist covered wafer and a UV light source.

When patterning metal, Lift-off is a method that avoids etching of the metal. This process masks a negative photoresist pattern, which is then covered with the metal. The photoresist is then removed with a solvent, together with the metal on top of the photoresist. One issue with photoresist masking is that some photoresist development solvents can etch exposed metal surfaces on the wafer.

Another masking technique is when a shadow mask is used. The first shadow mask was presented in 1986 [1]. A shadow mask is a mechanical mask that can be used for metal patterning [2, 3, 4] similar to lift-off or as a hard mask during plasma etching [5], and also for polymer spray coating [6]. They can be active, that is that the aperture openings can be adjusted, or passive with fixed aperture sizes. Metal shadow masks made by milling are suitable for larger aperture sizes, laser cutting can be used to create smaller aperture sizes. However, when high resolution micro-aperture sizes are needed, microfabricated shadow masks are used. Microfabricated shadow masks are often made out of Si [2, 3, 4, 5], where a Si wafer is patterned with photolithography and the aperture is etched either by wet or dry etching. These shadow masks are then clamped [2, 5] or taped with heat resistive tape. In [6] a different kind of shadow mask made out of Ni with a novel magnetic clamping method was presented. However, all of the presented shadow masks still needs a photolithography step and an etch step to create the shadow mask.

3D printing can produce shadow masks rapidly and at a low cost compared to standard shadow masks, partly because no photolithography and etch step is needed, and mostly because the printing time is within minutes and material cost is low. In [7, 8] 3D printed shadow masks are used. However these shadow masks still needs to be attached to the substrate with either heat resistive tape or with an alignment fixture. If the shadow mask is not in full contact with the substrate and there is an air gap between the shadow mask and the substrate, the transferred pattern can become much wider than wanted, and deformed in the deposition chamber.

In this paper we present a 3D printed shadow lift-off mask, printed directly onto the Si wafer. This way the shadow mask is in full contact with the wafer and there are no air gaps between the shadow mask and the wafer. The ABS shadow mask is released with acetone which does not etch any metals. The printing process is done at a low cost, low temperature and at a rapid pace.

2. Shadow mask design

The shadow mask design was done with google sketchup and then exported to the 3D printer. The shadow mask has 4 x 6 apertures. The first row consist of six squares with the widths (from left to right) 2 *mm*, 1.8 *mm*, 1.6 *mm*, 1.4 *mm*, 1.2 *mm* and 1 *mm*, the second rows consists of circles with the same sized dimensions, and these rows are then repeated once more see Fig. 1.

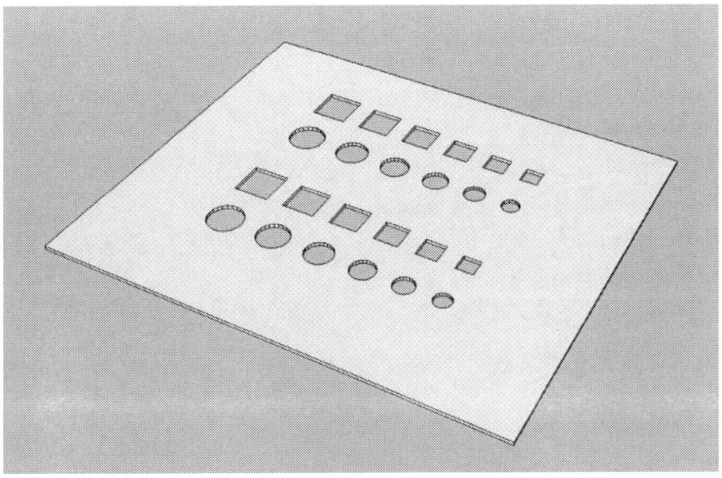

Figure 1: The mask design in google sketchup.

3. Development of 3D shadow mask

The 3D printer used was a Fused Deposition Modelling (FDM) in house built Delta 3D printer. The FDM 3D printer uses hot liquefied thermoplastic that quickly cools down and solidifies, making it possible to print the next layer on top of the previous one. The first layer can be printed on most types of materials and the adhesive properties can be improved by heating up the material. The Si wafer can be placed on the heated bed directly under the nozzle. 3D printing systems such as Selective Laser Sintering (SLS), which are powder based, are expensive and the 3D print will not attach to the Si wafer. Direct light processing (DLP) 3D printers have a limited amount of materials that can be used and the printing is done by printing the layers under the previous cured layers in a liquid. This method is not suitable since the Si wafer cannot be placed anywhere in the 3D printer.

With the in house built FDM 3D printer the x-y-z movements are smooth due to non-backlash magnetic joints and 2.5 μm step size on the stepper motors. What affects the resolution of the print is mainly the nozzle size, the nozzle temperature, the feed rate of the polymer and the speed of the nozzle head. A limitation with the FDM printer is that if the surface has high aspect ratio structures, the printer nozzle might not be able to print evenly without maybe damaging the structures on the Si wafer.

The polymer used with the FDM 3D printer was Acrylonitrile Butadiene Styrene (ABS). Available nozzle opening sizes range between 0.1 *mm* and 0.4 *mm* in diameter. The chosen nozzle opening was 0.4 *mm* in diameter. The smaller the nozzle, the thinner each printed layer would be, which will increase the printing time. The 3D printer was first calibrated without the Si wafer. Thereafter the Si wafer was placed onto the printer bed, and the software was adjusted to start printing at a height of 500 μm (the height of the Si wafer). The wafer was then sprayed with hair spray that leaves a thin layer of water soluble polymer Polyvinyl alcohol (PVA) on the wafer. The PVA polymer enhances adhesion of the ABS to the wafer and also protects the apertures from ABS, since any ABS residues can be washed away with water. The shadow mask thickness was 100 μm and a total of 60 mg of ABS was used. The total cost of the ABS plastic and the running time of the 3D printer was less than one euro cent. The printing time was 4 min and the bed temperature was 70°C.

4. Masking with 3D shadow mask

The Si wafer with the 3D printed shadow mask were rinsed with IPA, followed by a water rinse. The wafer was placed in the FHR MS 150 sputtering tool. A layer of 44 nm Ti and 200 nm Au were sputtered on top (Fig. 2a and b). The wafer with the shadow mask was immersed in an acetone bath and agitated for one hour. The ABS shadow mask was then released, Fig. 2c.

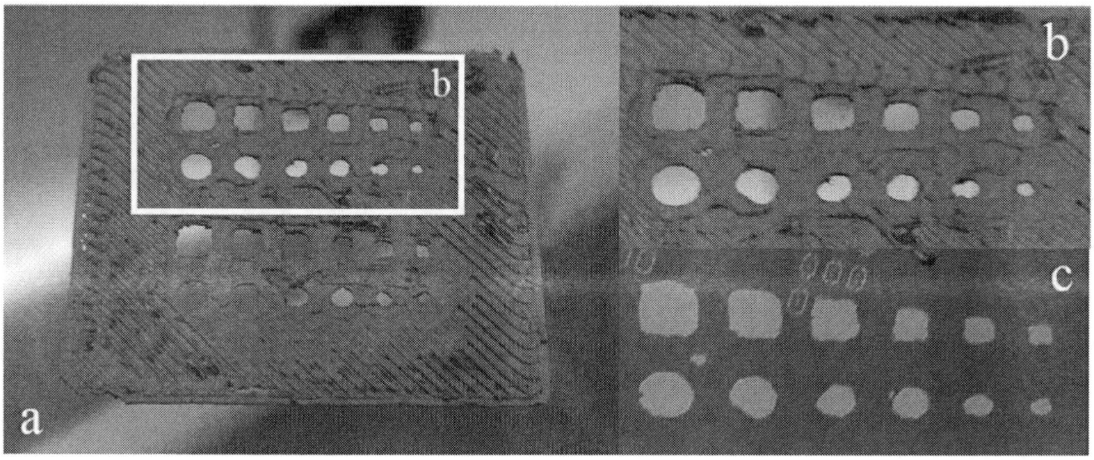

Figure 2: a) The 3D printed shadow mask on the Si wafer after Ti/Au sputtering. b) A close-up of the apertures in a). c) The Ti/Au patterns after the 3D printed shadow mask was released.

5. Evaluation of pattern transfer to the Si substrate

Fig. 3 shows the printed aperture before Ti/Au sputtering, and the Au patch generated from that aperture after lift-off. The widths of the Au patches were measured and compared to the design widths and the printed widths, see Tab. 1. The shadow mask aperture widths are overall smaller than the design, however the Au patches are bigger than the shadow mask aperture widths. The biggest Au patches were between -14% and $+2\%$ of the designed target width, and the smallest Au patches were between -32% and -20% of the target width.

6. Discussion

The printed aperture widths were smaller than the design, probably due to smudging of the ABS during printing. However, the Au patterns transferred to Si wafer were larger than the printed aperture widths thus compensating partly this. The largest Au patch were between -14% and $+2\%$ of the target width. The smallest Au patches were between -32% and -20% of the target width. The 3D printer prints more accurate for the larger structures than the smaller structures. This however can be adjusted by

Table 1: The average widths and standard deviation measured on the printed mask and the patterned Au are compared to the design width.

Design width (mm)	2.0	1.8	1.6	1.4	1.2	1
Printed width (mm)	1.79±0.15	1.61±0.16	1.37±0.17	1.18±0.03	1.01±0.19	0.76±0.17
Au width (mm)	1.87±0.15	1.66±0.14	1.41±0.16	1.24±0.09	1.04±0.12	0.74±0.06

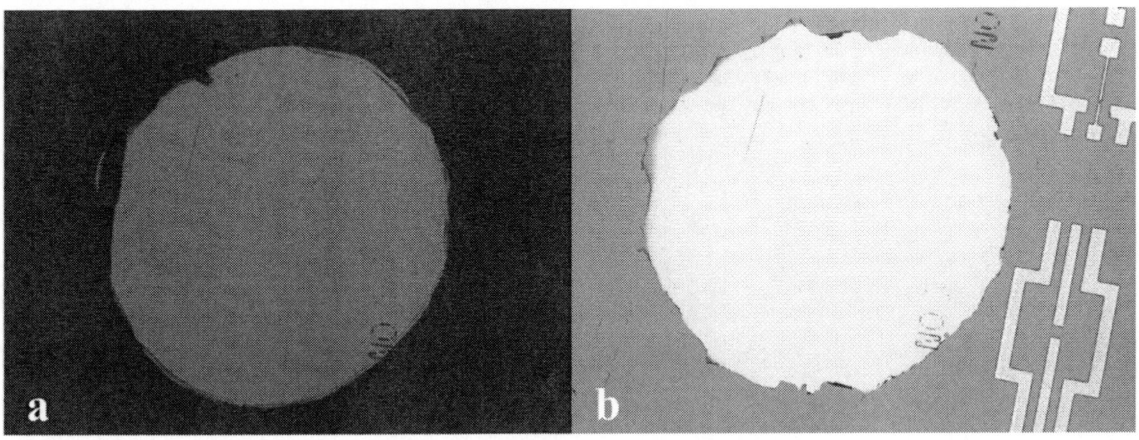

Figure 3: a) Printed aperture. b) Au patch produced from that aperture.

using a one of the smaller nozzle diameter. A smaller nozzle diameter will reduce the flow of ABS, thus resulting in less smudging during printing. A smaller nozzle diameter will also give a better resolution and thinner printing layers, which will also reduce smudging. A 0.25 *mm* nozzle diameter will print the same structure in about 9 *min*. The deviation from the target width can also be compensated for in the design.

7. Conclusion

A 3D printed ABS shadow mask prototype was printed directly on the Si wafer. The shadow mask and Si wafer were sputtered with Ti and Au and the shadow mask was released from the Si wafer by immersion into an acetone bath. The Au patches were closer to the target for the wider patches and more off target for the smaller patches. However, this can be adjusted with a smaller nozzle and a compensated design. The process is a low temperature process suitable for electronic structures not needing very high resolutions, the printing time is fast and the cost of the shadow mask is less than one euro cent.

Acknowledgments

The authors would like to acknowledge the Swedish Research Council VR Grant Number C0466101, the Swedish energy council, and the Chalmers production area of advanced science for funding of the reported work.

References

[1] Dohler G H, Hasnain G and Miller J N 1986 *Applied Physics Letters* **49** 704 ISSN 00036951 URL http://scitation.aip.org/content/aip/journal/apl/49/12/10.1063/1.97573
[2] Brugger J, Andreoli C, Despont M, Drechsler U, Rothuizen H and Vettiger P 1999 *Sensors and Actuators A: Physical* **76** 329–334 ISSN 09244247
[3] Tixier A, Mita Y, Gouy J P and Fujita H 2000 *Journal of Micromechanics and Microengineering* **10** 157–162 ISSN 0960-1317 URL http://stacks.iop.org/0960-1317/10/i=2/a=310?key=crossref.a132af2290fa5324ea79da2d297e0ee8
[4] Apanius M, Kaul P B and Abramson A R 2007 *Sensors and Actuators, A: Physical* **140** 168–175 ISSN 09244247
[5] Kolari K 2008 *Sensors and Actuators A: Physical* **141** 677–684 ISSN 09244247
[6] Keller S S, Bosco F G and Boisen A 2013 *Microelectronic Engineering* **110** 427–431 ISSN 01679317
[7] Sowmya N S, Oraon N, Sen S and Rao M 2015 *2015 IEEE International Conference on Electronics, Computing and Communication Technologies (CONECCT)* (IEEE) pp 1–5 ISBN 978-1-4799-9985-9 URL http://ieeexplore.ieee.org/lpdocs/epic03/wrapper.htm?arnumber=7383925
[8] Khan S, Dang W, Lorenzelli L and Dahiya R 2015 *2015 XVIII AISEM Annual Conference* (IEEE) pp 1–4 ISBN 978-1-4799-8591-3 URL http://ieeexplore.ieee.org/lpdocs/epic03/wrapper.htm?arnumber=7066786

Process flow to integrate nanostructures on silicon grass in surface micromachined systems

H Mehner[1], L Müller[1], S Biermann[2], F Hänschke[3], M Hoffmann[1]

[1] Micromechanical Systems Group, IMN MacroNano®, Technische Universität Ilmenau, 98693 Ilmenau, Germany

[2] Micro-Hybrid Electronic GmbH, 07629 Hermsdorf, Germany

[3] Leibniz-Institut für Photonische Technologien e.V., 07702 Jena, Germany

hannes.mehner@tu-ilmenau.de

Abstract. The process flow to integrate metallic nanostructures in surface micromachining processes is presented. The nanostructures are generated by evaporation of microstructured silicon grass with metal. The process flow is based on the lift-off of a thin amorphous silicon layer deposited using a CVD process. All steps feature a low temperature load beneath 120 °C and high compatibility with many materials as only well-established chemicals are used. As a result metallic nanostructures usable for optical applications can be generated as part of multi-layered microsystems fabricated in surface micromachining.

1. Introduction

Silicon grass as a non-periodical needle-like microstructure shows high potential for numerous applications, such as controlling wetting behaviour of surfaces or for chip assembly [1, 2]. The additional vapour deposition of metal results in nanostructured metallic needles on top of the microstructured silicon grass (see Fig. 1).

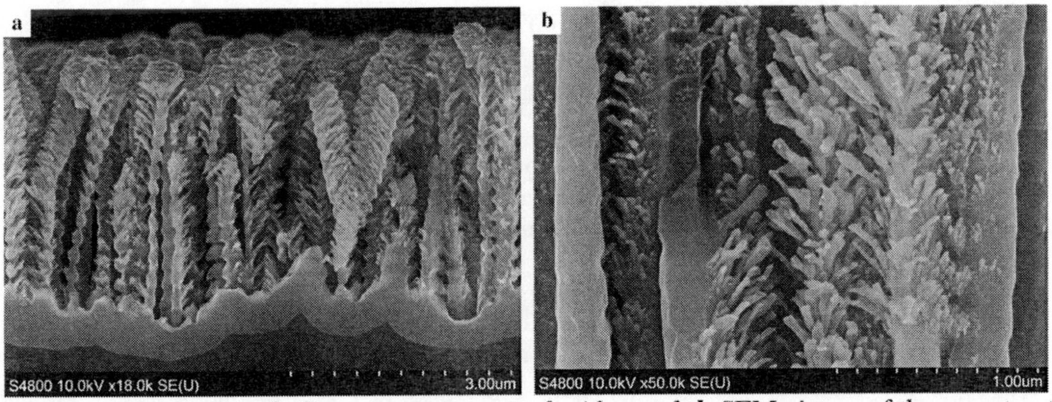

Figure 1. a SEM picture of silicon grass evaporated with metal; **b** SEM picture of the nanostructured needle-like metal grown on the sidewalls of silicon grass

These structures can be used for optical applications as for high emittance infrared sources or absorbers [3]. As described in [3], the silicon grass is typically etched into the bulk of a silicon wafer, which limits the position of the nanostructures in the microsystem and moreover restricts the fabrication flow.
Here we present a process flow to generate silicon grass using surface micromachining processes only. The presented technology enables the integration of silicon grass with vapour deposited nanostructures at the point of interest and in an arbitrary position of a multi-layered microsystem; such as a free-standing membrane that features thermocouples and shall be used as a high efficient infrared absorber.

2. Fabrication Flow

The process flow starts on a thermally oxidized silicon wafer (Fig. 2 **a**). Subsequently, a patterned thick photoresist for the lift-off is lithographically generated (Fig. 2 **b**). An amorphous silicon layer with a thickness of about 5 µm is deposited using an ICP CVD process (Fig. 2 **c**). This step causes the maximum temperature load of about 120 °C for all process steps. Afterwards the microstructured silicon grass is etched into the thin amorphous silicon layer using a fluorine based dry-etch process. Here, the thermal SiO_2 layer is used as an etch stop layer. The self-masking mechanism which is crucial for the generation of silicon grass during the dry-etch process is extensively described in [4].

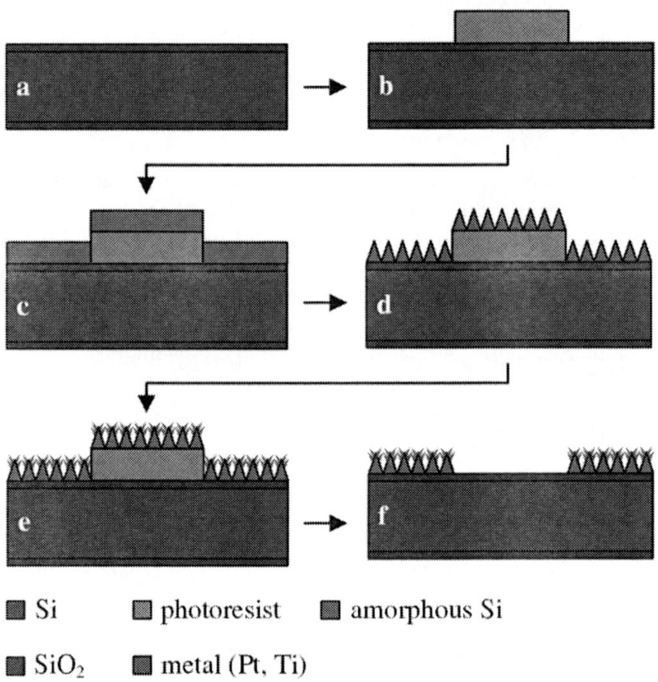

Figure 2. Process flow utilizing surface micromachining processes only, for integration of metallic nanostructures on top of silicon grass in surface MEMS

Afterwards, metal is vapour deposited on top of the silicon grass in order to form needle-like nanostructures (Fig. 2 **e**). In the last step the lift-off is realized utilizing a conventional remover product that is compatible with a lot of metallic, dielectric and semiconductive materials.

3. Process results

Dry-etching of silicon grass depends on the load. The lift-off process whereat the complete substrate is covered with an amorphous silicon layer causes a hundred percent load always. Therefore a stable process result that is independent of the current design is achieved. Process results after etching the silicon grass nearby the photoresist are shown in figure 3.

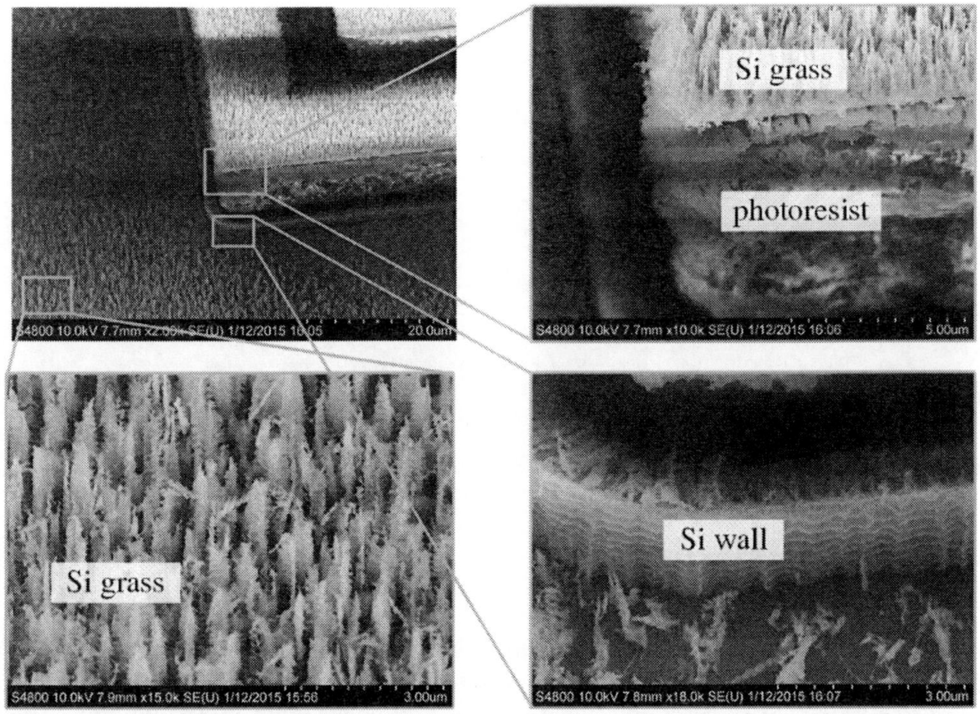

Figure 3. SEM pictures of silicon grass dry-etched in a thin amorphous silicon layer

Silicon grass is formed on top of the photoresist, too. In the close environment of the photoresist no silicon grass is generated. It is likely, that ion reflections at the photoresist sidewalls cause an enhanced passivation etching. Residual passivation clusters on the silicon are elsewhere responsible for the forming of silicon grass. Without these clusters, the silicon layer is removed completely. Directly aside the photoresist no etching occurs and a solid silicon wall is formed showing the typical etch ripples.

Figure 4. SEM pictures of the lift-off result for patterning silicon grass evaporated with metal

Figure 4 shows the process result after the lift-off process was executed. A typical non-periodic nanostructured surface is formed. The metal grows on top of the silicon grass as well as on the scallops of the etched silicon grass and forms fan-shaped crystalline nano-needles.

The developed process chain was used for integration a patterned silicon grass area evaporated with metal into a microsensor for detecting infrared radiation based on the Seebeck-effect (see Fig. 5). The structures are known to augment infrared absorption which therefore is expected to increase the sensitivity and efficiency of the microsensor.

Figure 5. SEM pictures of a patterned infrared absorbing area consisting out of silicon grass evaporated with metal; the absorber area is integrated on a multi-layered microsensor for measuring infrared radiation using the Seebeck effect.

4. Conclusions

Silicon grass is generated by dry-etching a thin amorphous silicon layer that is deposited in a CVD process. Cascaded metallic nanostructures generated by vapour deposition on silicon grass can be fabricated at the point-of-interest by using surface micromachining processes only. The presented process flow exhibits a very low temperature load (<120 °C) so that the integration of such structures in temperature-sensitive systems is possible, e.g. on thermally stable polymers.

Acknowledgements

This work was funded by the Bundesministerium für Wirtschaft und Energie (BMWi) within the project PyTOM (ZF4030901).

References

[1] Stubenrauch M et al 2006 *J. of Micromech. and Microeng.* **16**
[2] Kim J and Kim C-J 2002 *IEEE Int. Conf. MEMS*
[3] Müller M et al 2014 *J. of Micromech. And Microeng.* **24**
[4] Leopold S et al 2011 *J. Vac. Sci. Technol. B* **29**

MEMS gas ionization sensor with palladium nanostructures for use at ambient pressure

L Müller, H Mehner, M Hoffmann

Micromechanical Systems Group, IMN MacroNano®, Technische Universität Ilmenau, 98693 Ilmenau, Germany

l.mueller@tu-ilmenau.de

Abstract. A microfabricated gas ionization sensor with integrated field enhancing silicon-palladium nanostructures is presented. The sensor can be used to determine the ambient gas by its specific breakdown voltage. Technological details of the fabrication and the assembly of the sensor are presented too.

1. Introduction

Gas ionization sensors are based on the different ionization properties of gases at high electrical field strength. In microsystems, a high electrical field can be achieved at moderate voltages due to small gaps of a few micrometers. Additionally the surface of one or both electrodes can be modified to locally enhance the field strength, e.g. by 1D-nanostructures such as carbon nanotubes (CNTs) [1-6], ZnO or Au nanowires [7-9]. These nanostructures can significantly improve the sensor performance, but there are also a few drawbacks. For the CNT based sensors, the oxidation at high electrical currents is an issue and the need for high temperature during the fabrication of those structures can also limit the use in MEMS.

In this contribution, we present the fabrication and integration of 1D silicon-palladium nanostructures into a microsystem using thin film MEMS technology at low temperature and show the potential for use in gas ionization sensors. In the following section the fabrication of the nanostructured electrode is described in detail, followed by the assembly of the gas sensor.

2. Experimental

2.1. Fabrication of the nanostructured electrode

At least one of the two electrodes used in the gas sensor has to have integrated field strength enhancing nanostructures. For the presented sensor, so-called 'silicon grass' is used as a template for growing of Pd-nanostructures. The silicon grass is a self-masked structure etched by deep reactive ion etching (DRIE) with slightly shifted process parameters compared to conventional silicon DRIE, which is described in [10].

The process flow for the nanostructured sensor electrode is given in figure 1. To ensure that the nanostructures protrude beyound the electrode's surface, at first a protruding area has to be defined by UV lithography (*a*) followed by silicon plasma etching of the surrounding surface (*b*). After a thin (200 nm) layer of Aluminum is deposited by electron beam evaporation (*c*) the photoresist is removed and the top Al layer is lifted (*d*). The remaining Al layer is used as a mask during plasma etching of the silicon microstructures ('silicon grass', *e*). Last step is the electron beam evaporation of the metal. It deposits in a kind of glancing angle deposition at the scallops of the silicon grass and thus forms reproducible nanostructures. Here, palladium is chosen because of the appearance of sharp metallic tips and the comparatively high resistance against oxidation.

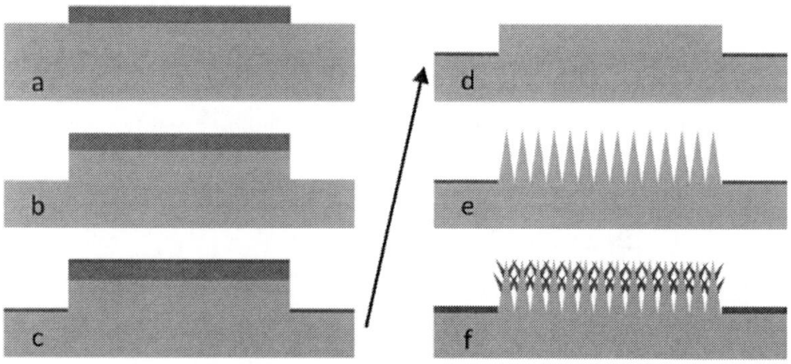

a lithography
b deep silicon plasma etching (DRIE)
c deposition of Aluminum (evaporation)
d lift off
e plasma etching of silicon grass (DRIE)
f deposition and growing of Pd nanostructures (evaporation)

Figure 1. Process flow for the fabrication of the nanostructured electrode.

It has to be highlighted that the growing of the sharp Pd-nanostructures is a result of the scalloped sidewall structure of the silicon grass only. No special tilting of the substrate during evaporation is needed, as it is the case of glancing angle deposition. For further details see [11].
Figures 2 and 3 show micrographs of the etched silicon grass, covered by the Pd nanostructures.

Figure 2. Tilted view (30° to the surface normal) of the silicon grass, covered by Pd-nanostructures

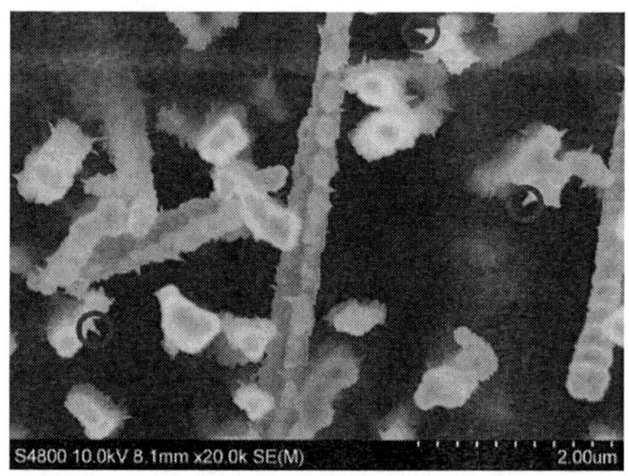

Figure 3. Top view of the structures of fig. 2 with higher magnification. The sharp tips of the Pd-nanostructures can be seen (red circles)

The size of the sensor chip is 6 x 10 mm^2 with an electrode area of about 4 x 4 mm^2. For the gas ionization sensor one electrode is fabricated as shown above and the other electrode has a plain metallic surface.

2.2. Assembly of the gas sensor

For use as a gas ionization sensor the two electrode chips have to be mounted face to face, with electrical contacts at each chip. Due to the need for relatively high voltages and low resulting ionization currents, proper isolation between the chips in the range of 10^{12} Ω is needed to avoid significant leakage currents that would otherwise reach the order of the ionization currents.

 doi:10.1088/1742-6596/757/1/012023

Fig. 4 shows the chosen approach to realize parallelism, low distance, electrical contacts and high isolation at once. The first step is to mount and glue the chips onto a PCB with subsequent wire bonding. The conductivity of the bulk silicon is high enough to ensure that the whole chip including the nanostructures has the same electrical potential. After preparing the two sensor parts a 'nano positioning and measuring machine' (NPMM) is used to place the two chips face to face with a distance of 40 µm between the nanostructures and the plain counter electrode. Parallelism is measured and iteratively fine-tuned with a resulting deviation of less than 1 µm over the 4 x 4 mm^2 area. After the adjustment is finished, the position of the two parts is fixed using highly viscous UV-curing glue. During the curing process the distance and parallelism is monitored and the shrinking of the glue results in an aberration of less than 3 µm.

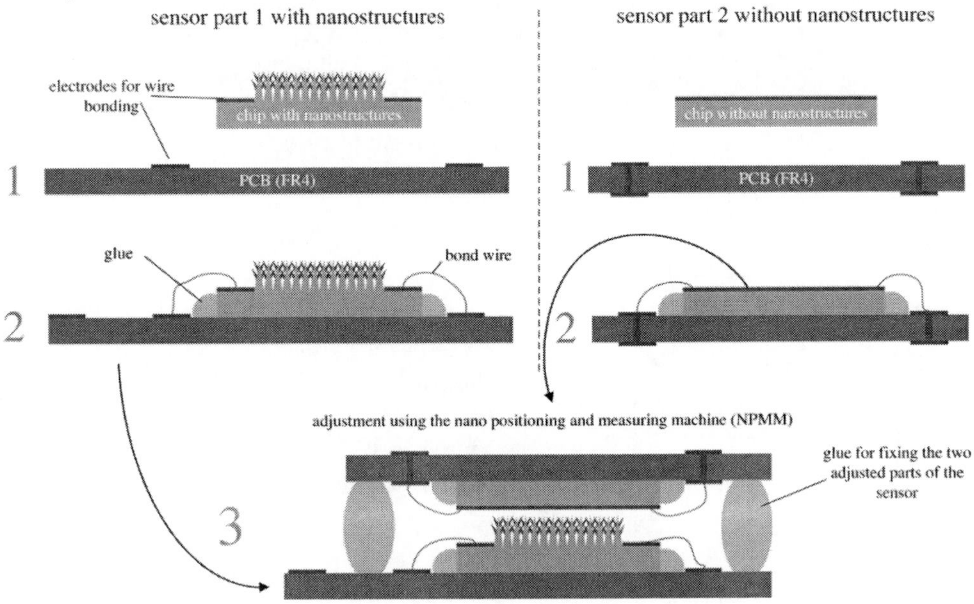

Figure 4. Assembly sequence.

To avoid the contact between the bond wires and the opposed chip, the upper chip is rotated by 90 ° in plane. So the rectangular shape of 6 x 10 mm^2 of each chip avoids electrical short circuit.

2.3. Measurements

To analyze the gas sensing properties of the gas sensor a test site was build (see fig. 5). The gas sensor itself is placed into a sealed PMMA measurement chamber with fluidic and electric feedthroughs. The gas flow is controlled by two mass flow controllers (MFCs) so that two different gases can be supplied independently at the same time with a flow up to 10 sccm. Vacuum generation within the measurement chamber is used to ensure quick and complete gas exchange. All electrical measurements are done at ambient pressure.

Electrical measurements are done using a Keithley 2290-5 high voltage source and 6485 picoammeter, also from Keithley. Fig. 6 shows the circuit diagram of the setup. The 100 MΩ resistor limits the discharge current to prevent a self-sustained breakdown resulting in an arc discharge. The two 1N3595 diodes protect the picoammeter against high voltages. The nanostructured electrode of the gas sensor was configured as the positive electrode.

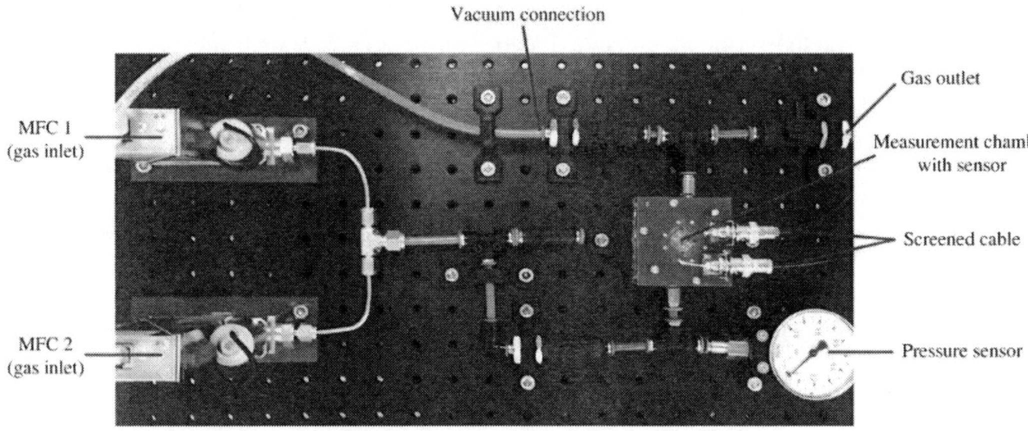

Figure 5. Gas measurement test side.

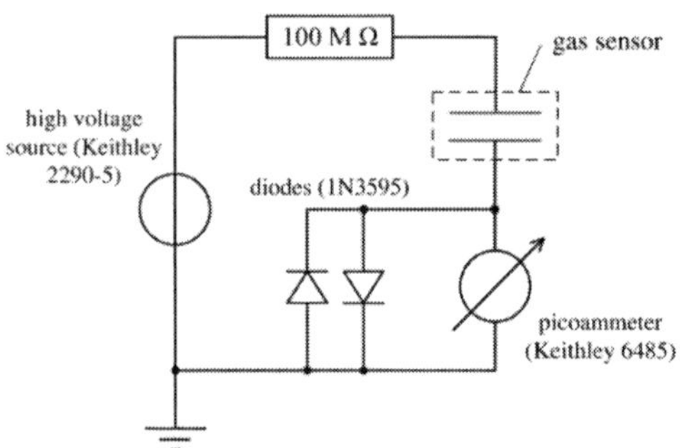

Figure 6. Circuit diagram of the electrical measurement setup, containing a high voltage source, a picoammeter, a current limiting resistor and protection diodes.

3. Results and discussion

The electrical response of the gas sensor was investigated with two different gases, argon and nitrogen have been chosen for the first tests.

During the measurements the gas flow was kept constant at 10 sccm Ar or N_2, respectively, and the source voltage was increased step-by-step until significant current flows. That point is defined as breakdown voltage, it significantly depends on the applied gas. Without the 100 MΩ resistor in series the current would increase dramatically due to the current-depending decreasing impedance of the ionized gas between the electrodes until a self-limiting arc discharge occurs. The 100 MΩ resistor limits the voltage between the electrodes as the voltage drop over the resistor increases with increasing current. As a result a stable glow discharge is generated without the destroying effect of an arc discharge.

In figure 7 the current-voltage graphs for the two different tested gases, Ar and N_2 are shown. Note, that the presented voltage is the actual voltage between the two electrodes calculated by $U = U_{source} - (R \cdot I)$, with R = 100 MΩ and I is the measured discharge current. The breakdown voltage of the two gases varies between 370 V for Ar and 700 V for N_2, respectively. With increasing *source* voltage (not shown here) the current increases, while the voltage between the electrodes decreases. With further increasing current in the range of 10^{-5} A the voltage begins to increase again, which is an indicator that the sensor is driven within the glow discharge regime.

From the measured breakdown voltages the corresponding electrical field strength can be calculated to 17.5 kV/mm for Nitrogen and 9.25 kV/mm for Argon, which is below the value for macroscopic breakdown field strength for these gases. This is an evidence to the field enhancement property of the silicon-palladium nanostructures.

The first experiments show a significant sensitivity for the breakdown voltage of the presented gas sensor. Further characterization experiments with more gases and gas mixtures will be done.

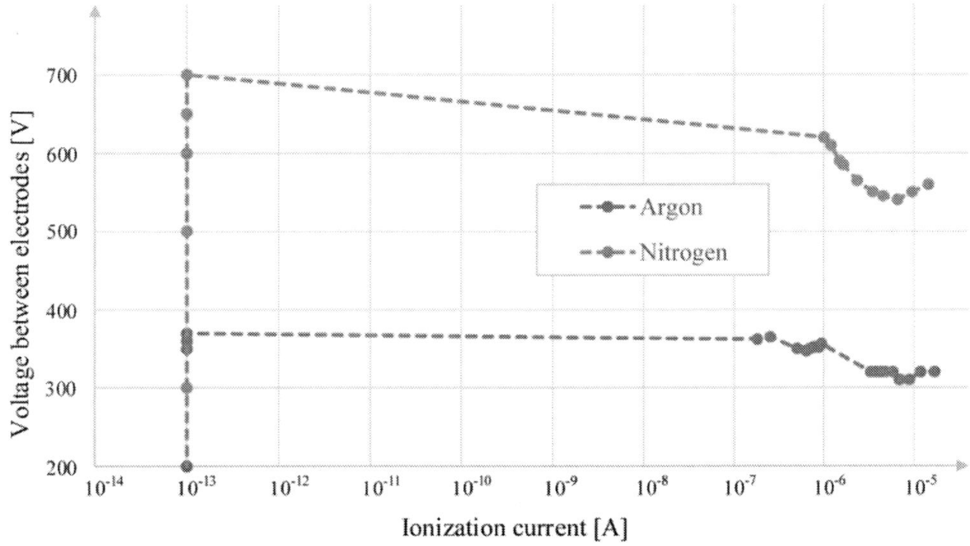

Figure 7. Current-Voltage graphs for the two different tested gases, Ar and N_2.

4. Conclusion
A gas ionization sensor based on the integration of field enhancing Pd nanostructures is realized. The breakdown voltage can be used to differentiate between the gases, which 1.9 times higher for N_2 than for Ar. Compared to other nanostructure based sensors, the presented device is fabricated using thin film technology at low temperatures (<100 °C) only. The assembly of the sensor chips onto separated PCB's is suitable to ensure high electrical resistance (>10^{15} Ω) and good parallelism. Further experiments on the sensitivity to other gases and the stability of the nanostructures will be done.

Acknowledgments
The authors would like to thank the German Federal Ministry of Education and Research (BMBF) for funding within the project 1D-SENSE (16ES0290).

References
[1] Modi A, Koratkar N, Lass E, Wei B Q and Ajayan P M 2003 *Nature* **424** 171–174.
[2] Zhang Y, Liu, Li X, Tang X J and Zhu C C 2005 *Sens. Actuators* A **125** 15–24.
[3] Hui G H, Wu L L, Pan M, Chen Y Q, Li T and Zhang X B 2006 *Meas. Sci. Technol.* **17** 2799–2805.
[4] Hou Z Y, Cai B C and Xu D 2008 *Appl. Phys. Lett.* **92** 223505.
[5] Wang M S, Peng L M, Wang J Y and Chen Q 2005 *J. Phys. Chem.* B **109** 110–113.
[6] Riley D J, Mann M, MacLaren D A, Dastoor P C, Allison W, Teo K B K, Amaratunga G A J and Milne W 2003 *Nano Letters* **3** 1455–1458.
[7] Liao L, Lu H B, Shuai M, Li J C, Liu Y L, Liu C, Shen Z X and Yu T 2008 *Nanotechnology* **19** 175501.
[8] Sadeghian R B and Kahrizi M 2007 *Sens. Actuators* A **137** 248–255.
[9] Sadeghian R B and Kahrizi M 2008 *IEEE Sens. J.* **8** 161–169.
[10] Leopold S, Kremin C, Ulbrich A, Krischok S and Hoffmann M 2011 *J. Vac. Sci. Technol.* B **29** 011002
[11] Müller L, Käpplinger I, Biermann S, Brode W and Hoffmann M 2014 *J. Micromech. Microeng.* **24** 035014

A 3D Polymer Based Printed Two-Dimensional Laser Scanner

H A Oyman[1,2], Y D Gokdel[2], O Ferhanoglu[3] and A D Yalcinkaya[1]

[1] Boğaziçi University, Department of Electrical and Electronics Engineering, Bebek, 34342, İstanbul, Turkey.
[2] İstanbul Bilgi University, Department of Electrical and Electronics Engineering, Eyüp, 34060, İstanbul, Turkey
[3] Istanbul Technical University, Department of Electronics and Communication Engineering , Maslak, 34469, İstanbul, Turkey

E-mail: artun.oyman@bilgi.edu.tr

Abstract. A two-dimensional (2D) polymer based scanning mirror with magnetic actuation is developed for imaging applications. Proposed device consists of a circular suspension holding a rectangular mirror and can generate a 2D scan pattern. Three dimensional (3D) printing technology which is used for implementation of the device, offers added flexibility in controlling the cross-sectional profile as well as the stress distribution compared to the traditional planar process technologies. The mirror device is developed to meet a portable, miniaturized confocal microscope application in mind, delivering 4.5 and 4.8 degrees of optical scan angles at 111 and 267 Hz, respectively. As a result of this mechanical performance, the resulting microscope incorporating the mirror is estimated to accomplish a field of view (FOV) of 350 μm x 350 μm.

1. Introduction

Miniaturization of the laser scanning microscopes to endoscopic imagers and hand-held probes necessitate shrinkage of the size of the scanning unit. Although, galvanometers are preferred as the scanning units in early and conventional laser scanning microscopes (LSM) [1], nowadays, the scanning units are getting replaced by micro-scanners in order to make the LSM more compact and portable [2, 3, 4].

These MEMS devices can be actuated using different kind of techniques such as electrothermal [5], electrostatic [6], piezoelectric [7], and electromagnetic [8] ones. MEMS based micro-scanners reported in handheld LSM system applications mainly use Si as the main fabrication material [9, 10].

The micro-scanners used in LSM systems have various topologies and been built using different kind of structural materials, in order to make them have better scanning angles, imaging and stress performance. The choice of fabrication material can also be used to make these micro-scanners as cheap as possible. Magnetically actuated Fire Resistant 4 (FR4) based micro-scanners that have been implemented for barcode scanning application are shown in literature [11]. Micro-scanners that use polymer as the structural material have been implemented for display applications [12].

Micro-scanner is located at the core of LSM and it is responsible for the two-dimensional orthogonal scanning. Even though various structural materials are present for fabrication, a great majority of the micro-scanners were implemented using Si planar microfabrication processes that restrict shapes of cross-sectional profiles to rectangular ones, in turn limiting the mechanical performance. In order to acquire a high optomechanical performance, i.e. low-cost, low power micro-scanner with low stress distribution and long life time, we resort to a native 3D fabrication process, namely polymer 3D printing.

In this study, 3D printing technology is used to fabricate polymer based micro-scanner which can carry-out two dimensional orthogonal scan movement. Section 2.1 reports optical design and ray tracing simulations. Overview of the micro-scanner is demonstrated in a CAD environment and FEM simulations are conducted in Section 3. Fabrication of the device is briefly explained in Section 4.1 and experimental results are outlined in Section 4.2. Conclusion and discussions are supplied in Section 5.

2. Device Operation

2.1. Optical Design

In order to determine mechanical design constraints of the targeted 3D printed micro-scanner, we performed optical ray tracing simulations as shown in Fig. 1 along with theoretical calculations. The micro-scanner is positioned with inclination 45 degrees to the optical axis and the objective lens. For the targeted confocal microscopy application, we aim to achieve i) approximately 1 μm resolution for sub-cellular imaging capability and ii) 350 μm x 350 μm FOV that is comparable to state-of-the-art benchtop laser scanning microscopes. [2, 13].

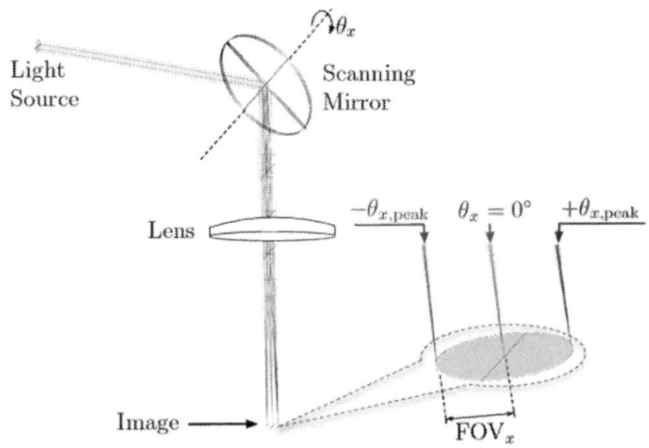

Figure 1. Optical ray tracing simulation for a confocal microscopy application.

Off-the shelf GRIN and miniaturized lenses to be used in LSM typically offer up to 0.5-0.6 NA [14] for the chosen wavelength of 650 nm, allowing tight focusing within limited volume. The choice of objective lens is made to accommodate an input beam size of 3.5 mm (in accordance with the mirror size of the micro-scanner) plus additional shift during scanning operation while providing the targeted 1 μm full-width-half-maximum (FWHM) lateral resolution. Therefore, we have considered an objective lens having 0.4 NA (large enough to produce desired resolution), 6.1 mm diameter (large enough to accommodate beam size and beam shift due to scanning) and 4.6 mm effective focal length (ensuring 0.4 NA). The lateral FWHM resolution is given [15]:

$$w = \frac{2\sqrt{\ln 2}\, 0.32\ \lambda}{NA} \tag{1}$$

For a 3.5 mm wide input beam with an effective NA of 0.175 that is underfilling the lens having 0.4 NA, we achieve a FWHM lateral resolution of 0.87 μm. We invoke the paraxial approximation in order to relate the field of view (FOV) to the zero-to-peak optical scan angle θ_o with the following equation.

$$FOV = 2\,\theta_o f \tag{2}$$

where f is the focal length of the objective. Therefore, in accordance with the desired FOV of 350 μm x 350 μm, one needs a zero-to-peak optical scan angle of 2.4 degrees, corresponding to a zero-to-peak mechanical scan angle of 1.2 degrees for the focal length of 4.6 mm. For the given optical architecture and beam size, the FWHM resolution and FOV were also simulated in a ray tracing environment and is observed to perfectly match analytical results as given in Fig. 1.

3. Mechanical Design & FEM Simulations

The device is completed satisfying the targeted constraints. It is constructed within a CAD tool. The cross-section of flexure is made circular to reduce the stress at the junction points. In order to use magnetic actuation, holes are opened to place small circular magnets having a diameter and thickness of 3 mm and 1 mm, respectively.

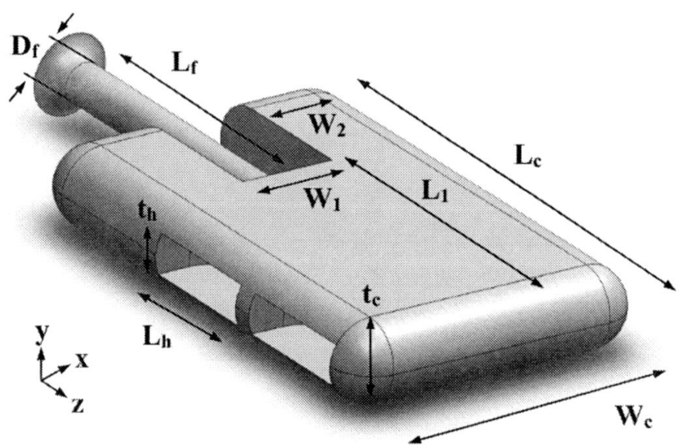

Figure 2. Drawing of the micro-scanner parametrically showing its dimensions.

The mode shapes and eigenfrequencies of the micro-scanner are obtained via finite element method simulations. Before the fabrication process, finite element simulation results give a rough idea about the resonant frequencies of the corresponding mode shapes of the micro-scanners.

Structural material of the micro-scanner device is VeroClear has a manufacturer specified density $\rho_v = 1045\,\mathrm{kg/m^3}$ and Young's Modulus $E_v = 2 - 3\,\mathrm{GPa}$ making it a suitable material for low-frequency resonant actuators. In order to model the micro-scanner in finite element simulations, E_v is set to 2.5 GPa and Poisson's ratio v_v is assumed to be 0.35 for VeroClear material, which lies in the natural range for polymers. Moreover, the magnets attached to the device are made of Neodymium which has a density (ρ_n) of $7500\,\mathrm{kg/m^3}$, Young's Modulus (E_n)

Table 1. Dimensions of the micro-scanner. (all in mm)

D_f	L_f	L_c	L_1	W_c	W_1	W_2	t_c	L_h	t_h
1	8	12	8	8	2,5	1,75	2	3	1

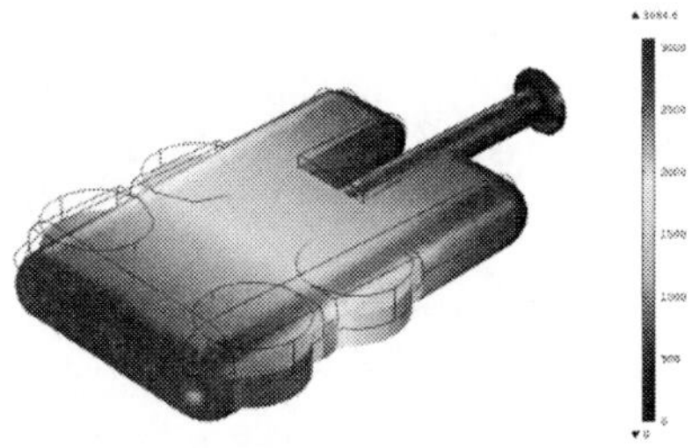

Figure 3. Out-of-plane bending at 137 Hz.

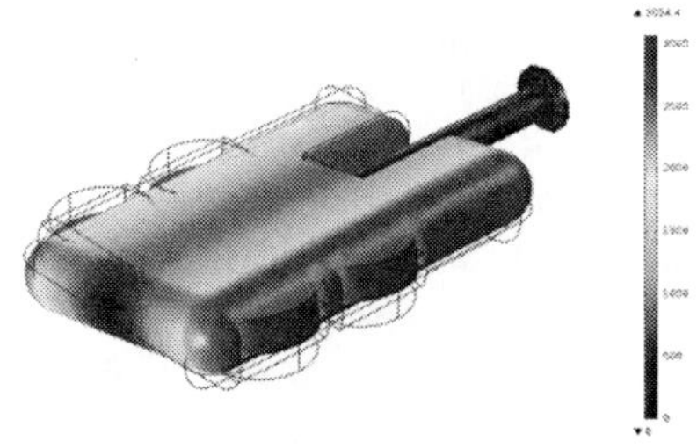

Figure 4. Torsion at 335 Hz.

of 160 GPa and a Poisson's ratio (v_n) of 0.24. The mode shapes for the movement of the micro-scanner are shown in Fig 3, 4 . The combination of the second and the third modes are used to get Lissajous patterns. The second mode of the micro-scanner is out-of-plane bending movement at 137 Hz (Fig 3). Torsion, which is the third mode, around the flexure is observed at 335 Hz (Fig 4).

4. Experimental Results

4.1. Fabrication

3D model designed in a computer-aided-design sofware is exported to the 3D prints. The next step is converting these solid structures into series of thin slices to be physically fabricated layer by layer. Liquid photopolymer material (VeroClear) is selectively exposed to a UV light source, causing it to cure into a solid object. Small neodymium magnets are sticked with epoxy resin as shown in Fig. 5.

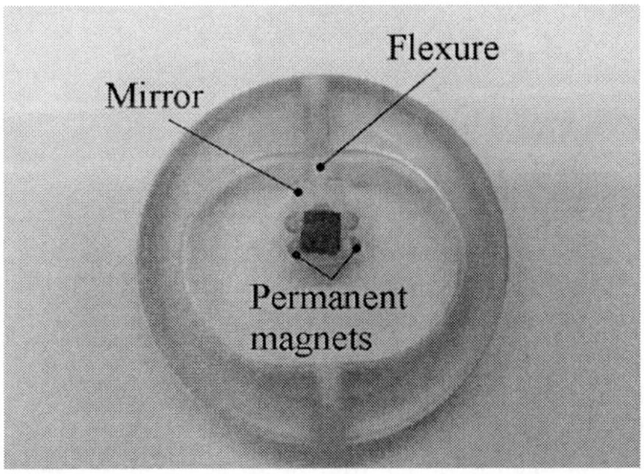

Figure 5. Polymer based 3D printed micro-scanner with attached permanent magnets.

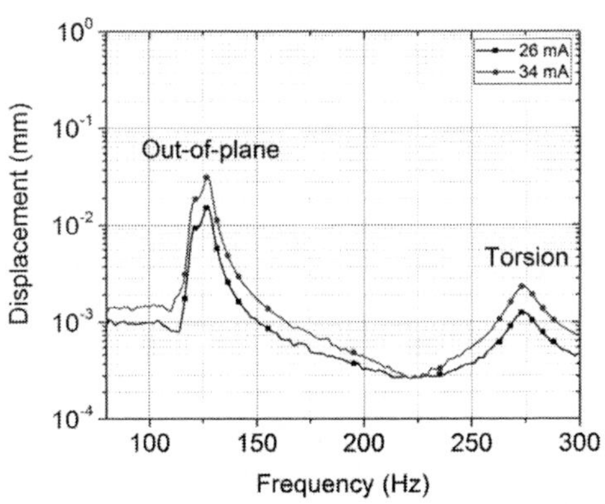

Figure 6. Frequency response of the micro-scanner for varying electrocoil current drives.

4.2. Measurement Results
A characterization setup, which includes an oscilloscope (Tektronix TDS 3032B), signal generator (Tektronix AFG 3101), laser doppler vibrometer (LDV) (Polytech OFV 2500), and PC is used to measure the frequency responses of the micro-scanner. The dynamic deflection of the micro-scanner as a function of frequency is shown in Fig. 6. The electro-coil is fed with a constant sinusoidal amplitude and by varying frequency of the signal, the vibration velocity values are collected by the LDV in ambient air.

As seen in Fig. 6 the slow resonance peaks of the micro-scanner corresponds to the out-of-plane bending mode are at 127 Hz. At this frequency, micro-scanner generate maximum deflections of 32 μm for a 34 mA current. Moreover, quality factor of the device is around 31 for the out-of-plane movement. Torsion around the flexure is observed at 274 Hz. At this frequency, micro-scanner generate maximum deflection around 2.4 μm. Furthermore, quality factor of the micro-scanner is around 25 for the torsion.

4.3. Experimental Lissajous Pattern
Laser source having a wavelength of 650 nm is directed toward the micro-scanner. Lissajous pattern has a length of 14 and 15 mm in vertical and horizontal axes while the coil is fed with a constant current of 64 mA as shown in Fig. 7. This image is occured at a distance of 358.2 mm away from the scanner, hence, TOSA of the scanner for the vertical and horizontal axes are 4.5 and 4.8 degress which meets the desired FOV.

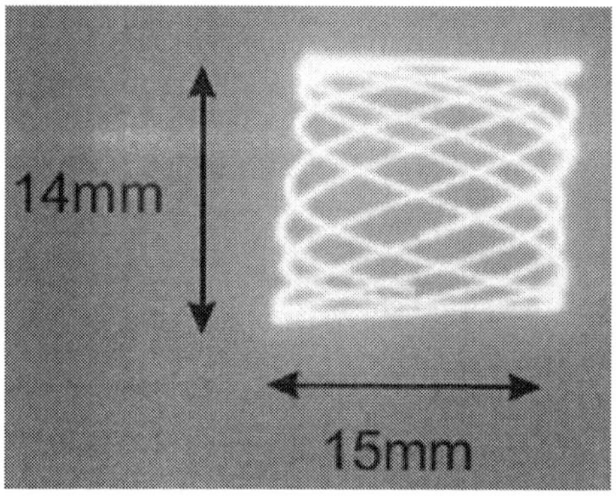

Figure 7. Lissajous patterns obtained at a distance of 358.2 mm from the scanner. The driving frequencies are 111 and 267 Hz for vertical and horizontal axes, respectively.

5. Conclusions and Discussions
First of all, optical ray tracing simulations of a micro-scanner is performed to determine the mechanical constraints to integrated this device in a confocal system. In order to have 1 μm resolution and 350 μm x 350 μm for a 3.5 mm input beam with an effective NA of 0.175 which fulfils an objective lens having 0.4 NA and focal length of 4.6 mm, one need a micro-scanner to utilize zero-to-peak mechanical scan angle of 1.2 degrees.

For this purpose, a 3D-printed micro-scanner is designed, simulated and characterized. Thanks to the 3D printing technology, flexure of the device is made circular to reduce the stress distribution on the flexure. To further optimize the micro-scanner, it is also possible to engineer the shape of the device in all three dimensions. In order to achieve magnetic actuation, small magnets are attached to the device. FEM simulations reveal that the designed structure makes out-of-plane bending movement and torsion at 137 and 335 Hz, respectively. Frequency response data of the micro-scanner is acquired via an LDV. The out-of-plane bending and the torsion mode is observed at 127 and 274 Hz, respectively. TOSA of the micro-scanner is measured for the vertical and horizontal axes and are found to be 4.5 and 4.8 degrees.

6. References

[1] Z. Xie, S. Jiao, H. F. Zhang, C. A. Puliafito, Laser-scanning optical-resolution photoacoustic microscopy, Optics letters 34 (12) (2009) 1771–1773.

[2] H. Ra, W. Piyawattanametha, Y. Taguchi, D. Lee, M. J. Mandella, O. Solgaard, Two-dimensional mems scanner for dual-axes confocal microscopy, Journal of Microelectromechanical systems 16 (4) (2007) 969.

[3] X. Mu, G. Zhou, H. Yu, Y. Du, H. Feng, J. M. L. Tsai, F. S. Chau, Compact mems-driven pyramidal polygon reflector for circumferential scanned endoscopic imaging probe, Optics express 20 (6) (2012) 6325–6339.

[4] K. Aljasem, L. Froehly, A. Seifert, H. Zappe, Scanning and tunable micro-optics for endoscopic optical coherence tomography, Microelectromechanical Systems, Journal of 20 (6) (2011) 1462–1472.

[5] Y. Pan, H. Xie, G. K. Fedder, Endoscopic optical coherence tomography based on a microelectromechanical mirror, Optics letters 26 (24) (2001) 1966–1968.

[6] W. Jung, J. Zhang, L. Wang, P. Wilder-Smith, Z. Chen, D. T. McCormick, N. C. Tien, Three-dimensional optical coherence tomography employing a 2-axis microelectromechanical scanning mirror, Selected Topics in Quantum Electronics, IEEE Journal of 11 (4) (2005) 806–810.

[7] J. M. Zara, P. E. Patterson, Polyimide amplified piezoelectric scanning mirror for spectral domain optical coherence tomography, Applied physics letters 89 (26) (2006) 263901.

[8] T. Mitsui, Y. Takahashi, Y. Watanabe, A 2-axis optical scanner driven nonresonantly by electromagnetic force for oct imaging, Journal of Micromechanics and Microengineering 16 (11) (2006) 2482.

[9] K. Kumar, R. Avritscher, Y. Wang, N. Lane, D. C. Madoff, T.-K. Yu, J. W. Uhr, X. Zhang, Handheld histology-equivalent sectioning laser-scanning confocal optical microscope for interventional imaging, Biomedical microdevices 12 (2) (2010) 223–233.

[10] Y. Wang, K. Kumar, L. Wang, X. Zhang, Monolithic integration of binary-phase fresnel zone plate objectives on 2-axis scanning micromirrors for compact microscopes, Optics express 20 (6) (2012) 6657–6668.

[11] A. D. Yalcinkaya, O. Ergeneman, H. Urey, Polymer magnetic scanners for bar code applications, Sensors and Actuators A: Physical 135 (1) (2007) 236–243.

[12] Y. D. Gokdel, A. O. Sevim, S. Mutlu, A. D. Yalcinkaya, Polymer-mems-based optoelectronic display, Electron Devices, IEEE Transactions on 57 (1) (2010) 145–152.

[13] T. D. Wang, C. H. Contag, M. J. Mandella, N. Chan, G. S. Kino, Dual-axes confocal microscopy with post-objective scanning and low-coherence heterodyne detection, Optics letters 28 (20) (2003) 1915–1917.

[14] J. Knittel, L. Schnieder, G. Buess, B. Messerschmidt, T. Possner, Endoscope-compatible confocal microscope using a gradient index-lens system, Optics Communications 188 (5) (2001) 267–273.

[15] W. R. Zipfel, R. M. Williams, W. W. Webb, Nonlinear magic: multiphoton microscopy in the biosciences, Nature biotechnology 21 (11) (2003) 1369–1377.

27th Micromechanics and Microsystems Europe Workshop IOP Publishing
Journal of Physics: Conference Series **757** (2016) 012025 doi:10.1088/1742-6596/757/1/012025

Alumina-based monopropellant microthruster with integrated heater, catalytic bed and temperature sensors

Zahra Khaji[1], Lena Klintberg[1], Dhananjay Barbade[2], Kristoffer Palmer[3], Greger Thornell[1,2]

[1] Div. of Microsystems Technology, Dept. of Engineering Sciences, Uppsala University, Uppsala, Sweden

[2] Ångström Space Technology Centre, Dept. of Engineering Sciences, Uppsala University, Uppsala, Sweden

[3] SSC Nanospace, Uppsala, Sweden

zahra.khaji@angstrom.uu.se

Abstract. A liquid propellant alumina microthruster with an integrated heater, catalytic bed and two temperature sensors has been developed and tested using 30 wt. % hydrogen peroxide. The temperature sensors and the catalytic bed were screen-printed using platinum paste on tapes of alumina that was stacked and laminated before sintering. In order to increase the surface of the catalytic bed, the platinum paste was mixed with a sacrificial paste that disappeared during sintering, leaving behind a porous and rough layer. Complete evaporation and combustion, resulting in only gas coming from the outlet, was achieved with powers above 3.7 W for a propellant flow of 50 μl/min. At this power, the catalytic bed reached a maximum temperature of 147°C. The component was successfully operated up to a temperature of 307°C, where it cracked.

1. Introduction

For precise attitude and positioning control of small satellites, scale appropriate propulsion systems that can provide thrust in the micro to milli Newton range are needed. A number of different principles, including electric microthrusters, solid propellant microthrusters, cold gas microthrusters, vaporizing liquid microthrusters, and monopropellant as well as bipropellant microthrusters, have been explored [1]. Cold gas microthrusters made in silicon have already been tested in space missions controlling the positions of CubeSats and microsatellites [2-5]. However, the thrust from cold gas microthrusters is limited to about 10 mN to 55 mN [1, 6], and there is a risk of gas leakage of the highly pressurized gas, especially during long missions, where also very small leakages can be fatal.

An attractive alternative that can give thrusts in the range of 10 to 200 mN [1] is to use liquid propellants that can decompose, such as hydrogen peroxide or hydrazine. The continuous operation temperature of silicon and Pyrex is limited, and wall temperatures of liquid propellant microthrusters can be as high as 630°C or above [7]. In addition, chemically active liquid propellants are also highly corrosive. Silicon, e.g., is attacked by hydrazine [8], making it necessary to use more sustainable materials like ceramics.

Compared with silicon, the thermal conductivity of ceramics like alumina is approximately 5 times lower, resulting in reduced thermal losses. 3-D shapes can be made using ceramic green tapes, i.e., flexible sheets made from ceramic powder mixed with a polymer binder. Embossing, laser cutting or milling are used to shape the sheets, and conductors are screen printed with metallic paste before the sheets are stacked, laminated, and finally sintered, making it possible to create multilayered structures with integrated conductors. This technique has been used to make different types of ceramic micro-thrusters, like vaporizing liquid propellant microthrusters, where water has been used as a propellant [9, 10], for a gaseous bi-propellants microthruster [11], and a solid propellant microthruster [12].

Although the corrosion resistance and high-temperature endurance make ceramics suitable for liquid propellant microthrusters that use chemically active propellants, there are few published examples. By integrating a catalytic bed initiating the decomposition, the chemical energy stored in reactive propellants like hydrogen peroxide and hydrazine can be released. Although, not showing any results from a realized component, Plumlee et al. [13] present simulations showing that the temperature in the walls of a ceramic thruster can exceed 800°C, using hydrogen peroxide as the propellant. Wu et al. have made a low-temperature co-fired ceramic thruster, where the decomposition of a hydroxylammonium nitrate liquid propellant was used [14]. There have also been attempts to use hydrogen peroxide in a silicon-based microthruster, but there have been difficulties of achieving complete chemical decomposition of the propellant within the thruster, probably because the temperature was not high enough because of thermal losses [15]. To overcome the excessive heat losses on small scale, a mixture of 90% hydrogen peroxide with an addition of ethanol in a glass microthruster has been evaluated, and it was shown that the stability of the average thrust was improved using the blended propellant [16].

In this paper, a liquid propellant microthruster made in alumina with an integrated platinum catalytic bed to decompose hydrogen peroxide is presented. A resistive heater and two resistive temperature sensors have been integrated into the thruster, making it possible to control and monitor the firing.

2. Design and manufacturing

The microthruster is comprised of five integrated parts: a platinum heater, a resistive platinum temperature sensor close to the heater, a catalytic bed, a catalytic chamber with a micronozzle, and a second temperature sensor placed above the catalytic chamber, figure 1. A divergence half angle of 28° was used for the nozzle since literature suggests a value in the range of 25°-30° [15, 17]. The catalytic bed and the heater have the same size, 7.5 × 4.3 mm before sintering, and overlap fully.

The thruster was built by stacking four layers of 2.5 × 2.5 cm large alumina tapes (150 μm thick, ESL 44007-150, Electro Science Laboratories, USA) that had been structured individually, using milling, and screen printed with platinum paste (ESL 5571, Electro Science Laboratories, USA). For the catalytic bed, the platinum paste was mixed with a sacrificial paste (ESL 4441, Electro Science Laboratories, USA) contributing to 10-15% of the volume, in order to increase the surface area of the structure after sintering. A graphite sacrificial insert that forms volatile products during sintering (ESL 49000, Electro Science Laboratories, USA) was placed in the catalytic chamber. A detailed description of the manufacturing procedure used has been given in [18].

As a last step before sintering, the contacts pads from the heater and the temperature sensor on the bottom of the components were connected to pads on the top by depositing platinum paste over the edge of the component. Some components lacking the top layer, thus leaving the catalytic bed exposed, were also made.

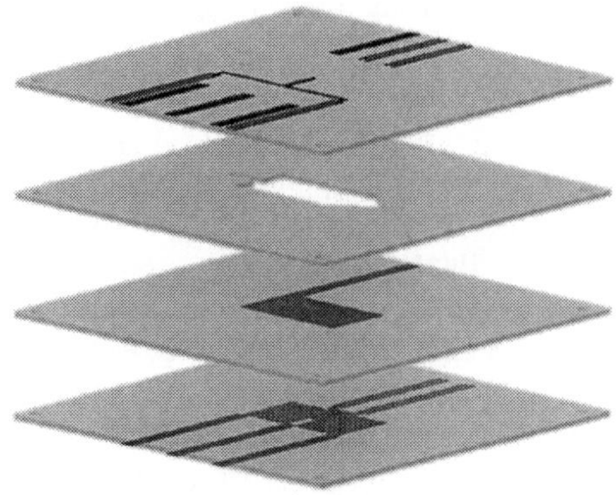

Figure 1. Exploded schematic of full microthruster chip, showing, bottom up: floor layer with heater and temperature sensor on backside, catalytic layer, chamber wall layer, and ceiling layer with leads and temperature sensor on top.

Before characterisation, the electrical contacts of the chip were connected to copper wires (200 μm in diameter), using conductive epoxy (CW2400, Chemtronics, USA), and a coned Nanoport (10-32 IDEX Health & Science, USA), was glued to the inlet using epoxy glue (EPO-TEK 353 ND, Epoxy Technology, USA).

3. Characterization

The complete thrusters were X-rayed (XT V 130, Nikon, Japan) before their temperature sensors were calibrated in an oven using a K-element thermocouple as a reference. MATLAB code was used to control the power supply (TTi QL 355TP) connected to the heater, as well as the digital multimeters (Keithley 2010 and HP 34401) that monitored the resistance of the temperature sensors through 4-point measurements.

The open components were used to study the catalytic bed during initial decomposition tests where the beds were loaded with 10 μl large drops of propellant using a micropipette. These components were also operated dry.

Using 1/16" PEEK tubing, the nozzle inlet of the complete component was connected to a gas tight 5 ml glass microsyringe (VWR International) mounted in a syringe pump (PHD 2000, Harvard Apparatus), and the nozzle exit was monitored using a Nikon SMZ 800 microscope. The syringe was filled with a hydrogen peroxide solution, (30 wt. %, Sigma-Aldrich), and a flow rate of 50 μl/min was maintained. The voltage used to heat the thruster was ramped from 0 V in steps of 0.5 V/min with dwells of 5 minutes at 13, 15, 17 and 19 V.

4. Results

Figure 2 a shows an X-ray picture of the thruster after sintering. The catalytic bed, which has the same dimensions as the heater, exhibits a coarse, porous platinum structure, figure 2 b, increasing the surface area, as the result of mixing the platinum paste with a sacrificial paste before screen printing. When hydrogen peroxide was added as droplets to the open version of the thruster, immediate decomposition was observed already at room temperature. The thruster outlet of the complete nozzle after sintering is shown in figure 2 c.

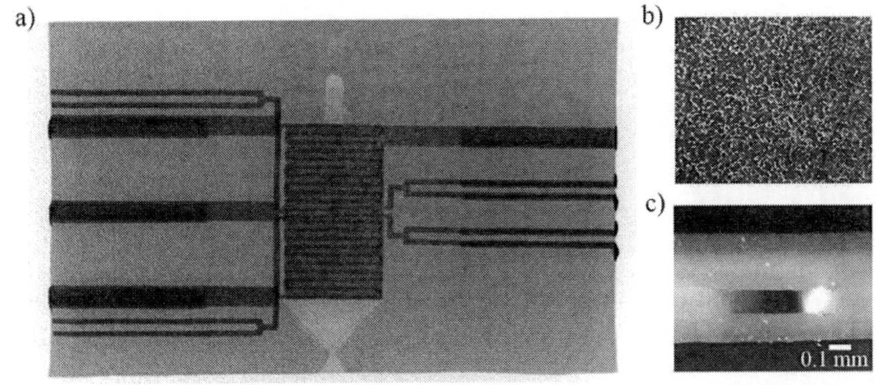

Figure 2. X-ray image of the full thruster chip (a), with the loops meeting at the centre of the meandering heater, being the temperature sensors. To the left, 3 wider leads, making it possible to drive only half the heater at a time, are seen. The porous structure of the catalytic bed (b). The nozzle after sintering has preserved its well-defined rectangular cross section (c).

For a thruster having the catalytic bed open and without feeding with any propellant, a voltage of 10 V, corresponding to a power of 2.7 W, resulted in a temperature of 230°C of the catalytic bed. Tests both with and without propellant flowing through thrusters showed that the temperature detected by the sensors on top and bottom of the thruster differ from each other with less than a few degrees. In order to reduce the wiring needed, only one of the sensors was therefor used for the actual firing experiments

When the applied voltage was ramped from 0 to 19 V at a propellant flow of 50 µl/min, the temperature in the component first rose slowly, and then, when the temperature was around 110°C, it reached a plateau corresponding to the phase transition from liquid to gas in the thruster, figure 3 a. At this stage, the component started to exhaust a mixture of liquid and gas, and the component spat fiercely, being calmer and less prone to spit as the temperature was further increased. During the dwell at 13 V, the temperature rose to 147°C, and only vapor came out from the nozzle. The temperature of the component rose during the dwells, but to less extent on increasing voltage. By holding a glass slide a couple of centimeter from the thruster, the condensation of the vapor could be detected, figure 3 b. The sample cracked during the dwell at 19 V, at 307°C. The crack cut through the center of the component, along the direction of the flow, splitting the component in halves. This was a failure mode seen for several components, usually occurring in the region of 250 to 307°C.

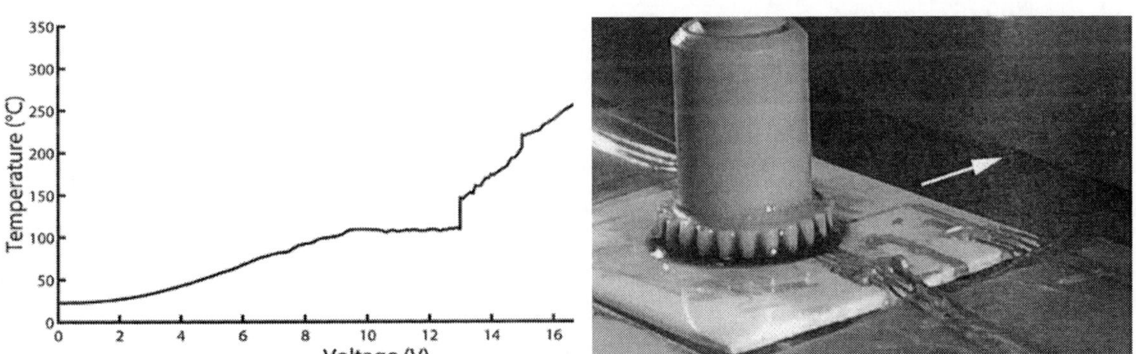

Figure 3. During the dwells at 13, 15, 17, and 19 V, the temperature increased (a). With the thruster activated using a heater voltage of 15 V, corresponding to a temperature of approximately 225 °C, no vapour is seen near the nozzle exit but condensation is observed at the glass slide. The arrow points in the direction of the ejected gas, toward the condensate at a glass slide (b).

During the startup of the thruster, where the decomposition of the hydrogen peroxide is not yet complete, it showed to be important to avoid letting the propellant come in contact with the silver epoxy that covered electrical pads since they then were chemically attacked. Also, the gluing of the Nanoport showed to be a weak point.

5. Discussion and Conclusions

Successful firing of a liquid propellant hydrogen peroxide driven microthruster, having a catalytic bed made of porous platinum, has been demonstrated at temperatures up to 307°C, where the component finally failed by cracking. This temperature is lower than the 409°C that the zirconia microthruster designed by Cheah et al. [10] survived in dry stage. However, and more important, this temperature is much higher than the temperature of 80°C the same authors reported as being the temperature where their component cracked when it was operated with liquid. This shows the potential in the design presented here.

When a ceramic microthruster is fed with liquid propellants, the large temperature difference between the liquid and the component itself is a source of large thermal stresses. This can cause failure also at temperatures far from what the ceramic material in itself can tolerate. Failure caused by thermal stresses in ceramic microthrusters has previously been reported by several authors [10, 14, 19] and is something that has to be addressed already on the design stage.

By integrating temperature sensors, precise control of the decomposition can be achieved, and here the temperature was showed to be more or less equal on both sides of the component. A heater voltage of 10 V corresponded to a component temperature of 230°C in the dry state for a component having the catalytic bed open, but during firing of a complete thruster with hydrogen peroxide, energy is consumed for the phase transition, and the temperature is only slightly above 100°C at the same driving voltage. Since heat is actually generated when the hydrogen peroxide is decomposed, the reaction can be self-sustained once started, if the thermal losses are low enough and high concentrations of hydrogen peroxide is used [13].

Although not being the limiting factor in this work, it is clear that in order to reach higher operating temperature, there is a need to use more robust ways to interface the component since temperatures close to functional limit of the epoxy glue used both for the Nanoport and the electrical wires have already been reached. It is not only the temperature that puts strain on the glue, but also the fact that hot hydrogen peroxide is extremely aggressive.

Future work will focus on reaching higher operational temperature of the component and improve the interfacing. Part of this, entails using a higher concentration of hydrogen peroxide.

Acknowledgements

The Swedish National Space Board and the Centre for Natural Disasters Science (CNDS) are acknowledged for project funding. The Knut and Alice Wallenberg foundation is acknowledged for funding the laboratory facilities.

References

[1] Rossi C 2002 Micropropulsion for space — A survey of MEMS-based micro thrusters and their solid propellant technology *Sensors Upd* **10(1)** 257-92
[2] Wright W P and Ferrer P 2015 Electric micropropulsion systems *Prog Aerosp Sci* **74** 48-61
[3] Guo J, Bouwmeester J and Gill E 2016 In-orbit results of Delfi-n3Xt: lessons learned and move forward *Acta Astronaut* **121** 39-50
[4] Davis T, Baker M T, Belchak T and Larsen W. 2003. XSS-10 micro-satellite flight demonstration program 17th Annual AIAA/USU Conference on Small Satellites
[5] Palmer K, Salaverri A, Sundqvist J, Li Z, Xia X, Kang B, Bai Y, Zhang C, Wu S and Grönland T-A 2016 On the development and demonstration of the NanoSpace cubesat propulsion module with closed-loop thrust control Space Propulsion 2016 Rome 3124889

[6] Cardin J, Coste K, Williamson D, and Gloyer P 2003 A cold gas micro-propulsion system for cubesats 17th Annual AIAA/USU Conference on Small Satellites

[7] Krejci D, Woschnak A, Scharlemann C and Ponweiser K 2011 Hydrogen peroxide decomposition for micro propulsion: simulation and experimental verification *AIAA Paper* 5855

[8] Kundu P, Sinha A K, Bhattacharyya T K and Das S 2013 Nanowire embedded hydrogen peroxide monopropellant MEMS thruster *J Microelectromech S* **22(2)** 406-417

[9] Karthikeyan K, Chou S K, Khoong L E, Tan Y M, Lu C W and Yang W M 2012 Low temperature co-fired ceramic vaporizing liquid microthruster for microspacecraft applications *Applied Energy* **97** 577-83

[10] Cheah K H and Low K S 2014 Fabrication and performance evaluation of a high temperature co-fired ceramic vaporizing liquid microthruster *J Micromech Microeng* **25(1)** 015013

[11] Wu M-H and Lin P-S 2010 Design, fabrication and characterization of a low-temperature co-fired ceramic gaseous bi-propellant microthruster *J Micromech Microeng* **20(8)** 085026

[12] Kaili Z, Chou S K and Ang S S 2005 Development of a low-temperature co-fired ceramic solid propellant microthruster *J Micromech Microeng* **15(5)** 944

[13] Plumlee D, Steciak J and Moll A 2007 Development and simulation of an embedded hydrogen peroxide catalyst chamber in low-temperature co-fired ceramics *Int J Appl Ceram Tec* **4(5)** 406-14

[14] Wu M-H and Yetter R A 2009 A novel electrolytic ignition monopropellant microthruster based on low temperature co-fired ceramic tape technology *Lab Chip* **9(7)** 910-16

[15] Hitt D L, Zakrzwski C M and Thomas M A 2001 MEMS-based satellite micropropulsion via catalyzed hydrogen peroxide decomposition *Smart Mater Struct* **10(6)** 1163

[16] Huh J and Kwon S 2014 Design, fabrication and thrust measurement of a micro liquid monopropellant thruster *J Micromech Microeng* **24(10)** 104001

[17] Kundu P, Bhattacharyya T K and Das S 2012 Design, fabrication and performance evaluation of a vaporizing liquid microthruster *J Micromech Microeng* **22(2)** 025016

[18] Khaji Z, Sturesson P, Klintberg L, Hjort K, and Thornell G 2015 Manufacturing and characterization of a ceramic microcombustor with integrated oxygen storage and release element *J Micromech Microeng* **25(10)** 104006

[19] Sun W, Cai G, Donadio N, Kuo B H, Lee J and Yetter R A 2012 Development of meso-scale co-fired ceramic tape axisymmetric combustors *Int J Appl Ceram Tec* **9(4)** 833-46

Journal of Physics: Conference Series **757** (2016) 012026 doi:10.1088/1742-6596/757/1/012026

Evaluation of 3D printed materials used to print WR10 horn antennas

Elof Köhler, Sofia Rahiminejad and Peter Enoksson

Chalmers University of Technology, Department of Micro and Nanotechnology, Sweden

E-mail: elof@chalmers.se

Abstract. A WR10 waveguide horn antenna is 3D printed with three different materials. The antennas are printed on a fusion deposition modeling delta 3D printer built in house at Chalmers University of Technology. The different plastic materials used are an electrically conductive Acrylonitrile butadiene styrene (ABS), a thermally conductive polylactic acid containing 35% copper, and a tough Amphora polymer containing at least 20% carbon fiber. The antennas are all printed with a 0.25 mm nozzle and 100 μm layer thickness and the software settings are tuned to give maximum quality for each material. The three 3D printed horn antennas are compared when it comes to cost, time and material properties.

1. Introduction

When fabricating prototypes or producing complex or hollow structures fast and at a low cost, not many fabrication techniques can offer this. 3D printing can be a very cost effective solution, it can reduce lead times, improve the design and/or lower the weight of the structure [1].

There are several different 3D printing techniques available all with both pros and cons. Stereolithography and DLP (Digital Light Processing) produces very accurate prints but the different types of materials available is limited. SLS (Selective Laser Sintering) can print metal and plastic without support material but it is far more expensive and not as fast as other techniques. FDM (Fused Deposition Modelling) is the most common 3D printer technology with an increasing amount of users and subsequently increasing amount of printable materials. The choice of 3D printer depends on the desired quality, material, cost and build speed.

Previous work has been done on 3D printed antennas using stereolithography [2]. In this project we are focusing on the FDM technique, mainly because of the amount of materials available. Several aspects need to be considered when printing with an FDM printer and the most prominent for this project will be explained below. The printed material will shrink when cooled down and depending on the material this can be quite extensive. This will give rise to warping, where the recently printed layer shrinks and slightly deforms the previous layers, hence forces the structure to curl upward. As this is a problem that effects the bottom layer the most this problem can be heavily reduced by having a heated bed, where the print surface is at elevated temperatures. Another problem is delamination which will occur if the layers do not bond properly. This occur if the temperature out of the nozzle is too low, if the layer height is too large or in some cases if the cooling of the previous layers are excessive. Cooling of previous layers are necessary when printing small structures to ensure that the current layer are printed on a solid surface, which is important for quality prints. The print quality is of course the main

Journal of Physics: Conference Series **757** (2016) 012026 doi:10.1088/1742-6596/757/1/012026

Table 1. Maximum operating temperature, Flexular Modulus and Density for the different materials

	Maximum Temperature (°C)	Flexular Modulus (GPa)	Density (g/cm^3)
Conductive ABS	88	2	1.3
CopperFill	60	7	4.0
Carbon Fiber	80	6.2	1.4

aspect when 3D printing and print speed, bed temperature, nozzle temperature, layer height, nozzle size, cooling and several other aspects needs to be considered. The surface roughness can also have an impact on the performance of the horn antenna which of course needs to be taken into account [3].

Figure 1. In house built FDM 3D printer of delta type with heated bed.

Figure 2. Extruder with hardened steel nozzle for printing carbon fiber filament.

This paper presents a horn antenna printed with three different materials. The aim was to combine functionality with a low cost using a 3D printer built in house at Chalmers University of Technology. The 3D printer is an FDM delta printer capable of 350°C nozzle temperature and 130°C bed temperature, see Figure 1, 2. The different plastic materials used are electrically conductive Acrylonitrile butadiene styrene (ABS), ColorFabb CopperFill PLA, which is polylactic acid containing 35% copper, and ColorFabb XT-CF20, Amphora polymer containing at least 20% carbon fiber. The goal is to print a horn antenna that is mechanically robust, heat resistant and/or can be electroplated without the use of a seed metal. The electrically conductive ABS has a resistivity of 4.8 Ω·m, a flexible modulus of 2 GPa and a maximum operating temperature of 88°C, see Table 1. The high operating temperature is a nice feature but the thermal conductivity is very low at 0.17 W/mK. As a complement to the soft and thermally isolating ABS we have the CopperFill PLA that is very hard and with a thermal conductivity so high that it is problematic to print because of thermal creep. This material is however heavy and brittle. The carbon fiber material is neither electrically or thermally conductive, but it is mechanically hard and durable. The three printed antennas in this paper have been reprinted until they reached similar quality as the others, but they all have different settings.

2. Design

The horn antenna was designed in Google Sketchup and can be seen in Figure 3. The design is a rectangular horn, a round flange and a WR10 waveguide. The WR10 has an opening of 2.54 mm x 1.27 mm and the total length of the structure is 49.27 mm, the width 26.66 mm and the height 20.81 mm. The total volume of the design is 3.025 cm^3.

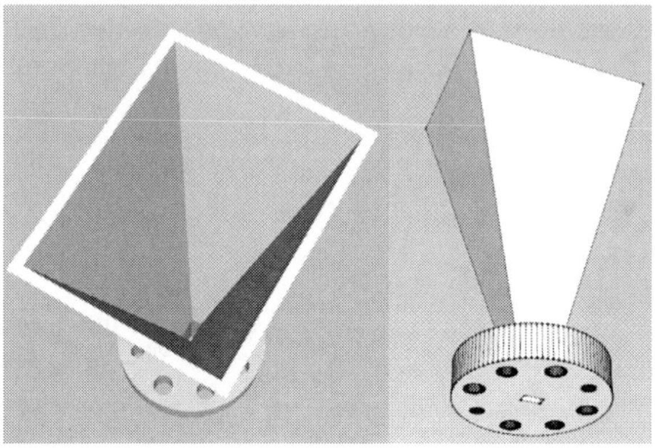

Figure 3. WR10 waveguide horn antenna. The total length is 49.27 mm, the width 26.66 mm and the height 20.81 mm.

3. Fabrication

The aim was to print all materials with the same quality, using a 0.25 mm nozzle diameter and 100 μm layer height.

Figure 4. Electrically conductive ABS horn antenna.

Figure 5. Thermally conductive PLA containing 35% copper.

The conductive ABS needs to be printed on a heated bed to avoid warping and the temperature needs to be high enough to ensure good lamination between the layers. The bed temperature was 70°C and the nozzle temp was 240°C. Cooling of extruded plastic is necessary to ensure high quality and even though the part was cooled by a fan each layer had to be slowed down to wait for the previous layer to cool down and solidify. The thinnest part of the structure had to be slowed to a speed of 15%. The final product weighted 4.15 g and measured 49.3 mm length, 26.8 mm width and 20.9 mm height, see Figure 4. The total print time was 3h36min with a material and electricity cost of approximately €0.13.

The CopperFill have less inclination to warp and does not require a heated bed to print. However, because of the high content of copper – 35% – the thermal conductivity of the material is high. This introduces a problem, if the filament feed rate is too low the filament temperature will rise further up in the extruder and soften the plastic with a clogged nozzle as a result. Without changing the nozzle diameter the only choice to reduce clogging was to increase the filament feed rate by increasing the print speed and to decrease the nozzle temperature. The upside of a thermally conductive plastic is that when printed the layers cool down very fast and can be printed on within a few seconds. The weight of the CopperFill antenna was measure to 12.10 g and the measured length was 49.3 mm, width of 26.7 mm and height of 20.8 mm, see Figure 5. The increased print speed gave a print time of only 1h18min but because of the higher price/kg and with the high density material the total cost reached €0.80.

ColorFabb XT-CF20 is a very durable and stiff material that needs not only a heated bed but also a hardened steel nozzle because of the abrasive nature of the carbon fiber. When printing this material it has a tendency to gather excess material on the side of the nozzle as a viscous drop that, when released, can destroy the print. To avoid this somewhat the print speed needs to be reduced. The carbon fiber antenna weighted 4.36 g and measured 49.3 mm length, 26.7 mm width and 20.8 mm height, see Figure 6. The total print time was 2h35min with a material and electricity cost of approximately €0.30.

From a visual inspection the quality of the three different antennas are similar and none of the antennas needed any post-processing.

Figure 6. the horn antenna from the Amphora polymer containing at least 20% carbon fiber.

4. Discussion

When considering the antenna application with these materials it is well worth to discuss quality and post-processing. There is a limitation on quality with the CopperFill PLA because of the thermal creep and the minimum required filament feedrate. It is however possible to print the ABS and the carbon fiber Amphora with 50 μm layer thickness, which would improve the quality quite a bit, this will however double the print time. It is also possible to reduce the nozzle size for ABS to 0.1 mm with 20 μm layer thickness, with even higher quality prints as a result. This would however give print times for the ABS antenna of 3-4 days.

For the post-processing it can be required for the carbon fiber antenna because of the tendency of drops to form on the nozzle, it was however not needed during these prints. If a smoother surface is needed the ABS is the best bet, as it can be printed at higher quality and also has the possibility to be treated in acetone vapor or acetone bath, which will melt and smooth out the rough parts.

The cost of the printed ABS antenna was 6 times less than the CopperFill antenna. However, the cost for all materials were below one Euro, including the cost of electricity and can be considered to be very low cost compared to the current options. The time to print the antennas are almost 3 times longer for the ABS antenna compared to the CopperFill antenna but also here this is a very short time compared to weeks of lead time.

5. Conclusion

Three different materials were used to print a WR10 horn antenna. From visual inspection the quality of the different materials were almost the same. The time to print the materials ranged from 1h18min for the CopperFill to 3h36min for the ABS with the carbon filled Amphora in the middle at 2h35min. The cost of the antennas including electricity to power the 3D printer was calculated to be a mere €0.13 for the ABS, €0.30 for the carbon fiber material and €0.80 for the CopperFill PLA. The different properties of the materials was noticed either by how difficult it was to print the PLA containing copper because of the high thermal conductivity, from the toughness of the carbon fiber antenna or the electrical conductivity of the ABS antenna. It was also clear that the antenna design could have been more complex as the 3D printing of these antennas did not pose any difficulties from a design point-of-view.

Acknowledgments

The authors would like to acknowledge the Swedish Research Council VR Grant Number C0466101, the Swedish energy council, and the Chalmers production area of advanced science for funding of the reported work.

References

[1] Gibson, I., Rosen, D.W., and Stucker, B. "Additive Manufacturing Technologies: Rapid Prototyping to Direct Digital Manufacturing", Springer-Verlag, USA, 2010

[2] M. van der Vorst and J. Gumpinger, "Applicability of 3D printing techniques for compact KU-band medium gain/high gain antennas", European conferance on Antennas and Propagation 2016

[3] C.R. Garcia, R.C. Rumpf, H.H. Tsang and J.H. Barton, "Effects of extreme surface roughness on 3D printed horn antenna", Electronics Letters, vol 49, no 12, 6th June 2013.

Journal of Physics: Conference Series **757** (2016) 012027 doi:10.1088/1742-6596/757/1/012027

Micro-Venturi injector: design, experimental and simulative examination

S Degenhardt[1], Y Cheriguen, T Geiling and M Hoffmann

Micromechanical Systems Group, IMN MacroNano®, Technische Universität Ilmenau 98694 Ilmenau, PF 100565, Germany

[1]author to whom any correspondence should be addressed

E-mail: sarah.degenhardt@tu-ilmenau.de

Abstract. This paper reports the first factorial design of a micro Venturi injector completed by a simulative investigation of the device. For the first time, a comprehensive correlation between the point of the maximum vacuum pressure generated by the Venturi nozzle and the variation of the inlet pressure is shown. The device reported in this contribution enables a new solution for robust low-pressure generation in parallel fluidic channels.

1. Introduction

The integration of microfluidic functionalities into conventional silicon MEMS devices has yielded growing interest especially with new technologies and products in biology and life sciences. Due to increasing complexity of devices, the parallel operation of several flow generators is required in some cases. Standard microfluidic components can be parallelized easily but miniaturization of effective pumps for flow or pressure generation remains difficult. Mostly, fluidic microsystems consist of miniaturized subsystems that are connected to a macroscopic pump. By using simple channels in parallel, the flow in each device depends on the flow in the parallel devices, i.e. if one channel blocks, the flow in the other channels will increase. For many applications this behavior is not desired. A complex solution to make systems more robust against failure of single channels might be the integration flow-regulated pumps or throttle valves. However, these solutions require additional sensors and active elements. In contrast, the integration of Venturi injectors offers a robust and easy solution to generate a flow in parallel microchannels without the need of any additional components.

The working principle of a Venturi injector is based on a fluid flow that causes low pressure at an inlet port. Thus, low pressure can be generated from a high pressure flow. A Venturi injector consists of pressure inlet, suction inlet, outlet and a convergent-divergent constriction. Low pressure is generated as air flows through the converging passage that gradually widens. The constriction causes a high velocity by which a low pressure at the suction inlet is caused. Thus, the fluid at the suction inlet is drawn into the nozzle and mixes with the inlet fluid.

Venturi injectors on a microscale have been reported as air-flow based tactile display [1], as a device for precise liquid aspiration [2] or as part of an optofluidic flow sensor [3].

The injector reported in this contribution is designed as integrated flow generator supplying an airflow to a particle detector. The low pressure generated at the suction inlet is utilized to produce a flow which transports particle-loaded air into a detection channel where the particles are counted and/or material and size are determined (see Figure 1). The Venturi injector has advantages over the use of a flow-regulated suction pump since multiple injectors can be operated in parallel using a constant inlet pressure, i.e. the flow of the single injectors depends only on the geometrical dimensions of the injector and the inlet pressure and not on the total number of injectors.

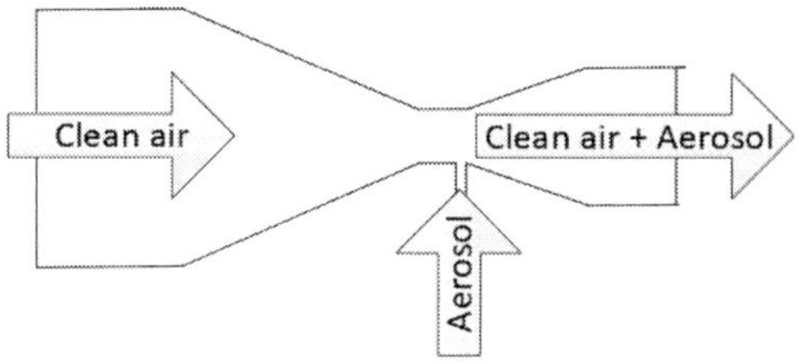

Figure 1. Working principle of the Venturi injector as part of a particle detector. Clean air flows through a constriction and thus aerosol from the suction inlet is soaked into the flow and mixes with the clean air.

2. Design and fabrication

The dependence of the vacuum pressure on the geometric dimensions at the suction inlet was investigated by using a design of experiments (DoE) approach. Aim of this approach is to observe the effect of design variables on the system response. In this case, the DoE was used to examine the impact of geometrical changes of the constriction on the low pressure generated at the suction inlet in order to find an optimum design of the Venturi injector.

For this purpose, a full factorial design with three factors was created. The variables considered to have the largest impact on the vacuum pressure are constriction width W, length L and the divergent angle Θ_{Div}. Table 1 lists the variable design parameters, the level of the single factors and the values of each level. According to the factorial design a total number of 16 devices with different geometries was fabricated. The depth of the constriction and the entire Venturi injector was 40 µm and the convergent angle was 30°. Both parameters were fixed for all designs. The design of the Venturi injector and in particular of the constriction section can be found in Figure 2.

Table 1: Design parameters of the design of experiments (DoE)

Factor	Level	Values
Length L	2	1500 µm; 3000 µm
Width W	2	100 µm; 250 µm
Divergent angle Θ_{Div}	4	20°; 30°; 40°; 50°

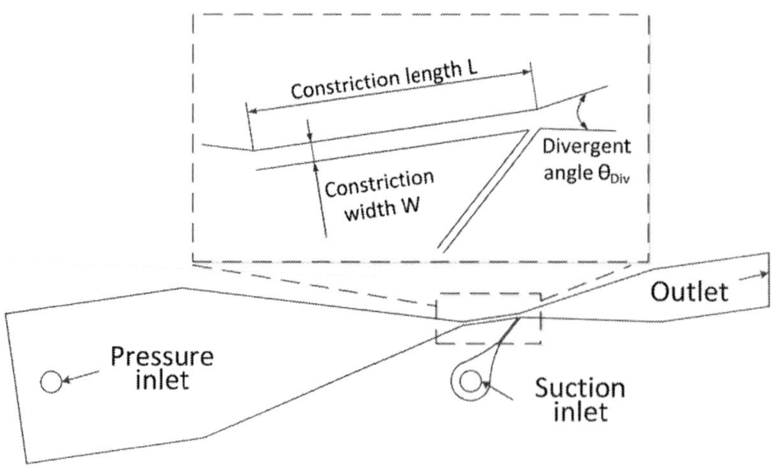

Figure 2. Dimensions of the micro-Venturi injector: The constriction length L, width W and the divergent angle are varied in the factorial design while the remaining geometrical parameters remain fixed.

The fabrication made use of standard microfabrication processes: The channels were etched in a DRIE process into silicon wafers and the openings for the fluid connection were also etched from the back side of the wafers. The outlet is on the edge of the chip (see Figure 3). The fluidic channels were covered by anodic bonding of a Borofloat® glass substrate. This manufacturing process enables simple integration of the Venturi structures into a large number of sensors, actuators and lab-on-chip devices.

To realize a connection to the experimental setup, NanoPort™ assemblies were used. These assemblies are specifically designed to provide a consistent fluidic connection for chip-based analysis. The NanoPort™ assemblies were bonded to the chips using an epoxy-based adhesive ring. They enable a stable and pressure tight connection up to pressures of >100 bar. Pictures of a fabricated chip device with attached NanoPorts™ can be found in Figure 3.

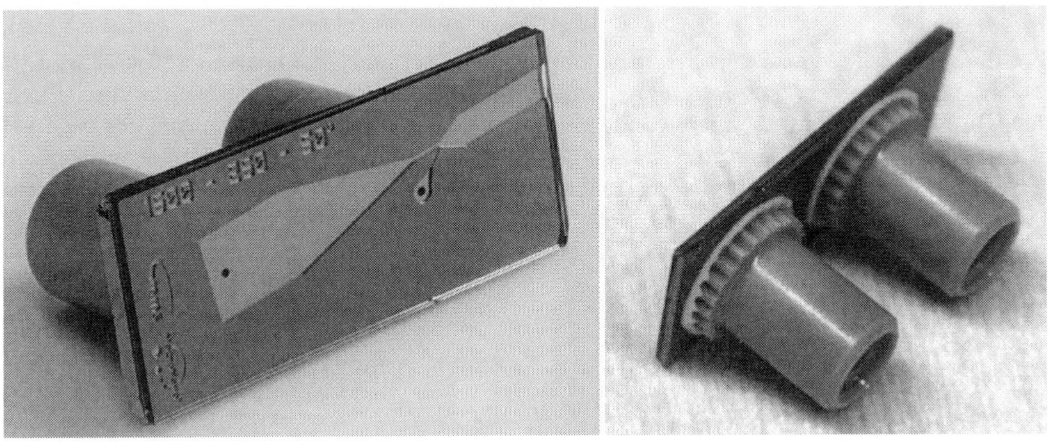

Figure 3. Photographs of the fabricated micro-Venturi injector. The microfluidic channels are etched into silicon and enclosed by a cover glass plate. The fluid connection to the chip is realized by NanoPort™ assemblies.

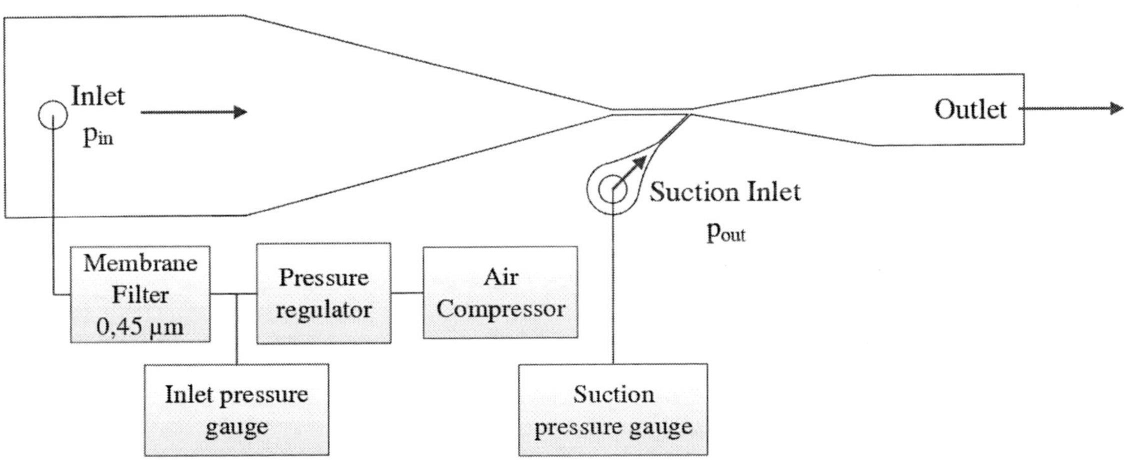

Figure 4. Experimental setup to measure the low pressure p_{out} at the suction inlet in dependence of the pressure at the pressure inlet p_{in}.

Journal of Physics: Conference Series **757** (2016) 012027 doi:10.1088/1742-6596/757/1/012027

3. Experimental

The devices were connected to a compressor with an intermediate pressure regulator and were tested with air at a pressure range from 200 mbar to 3000 mbar at the pressure inlet. The pressure at the suction inlet was measured with a pressure gauge. A schematic of the complete experimental setup can be found in Figure 4.

Figure 5 shows the measured outlet pressure p_{out} at the suction inlet vs. the inlet pressure at the pressure inlet p_{in} for 3 exemplary design variations. The designs depicted in the graph all have a constriction width of 100 µm, a constriction length of 3000 µm and the divergent angle varies from 20° to 40°. During the measurements it was found that the relative suction pressure decreases with rising inlet pressure, reaches a minimum and then increases. These results are comprehended by the simulation as shown in Figure 7. The location of the minimum pressure is not fixed as expected but it depends on the inlet pressure. This insight provides an explanation for the nonlinear behaviour of the Venturi injectors.

For the statistical evaluation of the measurement data the statistical analysis software Minitab was used. The maximum respective vacuum pressure $p_{vac,max}$ for each device served as the base for the evaluation. Figure 6 shows the main effect plot for $p_{vac,max}$. In this graph, the means for each value of the variables are plotted. The results show that the divergent angle has the greatest influence on the suction pressure, followed by the constriction width whilst the constriction length has the smallest impact.

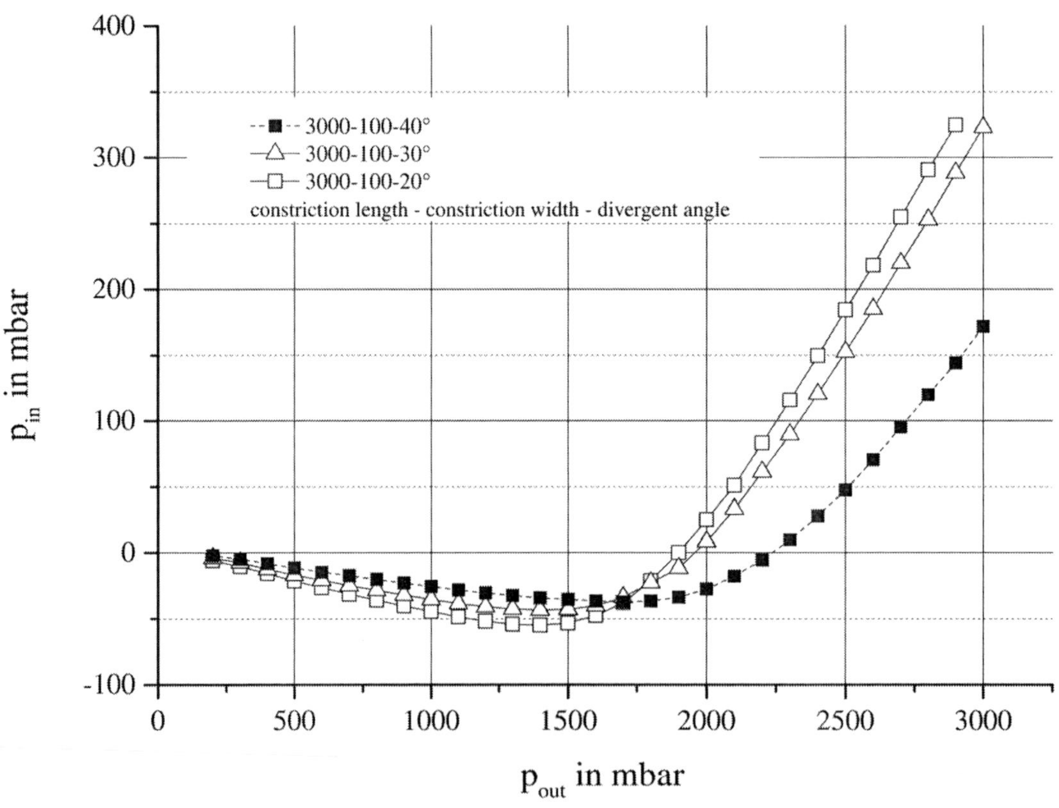

Figure 5. Measurement results of exemplary Venturi injector design variations.

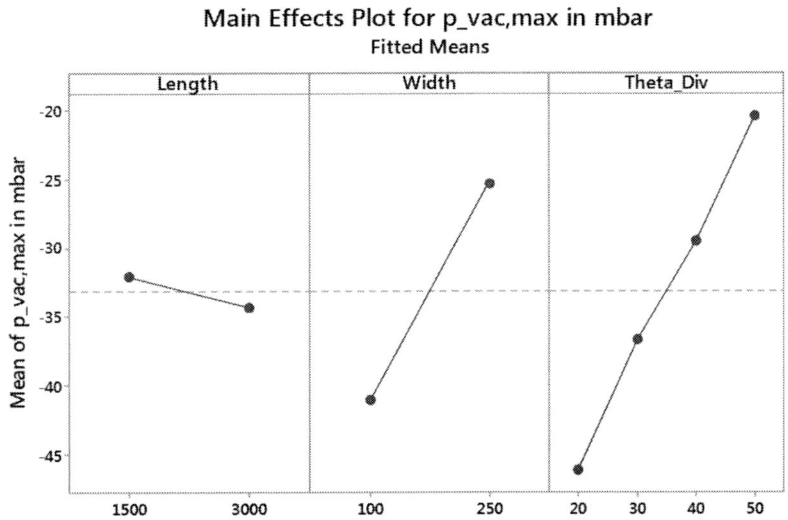

Figure 6. Results of the DoE: Dependence of the maximum vacuum pressure $p_{vac,max}$ measured at the suction inlet on the constriction length, width and divergent angle. As it turns out, the divergent angle has the greatest effect.

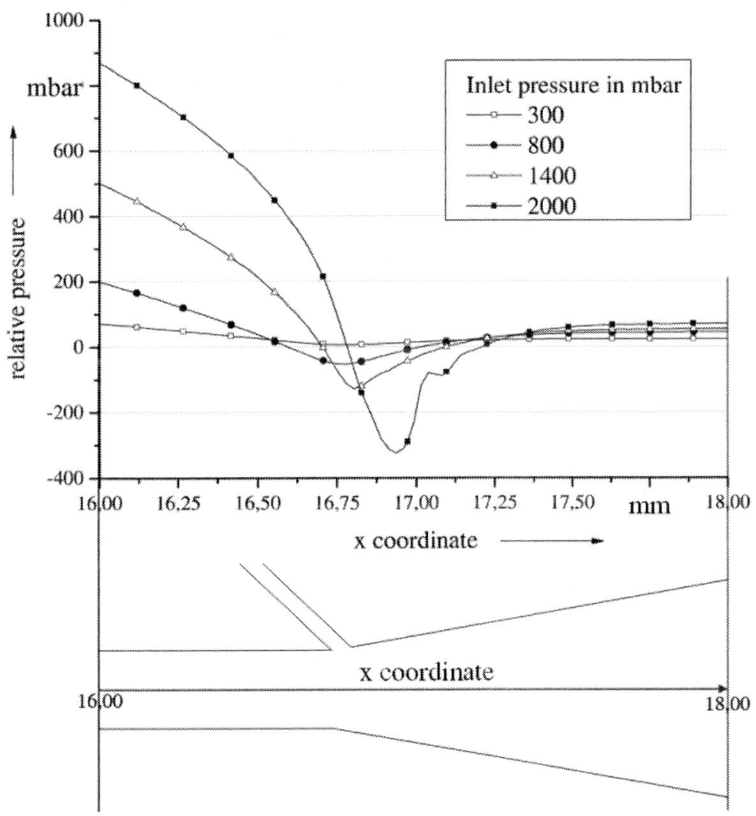

Figure 7. Results of the CFD simulation and relevant part of the Venturi device: The plot shows the relative pressure at the suction inlet in relation to the inlet pressure. The results show that the point of the maximum vacuum pressure is not fixed at the end of the constriction but changes with different inlet pressures.

4. Conclusions

This contribution reports the statistical examination of a Venturi injector based on the design of experiments (DoE) approach. The aim was to investigate the impact of the geometrical dimensions of the constriction on the maximum vacuum pressure at the suction inlet. Measurement results show that the divergent angle has the greatest influence on the suction pressure. The maximum vacuum pressure generated was -54.9 mbar at an inlet pressure of 1400 mbar with a constriction length of 3000 µm, a width of 100 µm and a divergent angle of 20°. The experimental results were supported by a CFD simulation using ANSYS Fluent.

Previous work [4, 5] described the design of a Venturi tube based on the geometrical parameters of the constriction. The experimental results presented in these publications showed that a vacuum pressure is generated along the entire range of applied inlet pressures. In contrast, our results show that the vacuum pressure collapses when the inlet pressure rises above a threshold inlet pressure. The measurements presented in this contribution clearly show a nonlinear correlation between inlet and outlet pressure. In previous work it was claimed that the greatest pressure drop occurs just before the divergent section [4]. The simulative and experimental results reported here could not support this claim and show that the location of the minimum pressure is not fixed, but it moves with changing inlet pressure.

Acknowledgments

This work was supported by the INTASENSE project, grant agreement number 285037, funded by the EC Seventh Framework Program theme FP7-ENVIRONMENT, Subprogram area EeB.ENV.2011.3.1.5-1.

References

[1] Arai M, Terao K, Suzuki T, Simokawa F, Oohira F and Takao H 2012 *Micro Electrical Mechanical Systems MEMS* (Paris, France) 148-51
[2] Perdigones F, Luque A and Quero J M 2010 *J. Microel. Eng.* **87** 2103-09
[3] Cheri M S, Shahraki H, Sadeghi J, Moghaddam M S and Latifi H 2014 *Biomicrofluidics* **8** 054123
[4] Yu H, Li D, Roberts R C, Xu K and Tien Ample N C 2012 *J. Micromech. Microeng.* **22** 035010
[5] Dachao L, Yongjie J, Haixia Y, Wenshuai L, Kexin X, 2013 *Nanotech. and Precision Eng.* **11** 410-17

Inertial focusing of microparticles and its limitations

FJ Cruz[1], **S Hooshmand Zadeh**[1], **ZG Wu**[1,2] **and K Hjort**[1]

[1]*Engineering Sciences, Uppsala University, Ångström Laboratoriet, Uppsala, Sweden*
[2]*State Key Laboratory of Digital Manufacturing Equipment and Technology, Huazhong University of Science and Technology, Wuhan, China*

javier.cruz@angstrom.uu.se

Abstract. Microfluidic devices are useful tools for healthcare, biological and chemical analysis and materials synthesis amongst fields that can benefit from the unique physics of these systems.

In this paper we studied inertial focusing as a tool for hydrodynamic sorting of particles by size. Theory and experimental results are provided as a background for a discussion on how to extend the technology to submicron particles. Different geometries and dimensions of microchannels were designed and simulation data was compared to the experimental results.

Keywords. Particle sorting, Inertial focusing, Micro-fluidic channel, PDMS.

1. Introduction

The field of microfluidics has experienced massive growth in the past two decades towards solving problems in healthcare, biological and chemical analysis, materials synthesis, and other emerging areas that can benefit from the scale, automation, or the unique physics of these systems. Microfluidic chips that focus, concentrate, order, separate, transfer, and mix particles and fluids have been demonstrated.
However, regarding sorting particles, not much progress has been made as we leave behind the microscale to approach the nano realm.

Inertial focusing is a phenomenon where suspended particles migrate across streamlines and focus at well-defined equilibrium points of the cross section. There is a necessity of large velocity gradients to influence particles and fluids while a laminar flow is maintained. Such combination of facts is only possible in micro scale, which is the reason for the lack of observation of similar behaviors in larger scales. Briefly, inertial focusing in straight channels is caused by the balance of two forces, figure 1.

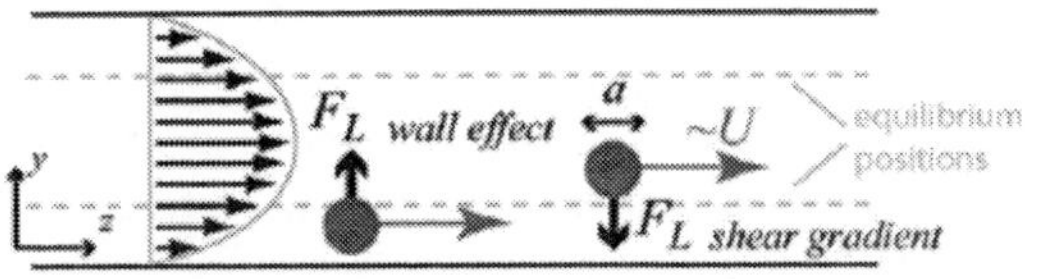

Figure 1. Main forces in a straight microchannel [1].

- a shear lifting force directed towards the walls of a channel due to the parabolic shape of the velocity profile.

$$F_{L,shear\ gradient} = \frac{2\rho U^2{}_f a_p{}^4}{D_h{}^2} \qquad \text{Eq. 1}$$

where ρ is the fluid density, U_f is the average flow velocity, a_p is the particle diameter and D_h the hydraulic diameter of the channel.

- a lifting force directed away from the wall due to interactions with the wall and a drag effect.

$$F_{L,wall\ effect} = 3\pi\mu a_p U_f \qquad \text{Eq. 2}$$

where μ is the dynamic viscosity of the fluid. Also, these forces need to be strong enough to achieve focusing, which will set the focus length. To keep the lifting forces balanced as we scale down the size of the particle, we must compensate by scaling up the fluid velocity and scaling down the hydraulic diameter to the same factor. If we keep a constant aspect ratio, the flow rate should be linearly decreased as well.

The addition of curvature to the channel induces an uneven centrifugal force (since such force is proportional to the square of the velocity) and enables the development of a secondary flow called Dean flow, which redistributes the velocity profile, enhancing the lateral motion of particles and modifying the equilibrium positions [2], figure 2.

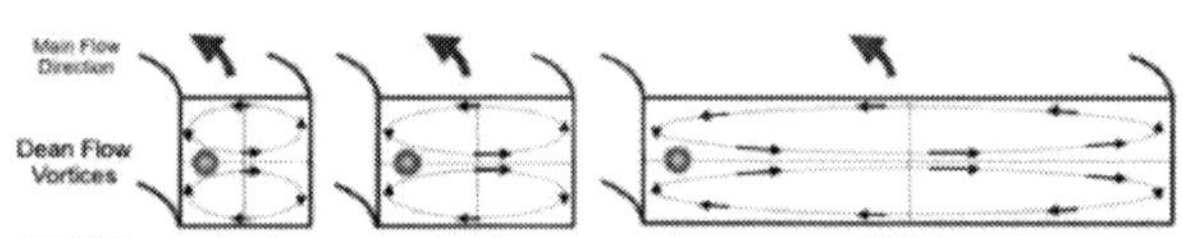

Figure 2. Secondary flow due to centrifugal forces [2].

Only one position remains stable and the enhancement of the lateral migration makes the particles reach such point within a shorter time than in straight systems.

A modification of the cross section with obstacles can also induce secondary flows that may be of advantage [3].

The aim of this paper is to provide experimental demonstrations on inertial focusing and a theoretical approach for the design of the channels and its limitations.

2. Experimental details

Microfluidic chips were fabricated with polydimethylsiloxane (PDMS, RT601, Wacker Chemie) and bonded to a glass slide.

PDMS was poured onto a SU-8 mold that was fabricated by UV lithography on a silicon wafer. It was cured for 1h at 70 °C, peeled off the mold and inlets and outlets were pierced with a biopsy punch.

To form the bond, the PDMS and glass surfaces were activated by plasma exposure with a corona discharger for 30 s prior to coming in contact. A hydrogen bond was formed, which turned covalent after the samples were put into the oven at 70 °C for 30 min.

Fluorescent polystyrene particles (Thermoscientific TM) with diameters of 10, 3 and 1 μm were suspended in deionized water at a concentration of 10^5 particles/ml.

A precision syringe pump (Harvard Instruments) was used to control the flow through the channels.

The experiments were carried out under an inverted microscope.

3. Results and discussion

Inertial focusing relies on the equilibrium of forces acting on the particles. Such forces are strongly dependent on the geometry of the channel, on its dimensions and on the shape and size of particles. This means that the channel should be tuned specifically for the targeted particle size and it will focus only a certain bandwidth of particles.

3.1. Experimental work

The phenomenon was first studied in straight microchannels with 10 µm particles, figure 3.
Equilibrium lines were achieved after a few millimeters from the inlet, in agreement to Dino Di Carlo [1], figure 4.

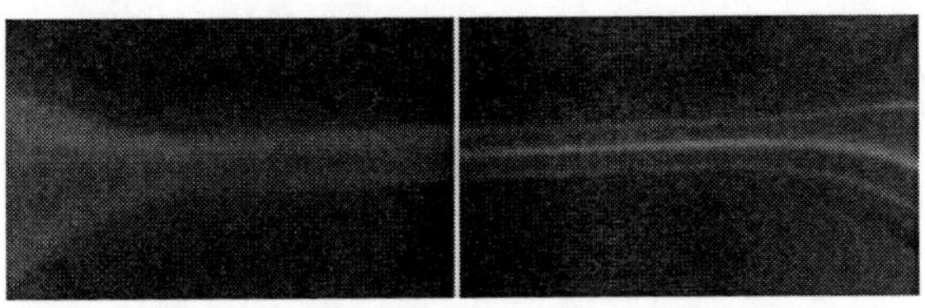

Figure 3. Equilibrium positions for 10µm particles in a straight microchannel, 60x100 µm (h x w) at 200 µl/min.

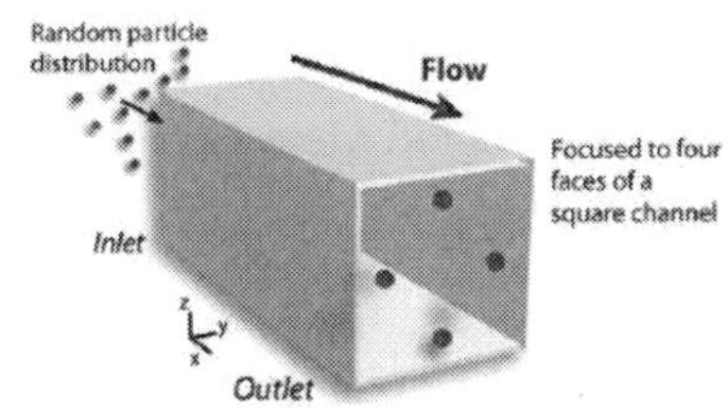

Figure 4. Theoretical equilibrium positions in a microchannel with rectangular cross section [1].

Adding curvature to the channels (spiral shape, growing radius from 0.7 mm up to 4.5 mm) not only led to a single equilibrium position but also faster alignment was achieved (after 1 loop, ~10 mm), figure 5. Changing the height of the channel, 3 µm particles were focused close to the inner wall while those of 10 µm were displaced to the center, figure 6.

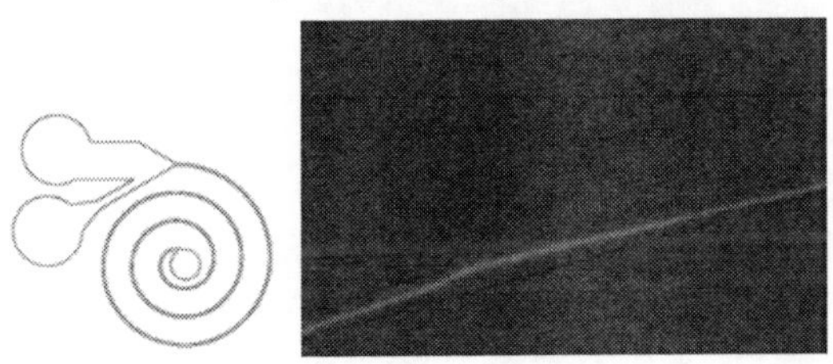

Figure 5. Equilibrium position for 10 µm particles in a curved microchannel, 80x100 µm (h x w) at 200 µl/min.

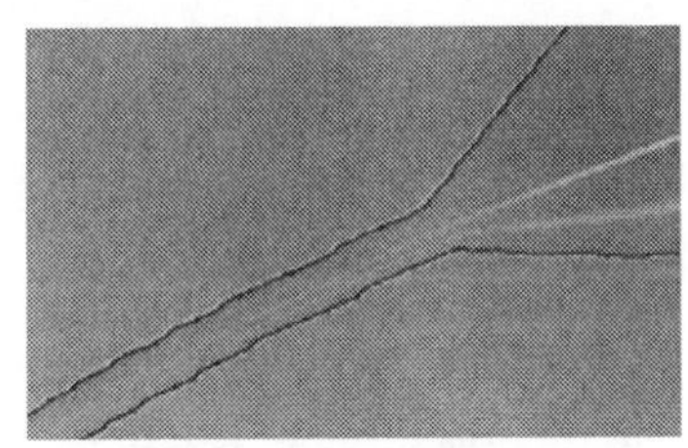

Figure 6. Equilibrium position for 10 µm and 3 µm particles (middle line and lateral line respectively) in a curved microchannel 30x100 µm (h x w) at 200 µl/min.

Finally, obstacles along the inner wall of a curved channel were studied to find if that would improve the performance [3]. In this case, 10 µm particles quickly focused close to the wall while 3 µm particles did not reach a line but gathered in a band, closer to the center, figure 7.

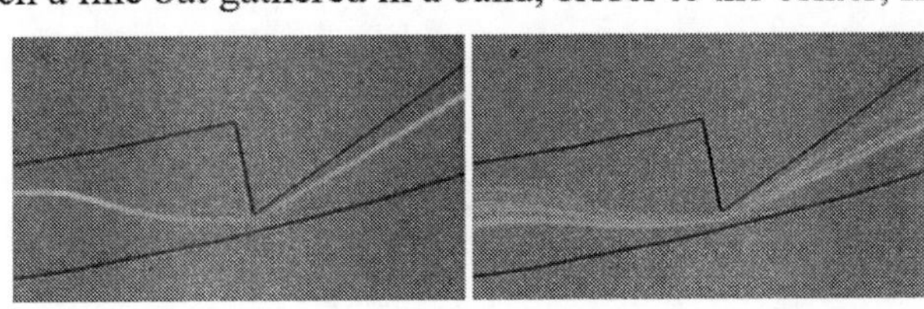

Figure 7. Equilibrium position for 10 µm and 3 µm particles (left and right respectively) in a curved microchannel with obstacles at the inner wall 40x100 µm (h x w) at 200 µl/min.

Obstacles were put also along the outer wall of a curved channel. The flow created by the obstacles dominated over the curvature and the particles were dragged to the outer wall.

No focusing was observed for 1 μm particles in any design. A channel with dimensions suitable for such particle size was tested - 10x30 μm (*h* x *w*). No result was obtained since the high pressure drop led to leakage of the system.

3.2. Theoretical discussion

As shown in eq. 1, the shear lifting force is strongly dependent to the size of the particle (to the power of four), to the flow velocity and to the dimensions of the channels (both to the power of two). On the other hand, the required channel length to achieve focus is proportional to the cube of the particle size (eq. 6). If nothing else is changed, alignment for a 10 times smaller particle requires a microchannel 1,000 times longer. To avoid such long channels, the cross section and the flow rate need to be tailored in order to focus a target particle size. It has the limitation that as particles decrease in size, higher velocities and smaller channels are needed, which turns into a large drop of pressure.
At the same time, such tailored conditions are valid for the target but not for a wide span around it; much larger particles will not fit in the channel and much smaller ones will not feel the lift force.

According to Di Carlo, the dimensions of the channel should be such that $a_p/h > 0.1$. The width should also be matched to the height in a relation of $h < w < 6h$ [1].
This agrees with our obtained results, where in a 30x100 μm channel the smallest particles we could focus were those of 3 μm since1 μm particles do not suffice the condition..

Following the formula recommended by Fuerstman et al. for the pressure drop in rectangular cross section channels [5], fulfillment of the conditions above mentioned will mean a growth of the pressure drop to the power of 3 with a linear shrinkage of the targeted particle size (if the same length was used).

$$\Delta P \approx \frac{Q\,12\,\mu\,L}{h^3\,w\,[1 - 0.630\frac{h}{w}]} \qquad \text{Eq. 3}$$

However, as we decrease the width of the channel, particles need to migrate less transversal distance. The required focus length depends on the slowest of the two velocities of lateral migration and Dean velocity. In our case, it will be the lateral migration velocity.

The required focus length (L_I):

$$L_I = \frac{U_f}{U_L}\,w \qquad \text{Eq. 4}$$

where U_L is the lateral migration velocity:

$$U_L = \frac{\rho C_L U_{max}^2 (a_p)^3}{3\pi\mu D_h^2} \qquad \text{Eq. 5}$$

and then:

$$L_I = \frac{24\pi\mu D_h^2}{\rho U_f \left(a_p\right)^3} w \qquad \text{Eq. 6}$$

Hence, the required focus length will scale linearly with the size of the particle, allowing for shorter channels and thus relieving some pressure drop.

Putting all together, the pressure drop needed to achieve focus of particles increases with the square of the decrease of particle size.

Some exemplifying conditions that will suffice the conditions to achieve focus of a targeted particle size are summarized in table 1. In the three last examples, the height is at its upper limit of ten times the size of the particle. A decrease in height would make the focusing easier but the pressure drop would increase.

Table 1. Example of conditions to achieve focus depending on the targeted particle size. With decreased particle size, the channel length can be decreased.

Particle Size (μm)	Channel length (mm)	Height (μm)	Width (μm)	Flow rate (μl/min)	Average speed (m/s)	ΔP (bar)
10	20	80	100	200	0.4	0.3
3	20	30	100	200	1.1	3.6
1	7	10	30	65	3.7	38
0.3	2	3	10	20	11	360

As stated earlier, when scaling the system with the aim of sorting 1 μm particles, the pressure drop was too high in our case of using PDMS bonded to a glass. The chips started to leak. For further studies aiming at sub-micron particles, the microchannels must be fabricated with more robust materials. A proper material would be glass, in which microfluidics have shown to stand pressures up to 690 bar [4] and which allows for observation of the channels thanks to its transparency.

4. Conclusions

Microfluidic designs, using inertial focusing for separation, offer a solution for sorting particles in the micro scale at high through-put ($Q > 60$ μl/min).

Focusing particles of 10 and 3μm was achieved with PDMS microchannels, while there was no success for those of 1μm. This is due to the high pressure drop necessary to reach inertial focusing, which requires more durable microfluidics, e.g., made of glass. Alas, to separate particles far smaller than 1 μm in size will require too high pressure drops also for more durable microfluidic systems.

5. Acknowledgements

This project has received funding from the European Union's Horizon 2020 research and innovation programme under grant agreement no. 644669.

Journal of Physics: Conference Series **757** (2016) 012028 doi:10.1088/1742-6596/757/1/012028

References

[1] H. Amini, W.H. Lee, and D. Di Carlo. "Inertial Microfluidic Physics." Lab Chip 14 (2014) 2739.
[2] J.M. Martel and M. Toner. "Inertial Focusing Dynamics in Spiral Microchannels." Phys. Fluids 24 (2012) 032001.
[3] H. Amini et al. "Engineering Fluid Flow Using Sequenced Microstructures." Nat. Comm. 4 (2013) 1826.
[4] R.M. Tiggelaar et al. "Fabrication, mechanical testing and application of high-pressure glass microreactor chips." Chem. Eng. J. 131 (2007) 163.
[5] Fuerstman et al. "The Pressure Drop along Rectangular Microchannels Containing Bubbles." Lab Chip 7 (2007) 1479.

Journal of Physics: Conference Series **757** (2016) 012029 doi:10.1088/1742-6596/757/1/012029

Extreme-temperature lab on a chip for optogalvanic spectroscopy of ultra small samples – key components and a first integration attempt

Martin Berglund[1,2], **Zahra Khaji**[1,2], **Lena Klintberg**[2], **Anders Persson**[1,2], **Peter Sturesson**[1-3], **Johan Söderberg Breivik**[1,2], **and Greger Thornell**[1,2]

[1] Ångström Space Technology Centre, Uppsala University, Sweden

[2] Div. of Microsystems Technology, Dept. of Engineering Sciences, Uppsala University, Sweden

[3] Dept. of Military Studies, The Swedish Defence University, Sweden

greger.thornell@angstrom.uu.se

Abstract. This is a short summary of the authors' recent R&D on valves, combustors, plasma sources, and pressure and temperature sensors, realized in high-temperature co-fired ceramics, and an account for the first attempt to monolithically integrate them to form a lab on a chip for sample administration, preparation and analysis, as a stage in optogalvanic spectroscopy.

1. Introduction

Despite being widely employed, neither *lab on a chip* (LoC), nor *high temperature* (HT), is really distinctly defined. The latter usually starts already at the "military hot", i.e., 125°C, and the former covers a really wide range of devices [1]. In fact, LoC could denote just a simple cavity, accessible to the naked eye, or, more often, an elaborate system of channels and chambers, where fluids are controlled and analyzed with auxiliary apparatus, or, more seldom, a rather complicated device with a high degree of functionality containing multiple integrated, active components and feed-throughs.

In this work, extremes in both respects are aimed for by integration of separately developed components into a sophisticated device intended for on-chip sample preparation reaching nearly 1000°C, and analysis, utilizing the optogalvanic effect (OGE) [2].

This LoC is intended for a range of applications, e.g., biomedicine, earth sciences, forensic science, and, ultimately, archeology, all of which benefit from accurate determination of carbon isotope ratios in samples so small and delicate that very little can be wasted on leaks and dead volumes in large instrument assemblies. This calls for a high degree of integration. Reported here, after a short overview of the system and its individual components, is the result from a recent attempt to attain such an integration to form a monolithic, ceramic lab on a chip, from a process perspective.

2. Device Operation and Design

A device like the one conceptualized here is a complex system of passive and active components arranged in line with a certain sequence, here mimicking parts of the macroscopic protocols described in [3]. Hence, the device under development here, starts with a minute amount of a solid organic sample, turns it into a conductive gas, and analyzes its carbon content with respect to isotope ratio – all on chip. Besides the two components of particularly importance for the sample phase transitions: the *combustor*, gasifying the sample, and the *plasma generator*, making it conductive, to support the OGE, there are valves, sealings, sensors, and channels for sample administration and monitoring.

With reference to figure 1, the sequence of steps from sample insertion to signal acquisition is:

1. insertion of (solid) sample in combustion chamber (to the right)
2. closing of inlet to this chamber (omitted in figure)
3. (gas) evacuation of device through two outlets (middle of rear long side)
4. pressure check through pressure sensors (along front long side)
5. sealing of outlets (omitted in figure)
6. combustion of sample while monitoring temperature and pressure sensor
7. opening of "breakable sealing" (near middle of chip)
8. pressure equalization as monitored through both pressure sensors (same as in (4))
9. temperature stabilization
10. ignition of plasma (in encircled gap of ring-shaped structure to the left)
11. OGE measurement (through wire probes not visible in figure)

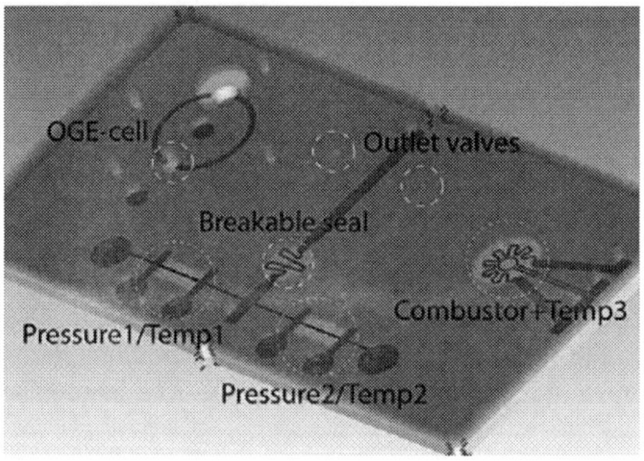

Figure 1. CAD rendering of device, showing all key components, which, in reality, are embedded and invisible. Sample is inserted to the right, signal is read from upper, left corner.

Based on the stacking of separately structured and patterned sheets, it was important to minimize process steps and to avoid accumulation of fabrication errors, such as misalignment, already in the design phase. Figure 2 indicates how embodiment of the above sequence uses vertical and horizontal routing of channels and leads to limit the number of metalized layers to four, and connecting the two ground planes of the stripline plasma source. (In fact, the layers were stepped out in a fashion requiring only one sheet to be metalized.)

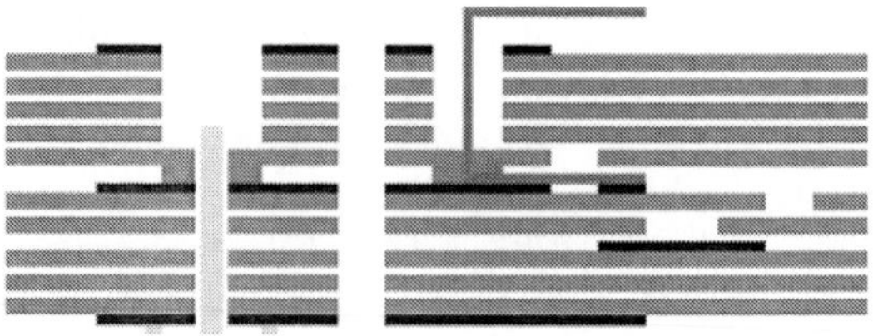

Figure 2. Schematic of the device's layers, with grey denoting individual or triplets of ceramic sheets, black metal patterns, green conductive adhesive, yellow an SMA connector, and orange metal wire.

3. Component Overview

With few exceptions, e.g., the innovative use of bond wires and internal electroplating, well-established High-Temperature Co-fired Ceramics (HTCC) technology was used for manufacturing of all components, both because of the ceramic's good high-temperature properties and its inertness, and because of the coming integration.

The combustor is a resistively heated cavity with the unique feature of having an integrated oxygen supply, formed by oxidizing a copper layer on the heater, and decomposing it by heating. As copper cannot be co-fired with the HTCC alumina, it was selectively electroplated on the platinum heater, subsequently, i.e. inside the cavity. An X-ray image of the final combustor, showing the vias to the heater and temperature sensor leads is shown in figure 3 together with the as-plated copper in an open combustor. Thermogravimetric analysis showed that the oxygen taken up during oxidation was released when heated, starting at around 750°C, and completing at around 900°C. Thermography, again on an open combustor, revealed an increasingly uniform heating of the combustor floor, when the heating power was increased, with a small but noticeable cooling from its leads. Residual gas analysis conducted during the combustion of a carbon-rich sample, together with microscopy, showed that oxygen released from the copper oxide reacted with the carbon, and formed carbon dioxide. [4]

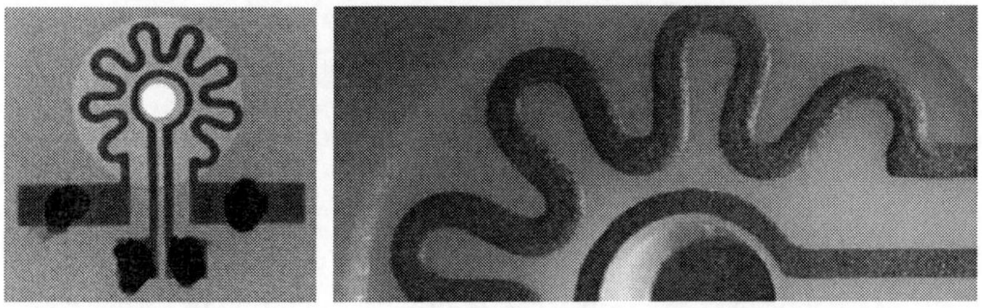

Figure 3. X-ray image of full-fledged combustor with a combined heater and oxygen supply meandering around the circular temperature sensor, inside which the sample inlet and gas outlet combine to make a bright circle (left), and microscope close-up of open reactor just after copper plating of the heater, not having affected the platinum temperature sensor (right).

The plasma source consists of a split-ring resonator, essentially a folded dipole antenna, in this case being sandwiched between ground planes to form a robust and interference-immune stripline device. Where the antenna's two prongs meet, there is a gap through which the sample gas is fed, and over which the electric field is strong enough to ionize it. Following on an extensive development of this kind of plasma sources [5-7], the one used here, was realized with HTCC technology, both for integrational purposes and because of the better thermal and chemical stability of the ceramic. Crucial to this transition, was the investigation of the materials' dielectric properties. From a variation of resonator dimensions, substrate thickness, and metal paste resistivity, it was found that the relative permittivity varied very little, whereas the dielectric loss varied a lot, probably as an effect of a non-uniform metallization thickness. Besides this, important knowledge was gained on how machining of ceramic sheets affects their lamination. In conclusion, a stripline microplasma source with a resonance frequency of 2.6 GHz, producing an intense and stable plasma, was developed. The final stand-alone component is shown in figure 4 together with a close up of the gap during operation. [8]

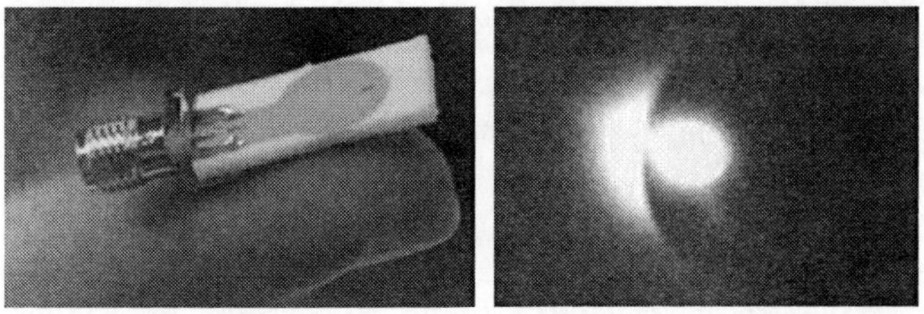

Figure 4. Completed HTCC-based plasma source with one ground plane and the sample and laser beam entrance to the gap of the split-ring resonator visible at top surface (left), and close-up when ignited in air at 10 Torr (right). Because of pixel bleeding and the translucency of the ceramic, the image is blurry and grainy, and the plasma appears much wider than it is.

The system contains two types of single-use valves. Although a second generation of lab chip will probably have a breakable sealing on its inlet to protect its interior from ambient contamination, the version here hasn't, and the first valve encountered in the sequence, is the sealable inlet. This operates by heating and melting a small amount of silver, which, so far, is either electroplated or manually deposited at the rim of the inlet, and having surface tension form a cap. Admittedly, and as evident from figure 5, this part of the device needs further development.

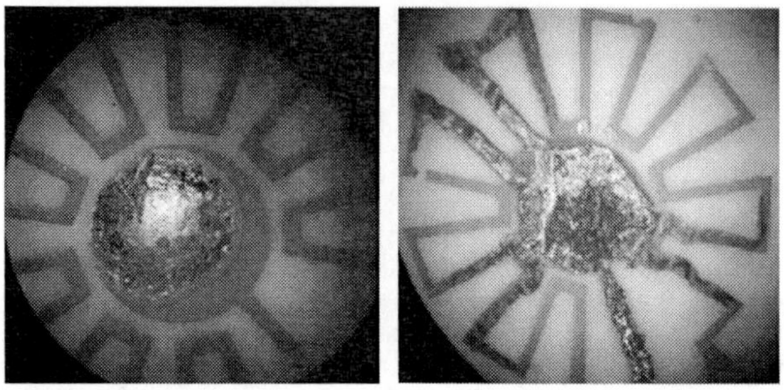

Figure 5. Well made sealing, with, here, manually deposited silver forming a half-sphere over the inlet (left), and the result of minute differences in metal adhesion and/or deposition where the heater meander has started to disadhere from the ceramic, and the molten silver has spread to it from its designated area at the rim of the inlet (right)

The next valve is the opposite, a breakable sealing membrane, which serves to clear the pathway to the analyzer part of the chip. The operation of this is based on causing localized thermal gradients and transients, too large for the brittle ceramic to handle, and controlling crack propagation. To find the right means, a thorough study on heater design, membrane thickness and membrane processing, was conducted. It was found that a few hundred mJ were enough to open most of the thin membranes, and that heater shape and processing scheme played a major role. The resulting openings, ranging from microcracks to loss of the entire membrane, figure 6, were quite reproducible [9].

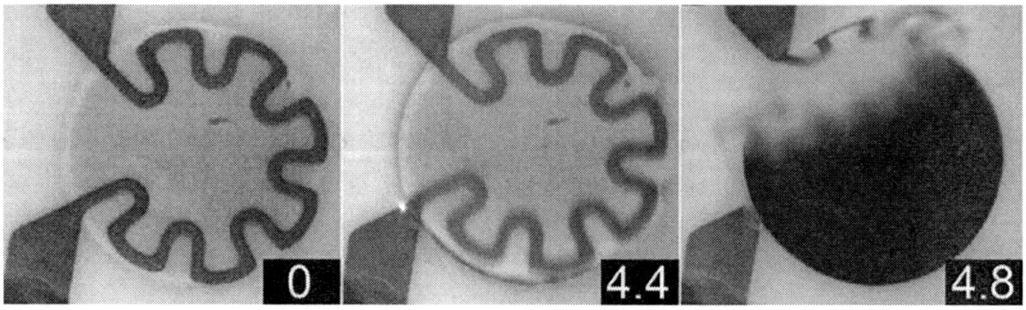

Figure 6. Frames from high-speed footage of a 40 μm thick alumina membrane cracking along its rim as a result of supplying the meander-shaped platinum heater with approx. 350 mJ. The number in each frame's corner is elapsed time in milliseconds from start of heating.

Finally, the chip is equipped with a range of sensors. Besides two plasma probes, there are three temperature sensors and two pressures sensors for housekeeping. Whereas a dedicated temperature sensor, located in the combustor (see figure 3), is made by screen printing platinum, and, hence, following standard HTCC processing, the rest are combined pressure and temperature sensors made by threading platinum bond wires in the stack of ceramic green sheets, prior to sintering. For temperature reading, simply the resistance of such a wire is measured. When heated, however, the wires, raised from their cavity floors as a result from densification of the green sheets on sintering, loose heat convectively in proportion to the surrounding gas pressure, allowing them to work like Pirani gauges, figure 7. Quite remarkably, the performance was satisfactory even on the few occasions the deflection of a wire made it hit the ceiling of its cavity [10].

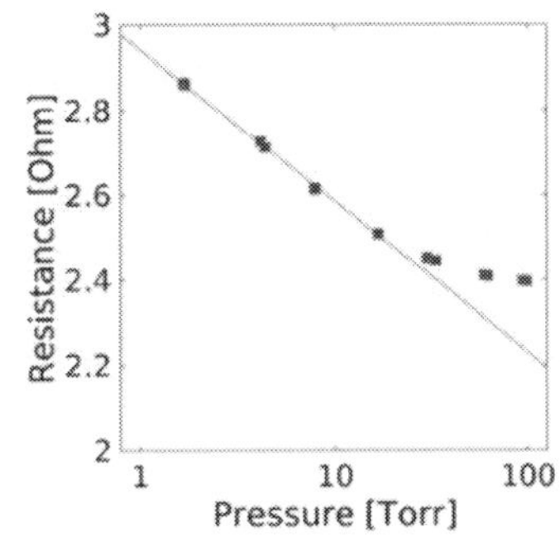

Figure 7. Resistance, being translatable to temperature, of a 25-μm diameter platinum wire vs. pressure in the enclosing cavity

For the plasma probes, which are located in the plasma source's gap, the above wire technology is used, but with a carefully tailored step to fuse them to form rounded tips allowing Langmuir-probe-like measurement of the plasma potential, which is the key to optogalvanic spectroscopy in this embodiment. Compared with the almost spherical tips made previously [11], the tips obtained in HTCC, figure 8, need further work.

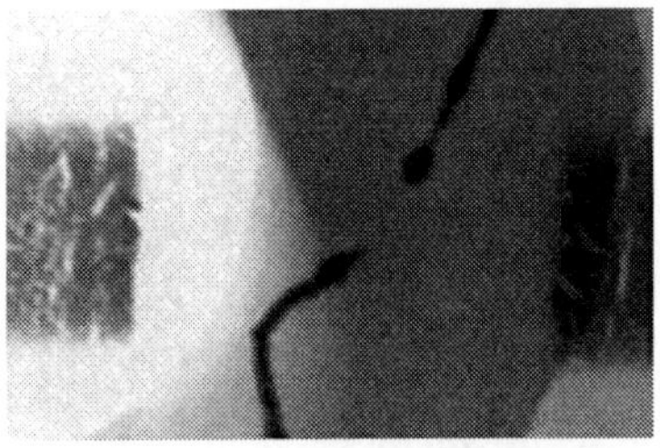

Figure 8. X-ray close-up of two fused plasma probes with semi-spherical tips revealing wire deflection and kinks from sintering.

4. Results at System Level

The integration of all the above components into a 50×25×1.6 mm chip was very successful. Despite the many layers, alignment errors were negligible, figure 9, and the curvature of the whole stack was just a few tenth of a mm. The only actual setback was the unfortunate cutting of the leads just outside the plasma probe cavities (not visible at the figure's magnification). The statistics doesn't allow for any certainty as of the cause, but a reasonable explanation is non-uniform compression of the chip during lamination because of the variation in support between solid green sheet areas and to-be cavities filled with fugitive material. Other than this, all leads and vias followed specifications. Thermography showed that components could be operated individually without severe thermal crosstalk.

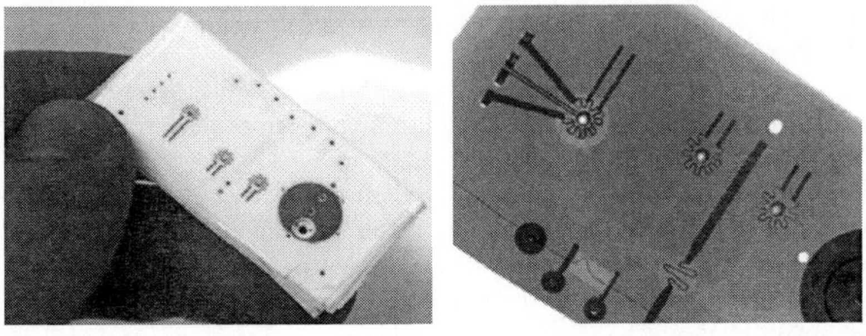

Figure 9. Device after sintering (left), with X-ray close-up of central part, showing alignment and printing quality, and the typical deformation of the sensor wires (right). Darker parts are screen printed platinum or wires, and the lighter shades within the chip are unfilled vias, cavities and channels.

5. Conclusions and Discussion

All inspections and tests conducted, show that integration was successful on the mechanical and thermal grounds defined. Regarding functionality, the damaged probe leads are of course a setback. Changing the locations of the components and routes in-between, may be a solution to this. Near-future work shall entail calibrating the LoC and inserting in the spectroscopy set up, and running the sequence given above. Beyond this, there are both design and process issues deserving further investigation. As for adding features, the above mentioned breakable sealing at the device's inlet is one. Another is a filter to prevent debris from membrane breaking to travel inwards. A third is a pressure reducer, for instance a relatively large cavity lined with getter material and evacuated already during sintering.

Acknowledgements

The Swedish National Space Board and the Centre for Natural Disasters Science are acknowledged for project funding. The Knut and Alice Wallenberg foundation is acknowledged for funding part of the laboratory facilities. The help and keenness of Jan Bohlin's and Anders Lund's at the Department of Chemistry, and of Anders Olsson's at the Department of Physics and Astronomy at Uppsala University, are highly appreciated.

References

[1] P. Abgrall and A.-M. Gué, "Lab-on-chip technologies: making a microfluidic network and coupling it into a complete microsystem – a review", *J. of Micromechanics and Microengineering*, 17(5), pp. R15-49, 2007.

[2] B. Barbieri, N. Beverini, and A. Sasso, "Optogalvanic spectroscopy", *Rev. Mod. Phys.*, 62, pp. 603-644, 1990.

[3] M. Salehpour, K. Håkansson, and G. Possnert, "Accelerator mass spectrometry of ultra-small samples with applications in the biosciences", *Nuclear Instruments and Methods in Physics Research B*, 294, pp. 97-103, 2013.

[4] Z. Khaji, P. Sturesson, K. Hjort, L. Klintberg, and G. Thornell, "Manufacturing and characterization of a ceramic microcombustor with integrated oxygen storage and release element," *J. Micromechanics Microengineering*, vol. 25, no. 10, 104006 (7 pp), 2015.

[5] M. Berglund, M. Grudén, G. Thornell, and A. Persson, "Evaluation of a microplasma source based on a stripline split-ring resonator," *Plasma Sources Sceinces and Technology*, 22(5), 055017, 2013.

[6] M. Berglund, G. Thornell, and A. Persson, "Microplasma source for optogalvanic spectroscopy of nanogram samples," *J. of Applied Physics,* 114(3), 033302, 2013.

[7] A. Persson, M. Berglund, and M. Salehpour, "Improved optogalvanic detection with voltage biased Langmuir probes," *J. of Applied Physics*, 116(24), pp. 243301, 2014.

[8] M. Berglund, A. Persson, and G. Thornell, "Evaluation of dielectric properties of HTCC alumina for realization of plasma sources," *J. of Electronic Materials*, vol. 44, Iss. 10, pp. 3654–3660, 2015.

[9] Z. Khaji, L. Klintberg, and G. Thornell, "Manufacturing and characterization of a ceramic single-use microvalve," accepted for publication in *J. Micromechanics Microengineering*, in May 2016.

[10] J. Söderberg Breivik, "Microscale ceramic pressure sensor eement for a carbon isotope analysis system for planetary exploration: – design, manufacturing and characterization," MSc Thesis, avialable through uu.diva-portal.org, 2015.

[11] M. Berglund, P. Sturesson, G. Thornell, and A. Persson, "Manufacturing miniature Langmuir probes by fusing platinum bond wires," *J. Micromechanics Microengineering*, vol. 25, no. 10, 105012 (7 pp), 2015.

Geometric optimization of magnetically actuated MEMS micromirrors

Francesco Pieri[1]

[1]Dipartimento di Ingegneria dell'Informazione, Università di Pisa, Pisa, Italy

f.pieri@iet.unipi.it

Abstract. In this paper we develop a set of self-consistent equations describing the static and resonance characteristics of a rectangular micromirror actuated by driving a current into a multi-coil inductor fabricated on the micromirror surface. Under fairly general assumptions, these equations can be solved analytically for the most common design targets, such as the driving current, deflection angle, resonance frequency, and overall dimensions. We use this solutions to optimize the micromirror geometry by minimization of a linear objective function. The trade-offs that emerge from optimization data allow simple quantitative design of low current, low size mirrors.

1. Introduction

MEMS micromirrors are intensely researched as essential components in several applications, largely aimed at the market of portable devices (mobile phones, tablets, etc.) [1]. Among the most important are picoprojectors for imaging, and optical scanners for 3D object reconstruction and motion tracking. Both applications typically require scanning of one (or more, i.e. for color image projection) laser beams in two orthogonal directions.

The 2D scanning can be obtained with a single, 2-axis torsional MEMS micromirror or with two separate torsional 1-axis micromirrors (figure 1). The requirements for scanning frequency and angle are dependent on the application. Raster scanning, where the vertical (i.e. column) axis is scanned at a much slower frequency that the horizontal (row) axis, is by far the most common scanning strategy. For this reason, the vertical axis is typically driven in quasi-static conditions, i.e. at a frequency much lower than the typical resonance frequency for the vertical resonance mode, while the horizontal axis is driven at its resonance frequency. A 2-axis solution has the obvious advantage of a greater miniaturization, at the cost of a more complex kinematics, both in gimbal-less [2, 3] and gimbaled [4, 5] solutions, undesired coupling between the two axes, so that a 2-mirror solution is typically simpler to develop.

From the point of view of actuation mechanisms, the ones most pursued are electrostatic, electromagnetic, and, to a lesser extent, piezoelectric and electrothermal. Electrostatic actuation is the most common, as the most simple technologically, but also requires of high actuation voltages. Electromagnetic actuation requires integration of a multiloop coil (or a magnet) on the mirror, works at low voltages, and is intrinsically linear. The main drawback is the high current required, which is an important limit for portable applications. As the actuation torque is linked to the mirror area, electromagnetic actuation is more suitable for large mirrors.

In this paper, we present the geometric optimization of an electromagnetic mirror suitable for low-frequency (i.e. quasi-static) applications, actuated through a coil created on the mirror plate. A magnetic field (typically provided by external magnets) is used to create a torque (and, consequently, an angular deflection) on the plate which is essentially proportional to the driving current.

2. Model

The structure of the mirror is shown in figure 2. The actuation coil has a rectangular spiral geometry. The external magnetic field is aligned along y. A number of loops in the form $n_c = (n + 1/2)$, with integer n, is assumed. The distance between adjacent metal lines g_l is set by the technology, while the metal width w_l is subject to optimization. Two torsional suspension springs, of rectangular cross-section (of width w_S and thickness t_S), provide the returning torque. The same thickness t_S is also assumed for the central plate. The material for the plate and springs is assumed to be silicon. Five design parameters, i.e. n_c, w_l, w_S, t_S, and the spring length l_S, completely define the mirror geometry. The design parameters and all the other dimensions and quantities are listed in table 1.

As the proposed magnetic micromirror is to be used as the slow mirror of a 2D scanner, an assumption of quasi-static actuation is made. Natural performance requirements for the mirror are: minimization of the actuation current at maximum angle, minimization of the mirror size, and a torsional resonance frequency much higher than the maximum driving frequency, to avoid undesired excitation of the resonance. While many loops result in a higher torque, and thus lower actuation current, they also increase the size of the mirror and its mass, which in turns lowers the resonance frequency. Several simultaneous trade-offs thus needs to be evaluated, and a generic optimization approach is consequently justified.

The geometrical, electromagnetic, and mechanical description of the mirror is contained in equations (1)-(4):

$$\begin{cases} p_l = g_l + w_l \\ l_x = d_x + (1 + 2n_c)p_l \\ l_y = d_y + 2n_c p_l \\ l_{Tx} = 2l_s + l_x \end{cases} \tag{1}$$

$$\begin{cases} T_x = 2k_\varphi \varphi_{max} \\ k_\varphi = G_{Si}\frac{J_{S\varphi}}{l_S} \\ J_{S\varphi} = w_S t_S^3\left(\frac{1}{3} - \frac{21}{100}\frac{t_S}{w_S}\right) \end{cases} \tag{2}$$

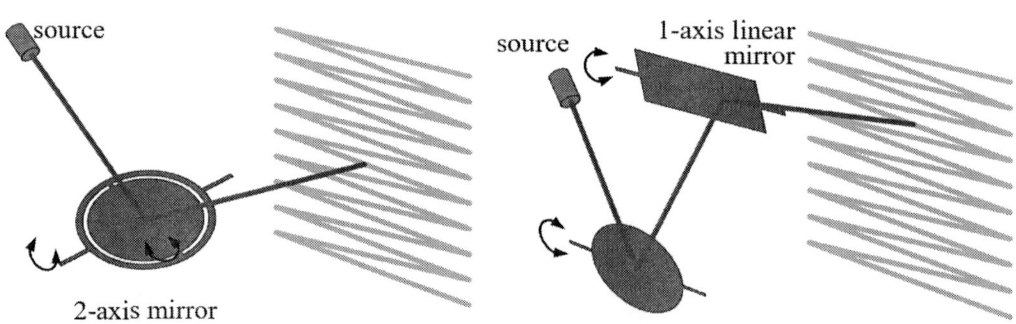

Figure 1. Schematic structure of a 1-mirror (left) and a 2-mirror raster scanner. Adapted from [1].

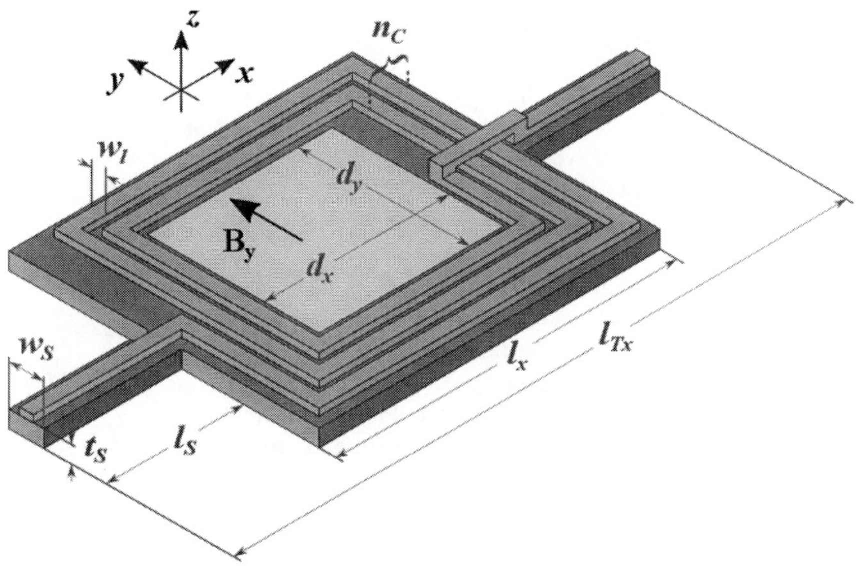

Figure 2. Structure of the mirror with the most relevant geometrical parameters. Design variables are marked in red. The value of n_c is 2 ½.

$$\begin{cases} f_\varphi^2 = \frac{2k_\varphi}{4\pi^2 J_{Px}} \\ J_{Px} = \frac{1}{12} l_y^2 M_P \\ M_P = l_x l_y t_S \mu_{Si} \end{cases} \tag{3}$$

$$\begin{cases} T_x = A_c B_y I_c \\ A_c = \sum_{k=1,5,9..} \frac{d_x+2kp_l+w_l}{2}\left(d_y + 2kp_l + w_l\right) + \\ \quad + \sum_{k=3,7,11..} \frac{d_x+(2k+1)p_l+w_l}{2}\left(d_y + 2kp_l + w_l\right) \end{cases} \tag{4}$$

Equations in (1) give obvious geometric constraints among the various geometric parameters; equations in (2) describe the torque-angle static characteristics of the mirror; equations in (3) give expression for the resonance frequency. Finally, equations in (4) define the relationship between torque and current. An assumption of linearity was made for the behavior of the suspending springs. The magnetic field was assumed uniform. Most equations are applications of basic physics of mechanical and electromagnetic systems and do not need special comments.

Additional remarks are only required for two of the equations. First, the sum for A_c, the equivalent area of the loop, is extended over the loop branches parallel to x, that are the only ones contributing to the torque. While it is not shown for simplicity, a closed form for this sum as a function of n_c exists. Secondly, a closed expression for the torsional constant for rectangular cross-sections does not exist, and approximate formulas are commonly given [6]. We chose a simplified expression which allows a closed solution for equations (1)-(4), but is only valid for $w_S > t_S$, and also has a reduced accuracy for aspect ratios t_S/w_S close to one.

3. Optimization

Because of the specific form used, equations (1)-(4) can be solved analytically to obtain a closed expression for any involved quantity as a function of the five design parameters n_c, w_l, w_S, t_S, and l_S. The solution was derived by using a software for algebraic manipulation, and is too lengthy to report here.

Table 1. Definition and values of the quantities used in the text.

Symbol	Parameter	Value
d_x	width of mirror surface (along main rotation axis x)	2 mm
d_y	width of mirror surface (along other axis y)	1 mm
l_x	total width of plate (including the coil) along x	-
l_y	total width of plate (including the coil) along y	-
w_l	metal width	-
g_l	distance between metal lines	20 µm
p_l	metal line pitch ($g_l + w_l$)	-
l_S	spring length (along x)	-
t_S	thickness of mirror central plate and springs	-
w_S	spring width (along y)	-
l_{Ty}	total occupancy along y	-
φ [φ_{max}]	[maximum] rotation along main axis x	[10° max]
n_c	number of coils (counting half-coil)	-
T_x	total torque on the plate	-
I_C	coil current	-
A_C	equivalent area of coil (i.e. area such that $T_x = B_y A_C I_C$)	-
B_y	magnetic field	100 mT
k_φ	torsional spring constant of a suspension spring	-
$J_{S\varphi}$	torsional constant of the spring cross-section	-
f_φ	torsional resonance frequency	
M_P	plate mass	-
J_{Px}	moment of inertia of the plate along rotation (x) axis	-
G_{Si}	shear modulus of silicon along x	79.5 GPa
μ_{Si}	density of silicon	2330 kg/m^3

The solution was used the starting point for subsequent optimization. As already mentioned, the three optimization targets were: a small current I_c at maximum deflection, a small total length l_{Tx}, and a high resonance angular frequency ω_φ. A maximum angular deflection φ_{max} of ±10°, compatible with a standard application, was chosen. A natural choice for the objective function to be minimized is then the linear combination:

$$f(n_c, w_l, t_S, l_S, w_S) = w_1 I_c - w_2 \omega_\varphi + w_3 l_{Tx} \tag{5}$$

where the w_i are positive weights to be tuned in dependence of the importance of each target variable. Without loss of generality, we can always set $w_1 = 0$. The objective function was minimized numerically with reasonable constraints for the design variables (shown in table 2) and for several different values of w_2 and w_3. Specifically, w_2 was varied in the range [10^{-5}, $2\cdot10^{-4}$] and w_2 in the range [50, 1000]. Larger ranges also led to possible optimal designs, but they were ignored because of the resulting unpractical values of the target quantities for real applications.

Table 2. Allowed optimization ranges for the design variables.

Parameter	min	max
n_c	1.5	79.5
w_l	20 µm	500 µm
t_S	30 µm	50 µm
l_S	0.3mm	2 mm
w_S	40 µm	200 µm

Interestingly but not unexpectedly, all optimum solutions led to the minimum possible cross-section for the suspension springs (t_S = 30 µm, w_S = 40 µm), so that the hypothesis of an aspect ratio higher than one is always verified. As the actual torsional constant of a rectangular cross-section is invariant for a swap of the two sides, the choice of the simplified expression for $J_{S\varphi}$ does not exclude possible optimal solutions.

Results for the simulations are shown in figure 3, which summarizes the results for a set of 36 different optimizations. Data points corresponding to the same weight given to the resonance frequency are given the same colour. Inside each colour group, each point corresponds to a different weight given to the mirror size (i.e. its total length l_{Tx}). The plot on the left, which shows the relationship between the current required to drive the mirror at maximum deflection and the size of the mirror, demonstrates a clear trade-off between the two. As a matter of fact, larger mirrors in the plot correspond to longer springs and a higher number of coils, which both reduce the required current. Our procedure ensures that the increase in occupancy caused by the coils and springs is optimally allocated between the two. The plot on the right shows the relationship between the current and the resonance frequency. Again, an evident trade-off exists between low currents and high frequencies. High resonance frequencies plainly require high stiffness of the springs, which in turn requires a high current. However, the data show that for practical applications, where the scanning frequency of the slow mirror is of a few tens of Hertz, the constraint on the frequency is not very important.

4. Conclusions

In this paper we started from the descriptive equations for the static and resonance behaviour of a torsional MEMS micromirror to develop a semi-analytical procedure to design mirrors with optimal performances. The optimization data allow the extraction of optimum frontiers between the requirement for low driving current and small size, and between current and resonance frequency. The same approach can be easily extended to evaluate the trade-offs between different target parameters, and adapted to different geometries. Further generalizations can include the size of external magnets in the occupancy of the device.

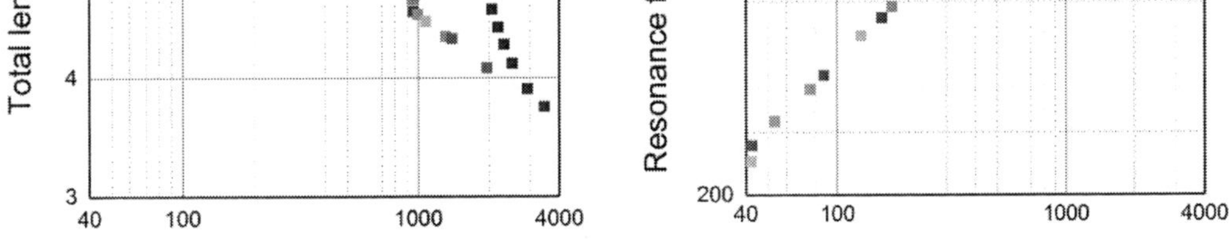

Figure 3. Total length of the mirror (left) and resonance frequency (right) as a function of the maximum driving current. Data are grouped by colour, each group corresponding to a different weight w_2 for the resonance frequency.

Acknowledgments
The author wishes to acknowledge the financial contribution by the ENIAC Joint Undertaking in the framework of project Lab4MEMS II under grant n. 621176-2.

References
[1] Holmstrom S T S, Baran U and Urey H 2014 MEMS Laser Scanners: A Review *J. Microelectromechanical Syst.* **23** 259–75
[2] Jun M-H, Moon S, Lee B H and Lee J-H 2014 A gimbal-less two-axis electrostatic scanner with tilted stationary vertical combs and serially connected springs via a microassembly process *J. Micromech. Microeng.* **24** 095008
[3] Tsai J and Wu M C 2005 Gimbal-less MEMS two-axis optical scanner array with high fill-factor *J. Microelectromechanical Syst.* **14** 1323–8
[4] Yanhui Bai, Yeow J T W, Constantinou P, Damaskinos S and Wilson B C 2010 A 2-D Micromachined SOI MEMS Mirror With Sidewall Electrodes for Biomedical Imaging *IEEE/ASME Trans. Mechatronics* **15** 501–10
[5] Jeong J-W, Kim S and Solgaard O 2012 Split-Frame Gimbaled Two-Dimensional MEMS Scanner for Miniature Dual-Axis Confocal Microendoscopes Fabricated by Front-Side Processing *J. Microelectromechanical Syst.* **21** 308–15
[6] Young W and Budynas R 2002 *Roark's Formulas for Stress and Strain* (New York: McGraw-Hill)

Magnetic Multilayer Fabrication Technology with Selective Activation of SU-8 Films

R Anthony[1, 2], C ÓMathúna[1, 2] and JF Rohan[1]
[1]Tyndall National Institute, Lee Maltings, University College, Cork, Cork, Ireland
[2]Electrical and Electronics Engineering, University College Cork, Cork, Ireland

Abstract. Developing magnetic multilayers are essential for reducing the core eddy current losses in the integrated power magnetic components (inductors/transformers). PVD based processes are typically used to achieve the multilayers with thin dielectric spacers. However, those processes are costly, and can be difficult to integrate. It is evident that cost effective alternative is needed. In recent years, electrochemical processes have been investigated to address these issues. One such method would be to successive metallization of insulating photoresists acting as spacer layer (such as SU-8) with soft magnetic films (such as Ni-Fe-Co alloys). This paper describes an experimental procedure to fabricate magnetic multilayers with a thin variant of SU-8 2 (< 1.5 μm) as inter-layers for integrated micro-inductors/transformers for power conversion applications.

1. Introduction

The power conversion circuitry is an integral part of any portable electronics. A major section of the circuitry is the energy storage and transfer passive components (inductors and transformers) which are often bulky and non-integratable. Increasing in the switching frequency of the DC-DC power converters in recent years has reduced inductance (and energy storage) requirements, thus these passives can now be fabricated on silicon with MEMS-based processes [1]. However, with the continuous reduction and increase in functionality of portable electronics, it is apparent that these components be further miniaturized without compromising on efficiency. To meet required device specification in a limited footprint is challenging. Although increase in switching frequencies have reduced footprint, increasing magnetic core flux density could also increase inductance density. This would potentially reduce the number of windings in the device and thus reduce the overall footprint area. Over the years electroplated soft magnetic films (Ni-Fe-Co alloys [2-6] with high flux density and low coercivities have been extensively studied (shown in figure-1). These metallic alloys have high saturation flux density, low coercivity and high conductivity. The magnetic core is restricted by the skin depth (δ) equation ($\delta = \sqrt{\rho/\pi f \mu}$). Where, ρ, f, μ represents conductivity (Ω-m), operation frequency (Hz) and permeability (H/m) of the magnetic film. For thickness > δ, the flux is contained at the surface of the material and exponentially decreases towards the centre due opposing field generated by the induced current. Recent developments in soft sputtered films such as amorphous CoZrTa and CoZrO films have high resistivity (ρ > 110 μΩ-cm) and saturation flux density (B_{Sat} > 160 A/m) [7]. However, these materials do not retain soft magnetic properties on thick sputter deposition and patterning could be challenging. Hence, electrochemical process remains the most widely accepted process to deposit magnetic films. Attempts have been made to achieve laminated core structures with electrochemical process [8-9]. An analytical study with $Ni_{81}Fe_{19}$ (ρ: 24 μOhm-cm; laminated racetrack micro-inductors has been investigated. Micro-inductor footprint area (1 mm^2) and dimensions (core length: 0.75 mm; inductor length: 1.44 mm and inductor width: 0.7 mm and core thickness and spacer: 0.8 μm) have been kept fixed. Although the eddy current losses reduces, the core hysteresis loss increases with laminations which along with winding loss then limits the efficiency for higher laminations (fig 2(a)). For high resistive soft-magnetic materials such as CZTB, lower number of laminations are needed compared to $Ni_{81}Fe_{19}$ to achieve same efficiency. Interestingly the micro-inductor operation frequency could also be increased on introducing core lamination (fig 2(b)). It is to be noted that the core relative permeability is considered frequency independent.

Authors have previously reported a surface activation method to metalize SU-8 50 dielectric [10]. In this work the authors exploited the process to fabricate magnetic multilayers by activating thinner SU-8 variant (SU-8 2). The experimental section will describe the SU-8 2 patterning, activation and metallization processes. The results and discussion will outline the characterization of laminated films for high-frequency applications and investigate the possible method of co-packaging the fabricated core in integrated microinductors and FEM analysis of the micro-inductor.

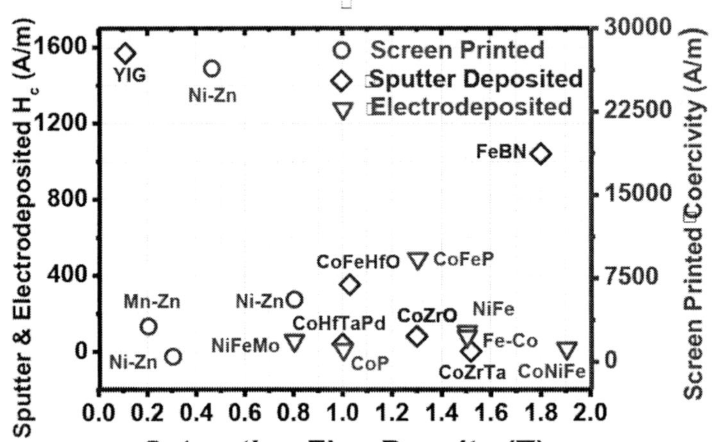

Figure 1. Coercivity vs saturation flux density of reported soft magnetic films.

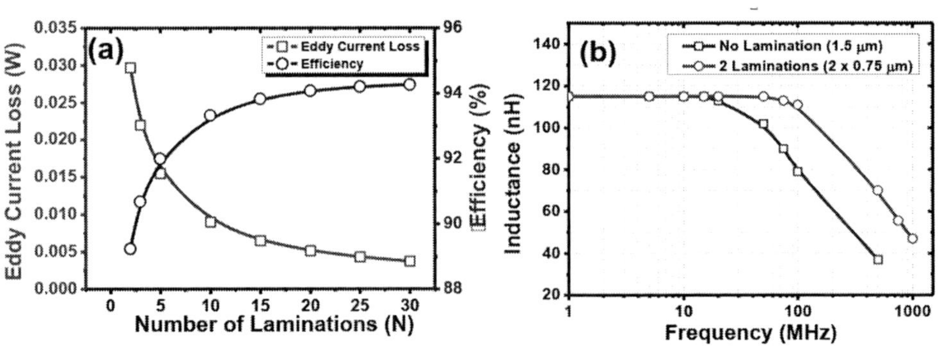

Figure 2. (a) Improvement in device efficiency and reduction in eddy current loss with number of lamination. (b) Inductance vs frequency curve with and without laminated micro-inductors.

2. Experimental Details

2.1. SU-8 2 Spin Coating

It is important to have thin and pin-hole free spacers in-between ferromagnetic (FM) layers to achieve laminated cores. In this case SU-8 2, a thinner variant of SU-8 epoxy based photoresist is used as spacer layer. Fig. 3 depicts the spin speed vs thickness curves for different SU-8 variants as reported in the datasheet [11]. To achieve thickness ~ 1 μm, the main spin speed of 3000 rpm / 40 s is selected. Table-

1 details the process steps. The photoresist exposure was performed under Canon PLA-501 FA optical mask aligner in soft contact mode and resist thickness of 1 μm -1.5 μm was achieved (as measured in Tencor Alpha Step-200 surface profiler). This is followed by a short post exposure bake essential for cross-linking of SU-8 molecule in ramp mode to 65 °C / 60 s and 95 °C/ 60 s. In case of patterned samples, the resist is developed in EC solvent for 60 s.

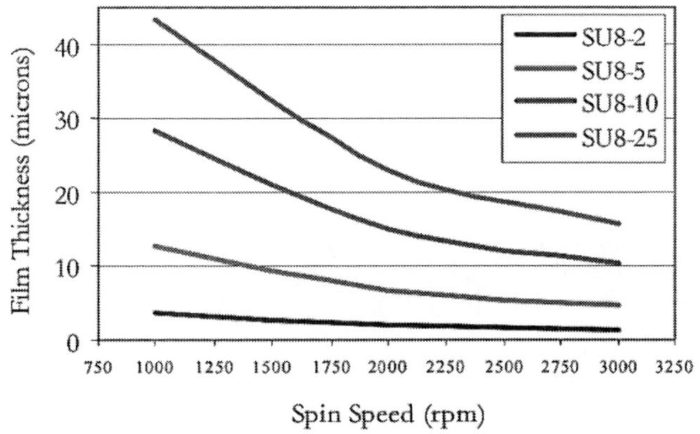

Figure 3. Spin speeds vs thickness curves for different variants of SU-8 (Obtained from datasheet [11]).

Table 1. SU-8 2 process steps

Process Steps	Quantity (mol-dm^{-3})
Spin profile	100 rpm/ 10 s +500 rpm /10 s +3000 rpm/40 secs
Soft Bake	Ramp from 30 °C to 95 °C/ 3 mins
Exposure Dose	100 mJ/cm^2
Post-exposure bake	65 °C / 60 s + 95 °C / 60 s
Development	60 s in EC-solvent

2.2. SU-8 2 Surface Activation and Electroless Deposition

The surface of the SU-8 is plasma treated in oxygen plasma asher (March Plamod GCM-200) at 350 mW for 60 s to remove organic contaminants. This is followed by 1800 s of immersion in (3-Aminopropyl)-triethoxysilane (APTES) solution. A monolayer is formed on the SU-8, selectively binding the APTES to the activated epoxy group of SU-8 surface in a covalent bond. It is then immersed in a solution of palladium ions to form palladium nuclei on the SAM layer which can be reduced and is essential for activation of successive electroless chemical deposition [10, 12-13]. This is followed by electroless deposition of $Ni_{81}Fe_{19}$ in in-house developed bath in absence of external magnetic film. Table-II lists the bath constituents and deposition condition for electroless deposition of $Ni_{81}Fe_{19}$. This process is repeated sequentially (fig.4) to achieve laminated films to achieve 2 and 3 layers of $Ni_{81}Fe_{19}$ of thicknesses (~ 1.7 μm to ~ 3 μm) and 1 μm thick SU-8 2 spacers. It is however observed that $Ni_{81}Fe_{19}$ films > 2 μm show high stress.

Table 2. Borane based $Ni_{81}Fe_{19}$ bath constituents and deposition conditions

Bath Contents	Quantity (mol-dm^{-3})
Di-ammonium citrate	0.027
Lactic acid	0.22
$NiSO_4.6H_2O$	0.060
$FeSO_4.7H_2O$	0.058
DMAB	0.034

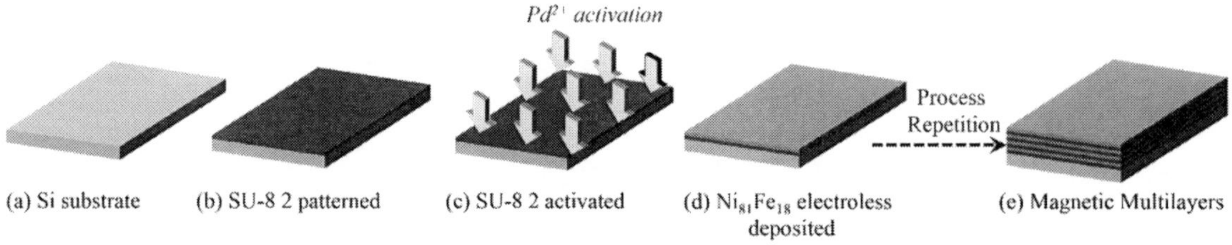

Figure 4. Spin speeds vs thickness curves for different variants of SU-8 (Obtained from datasheet

3. Results and Discussions

3.1. Characterizations

The magnetic characterization of electroless $Ni_{81}Fe_{19}$ were measured on surface activated silicon films. Fig. 5(a) shows the hysteresis curve for ~ 250 nm thick $Ni_{81}Fe_{19}$ measured in BH loop tracer at 10 Hz (SHB Instruments, USA, Model: MESA 200 HF). As observed from the figure, the coercivity is higher (365 A/m) compared to electroplated $Ni_{81}Fe_{19}$ (< 100 A/m), which is one of the drawbacks with the process. However, a high anisotropy field > 1397 A/m measured on surface activated deposition. The relative permeability spectra measured with wide-band complex permeameter (Ryowa electronics, Japan, Model: PMM 9G) suggested a uniform response up to 600 MHz at null field (fig. 5(b)) for the 250 nm thick deposit. The increase in cut-off frequency (f_c) could be attributed to the lower thickness (t) and relative permeability (μ_r) of the sample ($f_c \alpha \frac{1}{\mu_r t^2}$). Fig. 6 depicts the SEM micrograph of magnetic multilayers with sequential lithography-activation and deposition method. The shape of the sample (1mm x 3 mm) assists in uniform permeability, with high-frequency permeability characterization suggested uniform response up to 600 MHz, as observed in fig. 7.

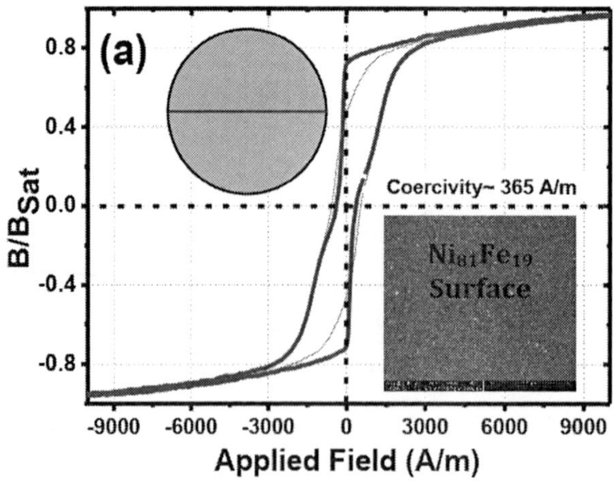

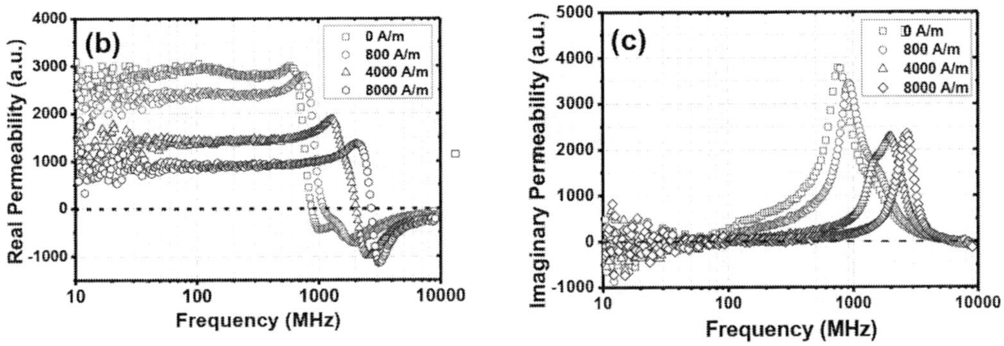

Figure 5. (a) Hysteresis loop of ~ 250 nm thick Permalloy films electrolessly deposited on surface activated silicon and its relative permeability vs frequency curves (3 mm x 3 mm).

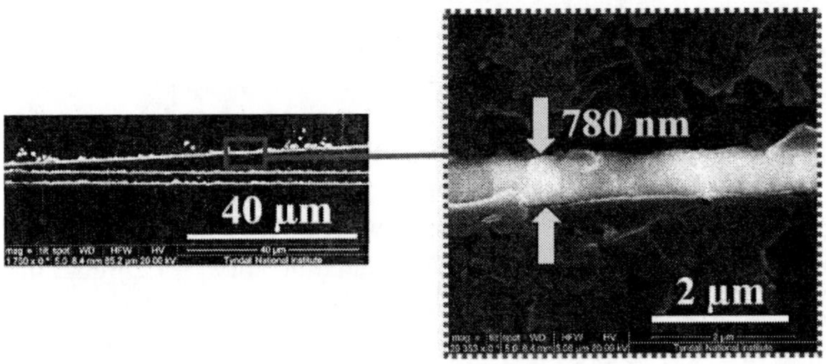

Figure 6. Cross-section SEM images of $Ni_{81}Fe_{19}$ mutli-layers with SU-8 2 spacers.

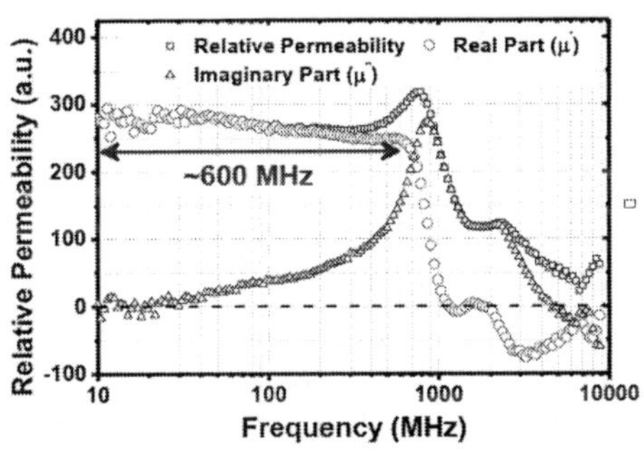

Figure 7. Initial permeability of 3 layers of $Ni_{81}Fe_{19}$ (~3 µm) and ~ 1 µm SU-8 spacer films (1 mm x 3 mm).

3.2. FEM Study of Solenoid Inductors with Co-Packaged Core

In this section a possible solenoid inductor with packaged core structure is demonstrated. The above mentioned multilayer core structure can be co-packaged into an integrated device structure with fabricated bottom and top windings. The silicon substrate of the core multilayer structure is thinned and packaged with adhesive mould, followed by resist patterned. This is followed by wire-bonding top windings (fig.8). Fig. 9 depicts the inductance vs number of lamination curve for the device simulated with Maxwell Ansoft finite element method (FEM) from air-core (no lamination) to 10 laminations. As expected, there is increase in inductance with number of laminations. As the net core thickness increases, this increase in inductance ceases. The future work involves optimizing the design and fabricating micro-inductors with co-packaged core.

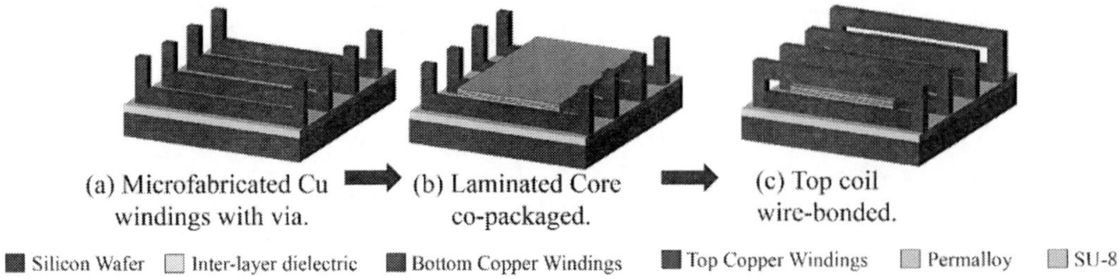

Figure 8. A schematic of fabricating solenoid inductors with co-packaged core.

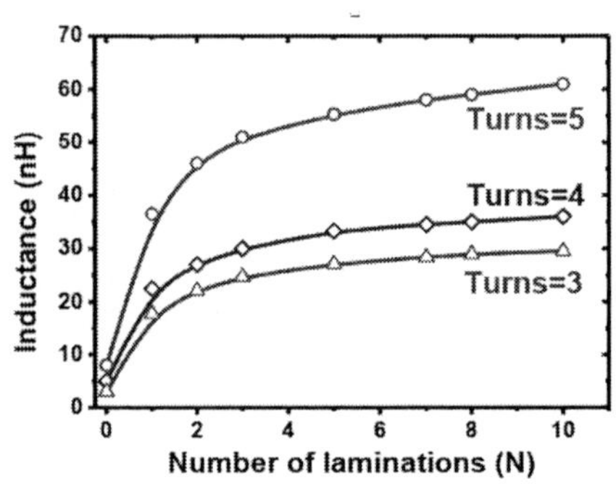

Figure 9. FEM simulated inductance vs lamination curve for 2 mm^2 laminated ($Ni_{81}Fe_{19}$: thickness: 0.8 μm separated by 1 μm SU-8 2) solenoid micro-inductor (3, 4 and 5 turns) with 5 mm core length.

4. Conclusions

We report a surface activation method to metallize SU-8 2 with soft magnetic material ($Ni_{81}Fe_{19}$). The magnetic measurements were conducted, which suggests a coercivity < 365 A/m. The high-frequency permeability tests showed uniform permeability response up to 600 MHz for 250 nm thick $Ni_{81}Fe_{19}$ on silicon. A sequential deposition of $Ni_{81}Fe_{19}$ on SU-8 2 to achieve magnetic multilayer is reported which showed improvement in frequency response with number of laminations. Finally, we propose a solenoid micro-inductor structure to with the co-packaged magnetic structure. FEM analysis suggested increase in inductance with laminations, which ceases with increase in net magnetic core thickness. The future work will involve details design study and optimization of the laminated device structure in terms of core shape and fabrication of solenoid micro-inductor.

References

[1] Meere R, O'Donnell T, Wang N, Achotte N, Kulkarni S and O'Mathuna SC 2009 *IEEE Trans. Magn.* **45** 4234.
[2] Roy S et. al., 2005 *J. Magn. Magn. Mater.* **290-291** 1524.
[3] Rhen MF, McCloskey P, O'Donnell T and Roy S 2008 *J. Magn. Magn. Mater.* **320** e819.
[4] McCloskey P et. al., *J. Magn. Magn. Mater.* **320** 2509.
[5] Park JY and Allen MG 1999 *5th IEEE Symp. Mag. Mater.* 389.
[6] Myung NV et. al. 2003 *J. Magn. Magn. Mater.* **265** 189.
[7] Gardner DS et. al. 2009 *IEEE T. Magn.* **45** 4790.
[8] Brunet M, O'Donnell T, Connell AM, McCloskey P and O'Mathuna SC 2006 *IEEE Trans. Micromech. Syst.* **15** 94.

[9] McCloskey P, Jamieson B, O'Donnell T, Gardner D, Morris MA and Roy S 2008 *J. Magn. Magn. Mater.* **320** 2509.
[10] Anthony R, Rohan JF and O'Mathuna C 2016 *Microelec. Engg.* **155** 33.
[11] SU-8 Datasheet (http://www.microchem.com/pdf/SU8_2-25.pdf).
[12] Anthony R, Shanahan BJ, Waldron F, O'Mathuna C and Rohan JF 2015 *Appl. Surf. Sci.* **357** 385.
[13] Anthony R, O'Mathuna C and Rohan JF 2015 *Proc. Phys.* **75** 1207.

Acknowledgments
The authors acknowledge European Union for funding the research work through project through FP-7 (Project: PowerSwipe) under grant agreement no.: 318529 for the support and Central Fabrication Facilities (CFF) for the support.

Journal of Physics: Conference Series **757** (2016) 012032 doi:10.1088/1742-6596/757/1/012032

Design and fabrication of ultrathin silicon-nitride membranes for use in UV-visible airgap-based MEMS optical filters

Mohammadamir Ghaderi and Reinoud F. Wolffenbuttel

Electronic Instrumentation Laboratory, Microelectronics Department, Faculty of EEMCS, Delft University of Technology, Delft, the Netherlands

E-mail: `m.ghaderi@tudelft.nl`

Abstract. MEMS-based airgap optical filters are composed of quarter-wave thick high-index dielectric membranes that are separated by airgaps. The main challenge in the fabrication of these filters is the intertwined optical and mechanical requirements. The thickness of the layers decreases with design wavelength, which makes the optical performance in the UV more susceptible to fabrication tolerances, such as thickness and composition of the deposited layers, while the ability to sustain a certain level of residual stress by the structural strength becomes more critical. Silicon-nitride has a comparatively high Young's modulus and good optical properties, which makes it a suitable candidate as the membrane material. However, both the mechanical and optical properties in a silicon-nitride film strongly depend on the specifics of the deposition process. A design trade-off is required between the mechanical strength and the index of refraction, by tuning the silicon content in the silicon-nitride film. However, also the benefit of a high index of refraction in a silicon-rich film should be weighed against the increased UV optical absorption. This work presents the design, fabrication, and preliminary characterization of one and three quarter-wave thick silicon-nitride membranes with a one-quarter airgap and designed to give a spectral reflectance at 400 nm. The PECVD silicon-nitride layers were initially characterized, and the data was used for the optical and mechanical design of the airgap filters. A CMOS compatible process based on polysilicon sacrificial layers was used for the fabrication of the membranes. Optical characterization results are presented.

1. Introduction

Optical interference filters based on a stack of alternating layers of two dielectric materials of different index of refraction provide significant potential for system integration of optical MEMS devices, such as microspectrometers and LEDs [1]. The availability of dielectric materials for the fabrication of an optical filter operating in the ultraviolet (UV) and visible range is, however, limited by several constraints. The most important limitation is the presence of absorption peaks of many common dielectric materials in the UV. CMOS process compatibility requirements further limit the list of materials that can be used for the fabrication of an optical filter. Using airgap-based interference filters provide a MEMS-compatible solution for the fabrication of high-performance optical filters in a CMOS-compatible fashion. Although UV and visible airgap-based optical filters using III-V materials have been investigated for LEDs [1], the silicon-based UV and visible MEMS airgap filters have only recently received attention [2].

The main challenge in the fabrication of airgap optical filters is maintaining optical flatness over the freestanding membrane [3]. Therefore, a tensile stressed membrane is required, while the stress gradient in the film has to be sufficiently low [3]. Filters composed of one or several layers with a thickness in the order of one Quarter-Wave Optical Thickness at the design wavelength (1 QWOT) have been considered, but it is extremely difficult to achieve a sufficient level of tensile stress without a stress gradient in such a thin layer (i.e. $\lambda_o/4n$, where n is the refractive index of the material; 70 nm for $\lambda_o = 400$ nm and $n \approx 1.5$) for a taut membrane without rupturing. Straining methods have been investigated to obtain optical flatness in silicon-oxide membranes, however, only a limited improvement was achieved [4].

An alternative approach is to use high Young's modulus materials, such as polysilicon and silicon-nitride, for the membrane material. Furthermore, the filters are composed of several layers with thicknesses in the order of multiples of the quarter-wavelength. The interference condition in the filters is generally satisfied by using (odd n) higher order designs (n-QWOT), thus thicker membranes could be considered, which would result in a similar spectral response.

The fabrication process and the preliminary results of ultrathin silicon-nitride membranes for airgap optical filter application have been investigated. Silicon-nitride thin-films were deposited using PECVD and subsequently annealed to obtain tensile stress. The residual stress and optical properties of the deposited thin films were characterized, and the results are presented in this paper. One-period ($N = 1$) airgap optical filters operating at 400 nm based on 1 QWOT and 3 QWOT silicon-nitride membrane were designed. The fabrication was carried out based on a CMOS-compatible fabrication process followed by a stiction-free drying process. Finally, the fabricated samples were optically characterized, and the results are discussed.

2. PECVD silicon-nitride as the membrane material in airgap optical filters

The excellent mechanical properties of silicon-nitride made it a suitable candidate for use in most MEMS devices. The Young's modulus of Silicon-nitride is about 200 GPa for the films deposited using PECVD [5]. This Young's modulus is significantly higher than the reported value for silicon-oxide (70 GPa [6]). The deformation energy and force in a membrane is inversely proportional to the Young's modulus of the film. Therefore, using silicon-nitride results in a decrease in membrane deformation, hence improving the flatness of the released membrane. In addition, the PECVD deposition of silicon-nitride films provides several advantages, such as the ability to tune the residual stress and optical properties of the deposited films. Therefore, the residual stress of the silicon-nitride can be tuned to obtain tensile stress films [5]. Finally, the etch rate of silicon-nitride in TMAH is very low [7], thus, can be used as a membrane material in combination with polysilicon sacrificial layers.

Table 1. Deposition parameters of PECVD silicon-nitride

SiH_4	0.28 slm
NH_3	1.8 slm
N_2	1 slm
HF/LF power	0.32/0.48 kW
Substrate Temperature	400 °C
Total pressure	2.8 Torr

In this research, the silicon-nitride films were deposited using PECVD and the deposition parameters are listed in Table 1. While the as-deposited residual stress of silicon-nitride films was measured to be equal to -200 MPa (compressive), the residual stress was increased to about 1 GPa (tensile) after an annealing cycle at $600°C$. This high tensile residual stress further ensures the flatness in the released membranes.

3. Optical design

3.1. Optical characterization of silicon-nitride

The optical properties of silicon nitride layers have been extensively studied (e.g. [8, 9]). In PECVD silicon-nitride layers, the gas phase composition of the plasma affects the silicon-to-nitrogen (Si/N) ratio in the films. As a result, the refractive index of the films varies linearly from 1.9 to 2.6 with an increasing Si content of the film (Si/N ratio from 0.7 to 1.7) [9]. Therefore, films with high refractive index can be deposited. The quality of the material combination used in a filter design is usually expressed in terms of their refractive index contrast. Therefore, silicon-rich nitride is in principle highly suitable for optical applications. However, the increased absorption of silicon-rich nitride is a limiting factor in optical applications. Increasing the silicon content in the deposited films results in a spectral shift of the absorption peak in the film toward the longer wavelengths. For instance, the absorption coefficient becomes significant (is higher than $\alpha = 10^3 cm^{-1}$) for stoichiometric Si_3N_4 for wavelengths shorter than about 270 nm ($k = 0.002$), while this limit increases to 800 nm ($k = 0.006$) for silicon-rich Si_9N_4 [8]. Therefore, the trade-off between the refractive index and absorption coefficient must be considered when designing a nitride-based interference optical filter.

The refractive index and extinction coefficient of the deposited nitride layer were measured using ellipsometry. Figure 1 shows the refractive index and absorption coefficient of the films. Although the extinction coefficient of the deposited silicon-nitride at 400 nm is significant ($k = 0.06$), the nitride-based designs are still highly interesting. This stems from the higher refractive index of silicon-nitride, as compared to the silicon-oxide, while the loss in one QWOT layers remains limited to $1 - exp(-\alpha d) = 0.08$. The refractive index contrast ($(n_H^2 - n_L^2)/2n_H^2$; a unitless quality factor between 0 and 0.5) is an important criterion for assessment of the thin films for the optical applications. The higher refractive index of nitride results in a considerable increase in the refractive index contrast (i.e. 43% increase in refractive index contrast as compared to the silicon-oxide with air pair). Furthermore, due to the high refractive index of silicon nitride, a QWOT of silicon-nitride is thinner, thus, the intensity loss due to the absorption is smaller.

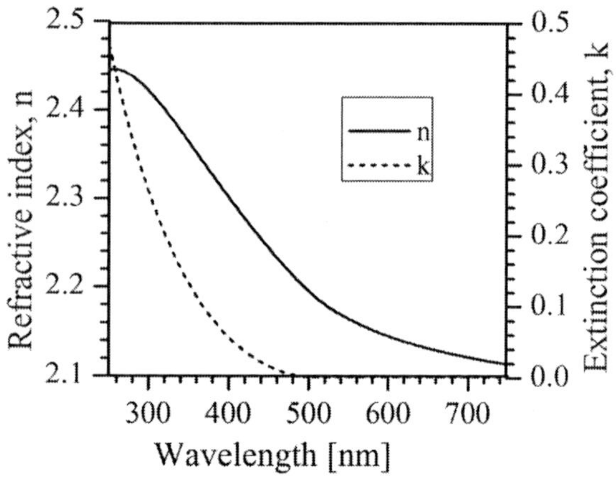

Figure 1. The refractive index and extinction coefficient of PECVD silicon nitride thin-films measured by Ellipsometry.

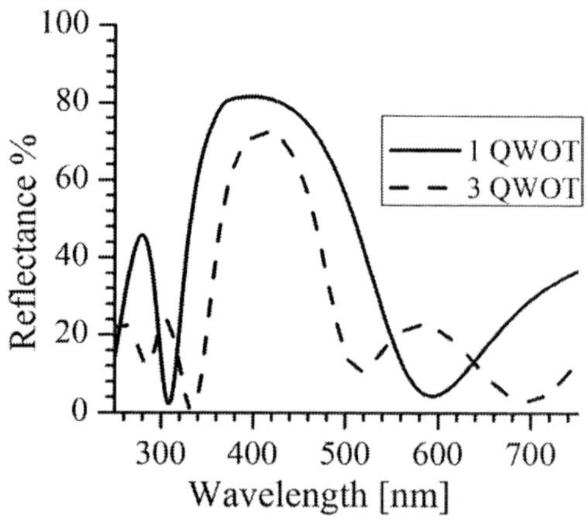

Figure 2. Expected reflectance of a single period ($N = 1$) DBR with 1 QWOT and 3 QWOT membrane simulated using transfer matrix method. The model is based on a silicon (substrate)/silicon-oxide (passivation)/airgap/silicon-nitride structure.

3.2. Optical design of a distributed Bragg grating
A distributed Bragg reflector (DBR) is one of the simplest forms of an optical filter and is composed of a repetitive and alternating high and low index quarter-wave thick layers. In general, any layer with a thickness equal to the odd-multiple of the quarter-wave also satisfies the interference condition and results in a similar spectral reflectance. In this research single period ($N = 1$) DBRs with single-quarter (1 QWOT) and three-quarter-wave thick (3 QWOT) layers, have been designed. Figure 2 shows the expected reflectance of the DBRs simulated using transfer matrix method (TFCalc).

As shown in Figure 2, a 1 QWOT membrane separated by a 1 QWOT airgap result in a peak reflectance of about 82% with a bandwidth of 192 nm. The 3 QWOT membrane separated by 1 QWOT airgap design, on the other hand, results in a peak reflectance of 72% and bandwidth of 121 nm. The reduced peak reflectance in the 3 QWOT design is due to the increase in the spectral absorption in the thicker membrane layer. Therefore, the 3 QWOT membranes are not very suitable for optical designs, especially when multiple periods of high and low index layers are required.

4. Fabrication
A CMOS-compatible process for fabrication of free-standing silicon-oxide membranes using polysilicon sacrificial layers was already introduced by our group [2, 10]. A similar fabrication process was used for the fabrication of silicon-nitride membranes. The fabrication was performed on prime 10 cm diameter silicon wafers. Figure 3 shows the schematic of the fabrication process. A repetitive deposition of LPCVD polysilicon and PECVD silicon-nitride was performed to build the layer stack according to the optical design. Polysilicon layers are used as the sacrificial layers, thus the thickness was chosen equal to the thickness of the airgap layers. After the deposition of the layer stack, a set of openings was patterned and etched through the deposited layers. An additional silicon nitride layer was deposited using PECVD to cover the openings. The reminder of this layer is then removed, leaving pin-shaped structures for anchoring the membranes after the release. A final patterning and etching of windows provide access for the sacrificial etching. Finally, wet etching in a TMAH-based solution followed by a stiction-free drying using CO_2 at its critical point was used to release the structures.

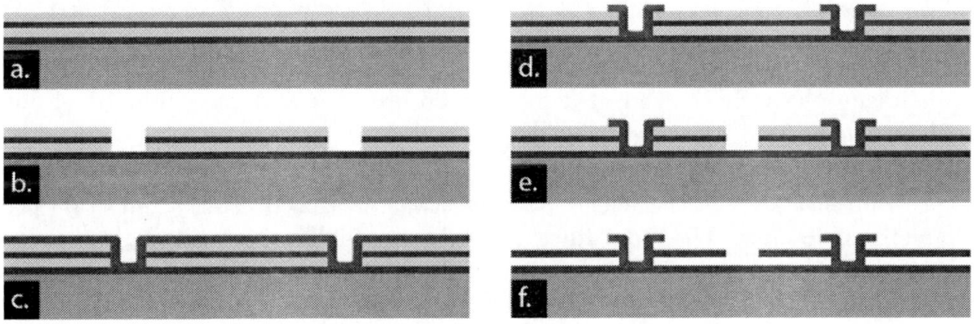

Figure 3. Schematics of the fabrication process.

5. Characterization
Several samples were partially under-etched and went through the critical point drying process. The etching was stopped before the complete removal of the sacrificial material to find out the maximum achievable membrane size, while maintaining a high yield. Figure 4a and 5a show the microscope images of the released samples. The fabricated 1 QWOT and 3 QWOT membranes

have a diameter-to-thickness ratio of about 500 and 150 respectively. Figure 4b and 5b show the show the 45° tilted SEM images of the partially released membranes.

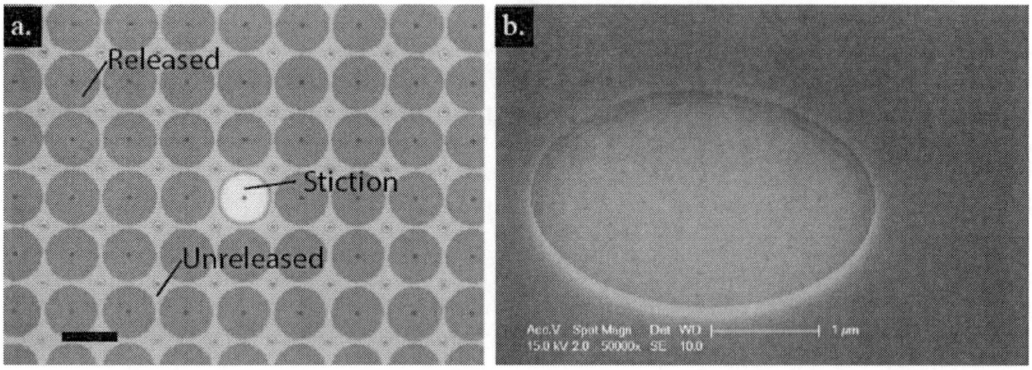

Figure 4. The 1 QWOT silicon nitride membrane separated by a 1 QWOT airgap: a) microscope image, the scale shown in the figure is 25 μm; b) 45° degree tilted SEM micrograph of a circular opening used for the sacrificial etching.

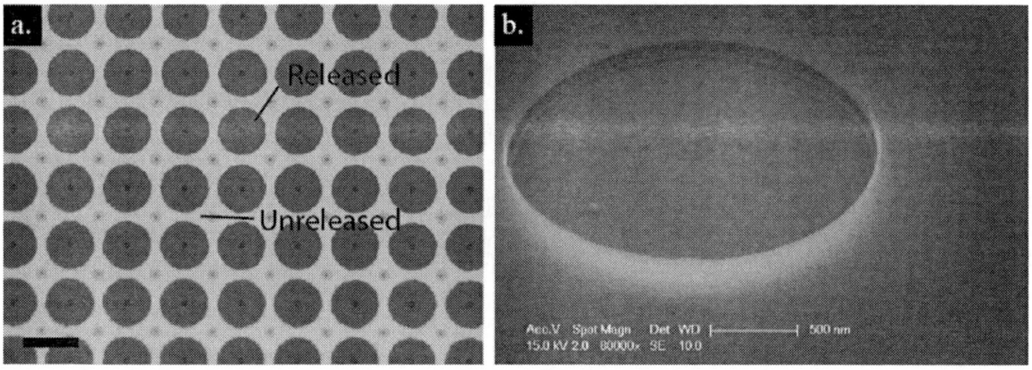

Figure 5. The 3 QWOT silicon nitride membrane separated by a 1 QWOT airgap: a) microscope image, the scale shown in the figure is 25 μm; b) 45° degree tilted SEM micrograph of a circular opening used for the sacrificial etching.

The spectral reflectance of the released samples was measured using a reflection probe with a core diameter of 200 μm. The input fibre channel was coupled to a white light source, and the output was fed to a spectrometer module (Flame, Ocean optics). The samples were placed about $1-2$ mm far from the fibre probe tip. Figures 6 a and b show the spectral reflectance of samples with a 1 QWOT and 3 QWOT silicon-nitride membrane over an area of about $2-3$ mm^2. Due to the large inspection area, several 100-s of filter elements are simultaneously illuminated, and the measured spectral reflectance is a combination of the spectral response from the released and the unreleased areas, while the posts and access holes are accounted for in the device area fill factor. The spectral reflectance reveals a peak at 400 nm, which is associated with the released membranes (about 90% fill factor), is present in Figure 6. Furthermore, the peak at longer wavelengths (about 600 nm) is due to the part of the area with the unreleased structures and is an indication of fabrication yield. An increase in the released area results in an increasing peak at 400 nm, while the peak at 600 nm decreases.

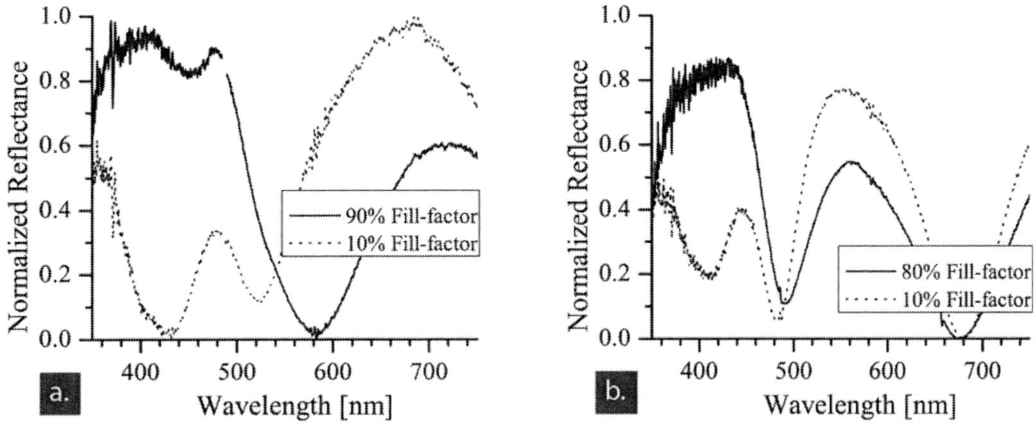

Figure 6. The large area spectral reflectance of a) the 1 QWOT membrane design and b) the 3 QWOT design.

6. Conclusion

In this paper, we present the design, fabrication, and preliminary results on several 1 QWOT and 3 QWOT membranes. Silicon-nitride is a suitable candidate for the large area free-standing membranes. Although silicon-nitride has many advantages, such as the high Young's modulus, ability to obtain a tensile stress, and high refractive index, the main disadvantage is its high extinction coefficient in the UV spectral range. Silicon-nitride films were deposited using PECVD and the mechanical and optical characteristics of the films were measured. A tensile stress of 1 GPa was achieved through annealing of the layer. Although the absorption coefficient in the deposited layers cannot be disregarded, a good spectral response can be achieved due to the high refractive index of the layers. The measured spectral reflection was also in good agreement with the expected response. On-going work is directed towards the fabrication advanced optical filters with higher periods ($N > 1$) and compound silicon-nitride/silicon-oxide membranes.

Acknowledgment

This work has been supported by the Dutch Technology Foundation STW under Grant DEL.11476. The process was carried out in the Else Kooi Laboratory (DIMES) and the Kavli nanolab both at Delft University of Technology.

References

[1] Chen D and Han J 2012 *Applied Physics Letters* **101** 221104
[2] Ghaderi M, Ayerden N P, de Graaf G and Wolffenbuttel R F 2014 *Procedia Engineering* **87** 1533 – 1536
[3] Ghaderi M, Karimi Shahmarvandi E, de Graaf G and Wolffenbuttel R F 2016 *Proc. Conf. on Micro-Optics* vol 9889 (SPIE) pp 98890A–98890A–9
[4] Ghaderi M, de Graaf G and Wolffenbuttel R F 2016 *Proc. Conf. on Micro-Optics* vol 9888 (SPIE) pp 98880R–98880R–8
[5] Zorman C A, Roberts R C and Chen L 2011 *MEMS Materials and Processes Handbook* (Boston, MA: Springer US) ISBN 978-0-387-47318-5
[6] Thurn J and Cook R F 2002 *Journal of Applied Physics* **91** 1988–1992
[7] Paranjape M, Pandy A, Brida S, Landsberger L, Kahrizi M and Zen M 2000 *Journal of Vacuum Science & Technology A* **18** 738–742
[8] Philipp H R 1973 *Journal of The Electrochemical Society* **120** 295–300
[9] Claassen W A P, Valkenburg W G J N, Habraken F H P M and Tamminga Y 1983 *Journal of The Electrochemical Society* **130** 2419–2423
[10] Ghaderi M, Ayerden N P, de Graaf G and Wolffenbuttel R F 2015 *Proc. Conf. on Micro-Optics* vol 9517 (SPIE) pp 95171M–95171M–6

Journal of Physics: Conference Series **757** (2016) 012033 doi:10.1088/1742-6596/757/1/012033

CMOS-compatible fabrication of metamaterial-based absorbers for the mid-IR spectral range

Ehsan Karimi Shahmarvandi, Mohammadamir Ghaderi and Reinoud F. Wolffenbuttel

Faculty EEMCS, Delft University of Technology, Mekelweg 4, 2628 CD Delft, Netherlands

E.karimishahmarvandi-1@tudelft.nl

Abstract. A CMOS-compatible approach is presented for the fabrication of a wideband mid-IR metamaterial-based absorber on top of a Si_3N_4 membrane, which contains poly-Si thermopiles. The application is in IR microspectrometers that are intended for implementation in portable microsystem for use in absorption spectroscopy. Although Au is the conventional material of choice, we demonstrate by simulation that near-perfect absorption can be achieved over a wider band when using the more CMOS-compatible Al. The absorber design is based on Al disk resonators and an Al backplane, which are separated by a SiO_2 layer. The fabrication process involves the deposition of Al and SiO_2 layers on top of a Si_3N_4 membrane, lithography and a lift-off process for patterning of the top Al layer.

1. Introduction

Since the introduction of metamaterial-based absorbers in 2008 by Landy *et al.* [1], these structures have been adapted to serve a wide variety of practical applications; including selective thermal emitters [2], wavelength-tunable microbolometers [3], refractive index sensing [4] and microspectrometers [5]. The typical structure of the metamaterial absorber consists of arrays of patterned metallic sub-wavelength structures on top of a dielectric spacer, which is backed by a thick metallic ground plane. The top periodic structure is responsible for the electrical response of the metamaterial absorber. The metallic backplane is thicker than the penetration depth of the incident wave to prevent any transmission. The coupling of two metallic layers results in the magnetic response. Therefore, by selecting the proper geometrical parameters, the effective permittivity ε and permeability μ can be tuned independently, resulting in impedance matching to the free space. Practical structures have been presented that exhibit near-perfect absorption of the incident wave a in certain frequency band can be obtained from microwaves [6], terahertz frequencies [7], the infrared spectral range [5] to the visible spectrum [8].

The essential characteristic of a structure that is to be used as a metamaterial is its sub-wavelength feature size. Consequently, metamaterials designed for microwave frequency range are relatively large and impose no significant requiremenst on the lithography. However for the fabrication of metamaterials operating in the infrared and visible ranges, techniques such as e-beam lithography or focused ion beam (FIB) methods should be used. These methods are capable of making structures with sizes that are in the order of tens of nanometers. However, unlike mask-based photolithography, these methods 'write' the pattern in a point-by-point serial process, which is a significant throughput problem in the fabrication of large-area metamaterials, as required in an absorber. In the part of the mid-IR spectrum that is suitable for absorption spectroscopy on combustible hydrocarbon gas (3-4 µm), the typical feature size of the metamaterial is just within the realms of a conventional lithography system. Such an approach is pursued here; however, the specifications in terms of resolution and step size of such a system on the performance of the metamaterials need to be investigated.

In this paper the fabrication process for a CMOS-compatible wideband metamaterial absorber on top of a Si_3N_4 membrane is presented. The application is the MEMS-based microspectrometer that was already reported in [5]. Three compatibility issues are addressed. The first is the choice of materials. Although Au is the conventional material of choice, near-perfect absorption can be achieved over a wider band when using the more CMOS-compatible Al. The second issue is the dimensions. The absorber design is based on four Al disk resonators in a sub-wavelength unit cell with dimensions in the 500-600 nm range. The third issue is the effect of the limitations of the lithography and other process tolerances on the absorber spectral performance. The fabrication process involves Al and SiO_2. The effects of the lithography and fabrication tolerances, such as the variation in the thickness of deposited dielectric layer, have been investigated using FDTD analysis. Based on the results of these investigations, a fabrication process for wideband absorbers is presented using masked lithography and a lift-off process for patterning of the top Al layer.

2. CMOS-compatibility

Most of the reported metamaterial absorber structures are based on Au, because of its higher electrical conductivity. Higher conductivity of Au results in stronger localized surface plasmon resonance (LSPR) and consequently in stronger magnetic resonance [5]. However, Au is poorly CMOS-compatible. Recently, we have proposed an IR metamaterial absorber using Al circular disks [5]. The FDTD analysis showed that a wide band absorption can be achieved using multiple disk resonators. Figure 1(a) shows the metal-dielectric-metal structure with a circular disk, while Figure 1(b) shows the comparison between the absorption peaks for with Al and Au with an identical geometry. Despite the higher loss of Al in the mid-IR range as compared to Au, the absorption peaks are spectrally shifted but almost identical shape, demonstrating that a near-perfect absorption can also be achieved when using the more CMOS-compatible Al as the plasmonic material.

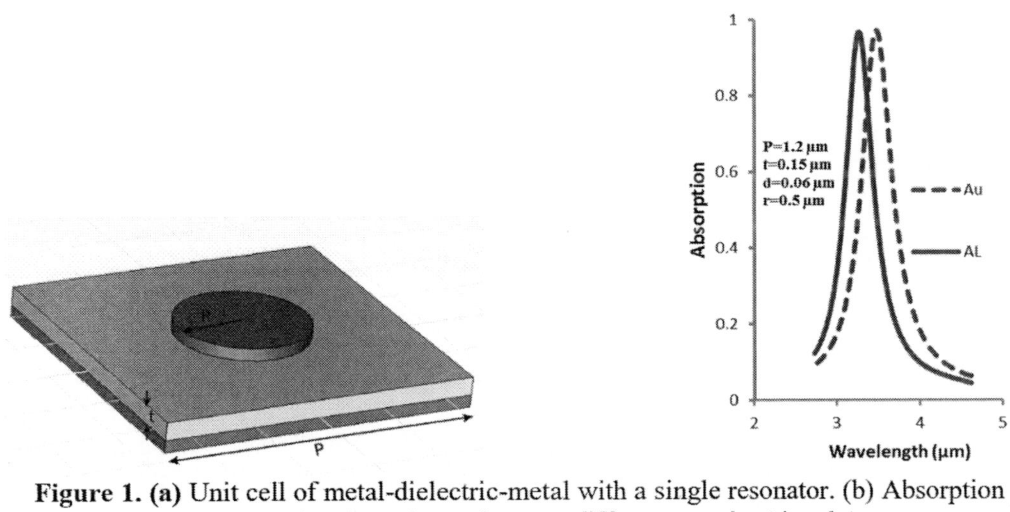

Figure 1. (a) Unit cell of metal-dielectric-metal with a single resonator. (b) Absorption peak for an identical single-resonator absorber when using two different metals; Al and Au.

2. Device operation and tolerances

Arranging differently sized Al disk resonators in a single unit cell results in a broadband absorption spectrum, if the different disks are tuned to different parts of the spectrum to be absorbed in such a way that the spectra do overlap. A design is presented here with four differently sized Al disks in one unit cell of the structure, as presented in Figure 2 (a). Wideband operation over the spectral range can be achieved by staggered tuning of four Al disk resonators in one 1.5 × 1.5 μm^2 unit cell. By using different values for the radius for the Al disks, between 0.50 μm and 0.63 μm, and tuning the thickness of the SiO_2 spacer (150 nm, in this design) wideband absorption over the 3-4 μm range was obtained (Figure 2). The simulations reveal an average absorption of about 95% with a ±4% ripple at normal incidence.

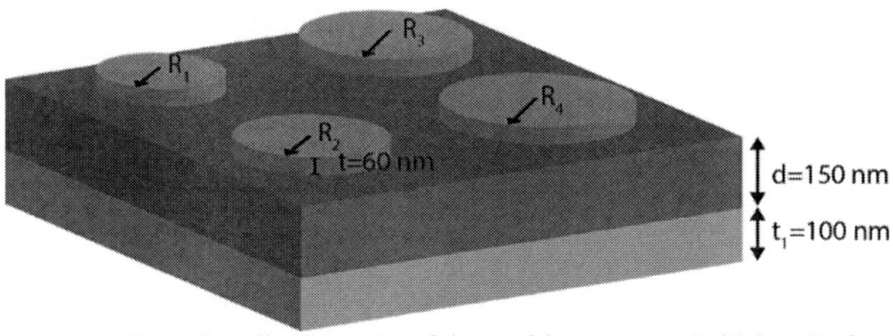

Figure 2. (a) The unit cell schematic of the multi-resonator (wideband) absorber. Periodicity (p) =1.5 μm. R_1=500 nm, R_2=540 nm, R_3=580 nm, R_4=630 nm

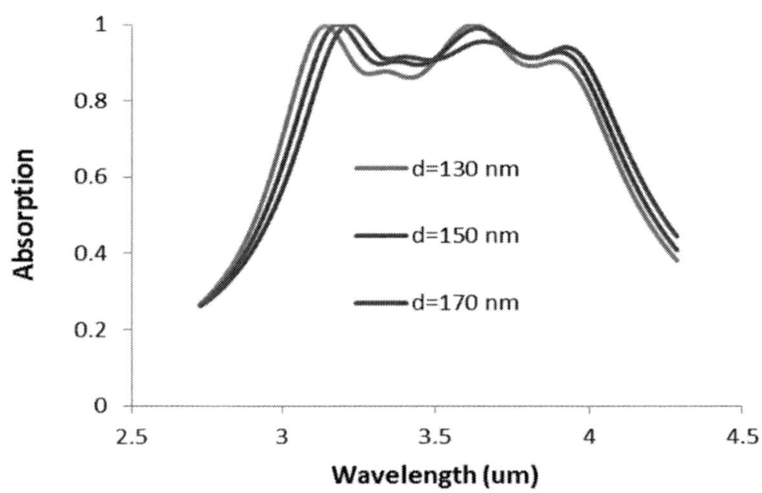

Figure 2. (b) Simulated absorption spectra of the multi-resonance absorber with dielectric thickness (d)=150 nm and ± 20 nm tolerance.

The tolerances in the thickness of the deposited SiO_2 could affect the absorber performance. Simulation results when considering a ±20 nm variation in the dielectric layer thickness are shown in Figure 2 (b). This result demonstrates that the spectral performance of absorber is not significantly affected.

The resolution of the e-beam lithography is in order of less than ten nanometers and would enable the precise definition of the Al disks in the 500-600 nm range. Therefore, e-beam has been used for the fabrication of reference structures, but the limitations of masked lithography for fabrication are also considered. Figure 3, shows the spectral performance in case of: octagonal structures, inner squares (inscribed polygon) and outer squares (circumscribed polygon), as compared to circular structures. This result confirms that the metamaterial is actually rather robust for deviations from the circular shape of the top structures. This can be explained by the magnetic plasmonic resonance mechanism of the absorption. The length or diameter of the top metallic pattern determines the resonance frequency, because it defines the LSPR. The results indicate a limited susceptibility to shape deformation. The resonance wavelengths vary within the band, but the wideband absorption remains acceptable. It should be noted that the inner square design is basically tuned to a significantly shorter plasmonic resonance wavelength.

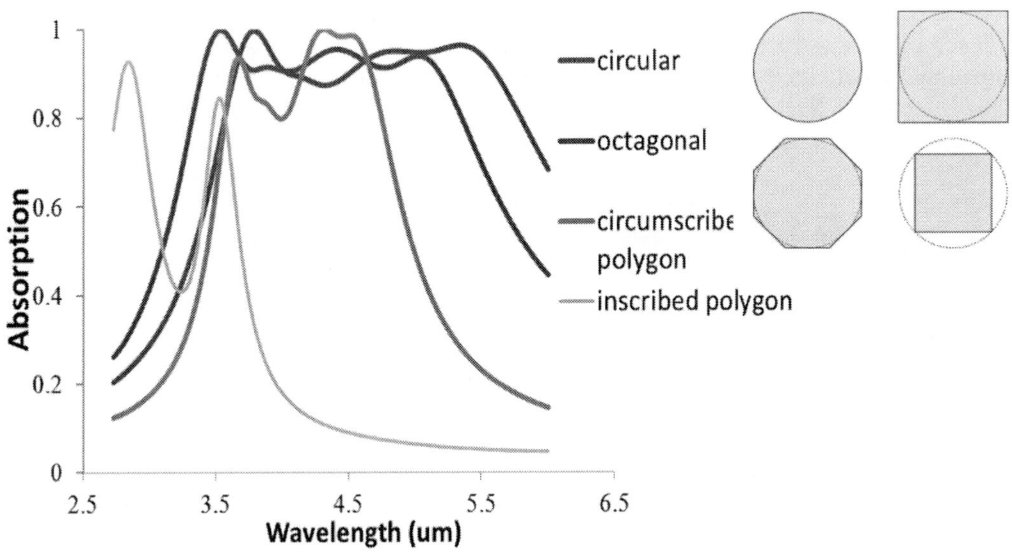

Figure 3. Simulated absorption spectra for different metallic patterns. The various shapes are shown in the inset.

3. Device Fabrication

This absorber is intended for use in combination with polysilicon thermopile based detector arrays fabricated on Si_3N_4 freestanding membranes as shown in Figure 4 [9].

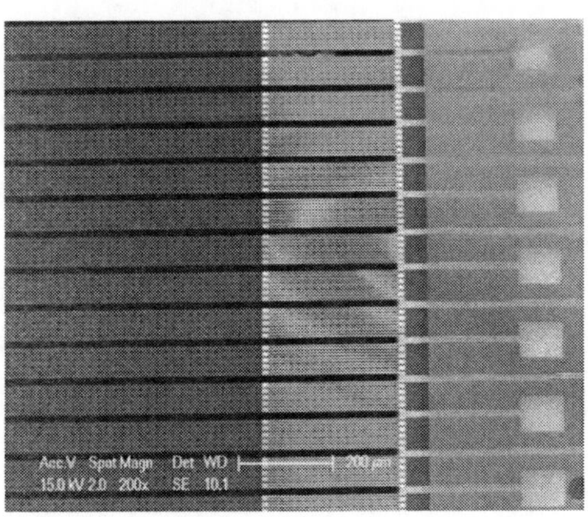

Figure 4. Released thermopile array without absorber.

After completion of the detector array fabrication and before the sacrificial release, a 100 nm thick Al film is sputtered on the detector array. This layer serves as the metallic back plane of the metamaterial absorber. Subsequently, a 150 nm thick SiO_2 layer is deposited on the Al layer by Atomic Layer Deposition (ALD). Several techniques are available for low-temperature SiO_2 deposition on top of the Al backplane, such as PECVD, sputtering and ALD. Althougth ALD is a very slow deposition technique it is the most sutaible technique for the kind of flexible expremementation that is needed at this stage. Afterwards, a layer of resist (PMMA) is spin-coated on the SiO_2 dielectric layer and lithography is used to define the pattern of the top Al structures. The technique that is considered for transferring the pattern is based on a lift-off process, which is a method of patterning using a sacrificial material (photoresist). Figure 5 shows the schematics of the fabrication process. After EBL exposure and development, an inverse pattern of the disks is made in the resist layer. Subsequently, the second Al layer is deposited on the samples. The remaining resist layer with the Al on top of it, are removed. The remaining structure layer will have the desired metallic pattern. Finally, a set of sacrificial openings is patterned and etched through the membrane and absorber layers. The sacrificial layer (TEOS) is etched in a high concentration HF solution (73% to avoid Al attack) and the samples were subjected to the supercritical CO_2 drying.

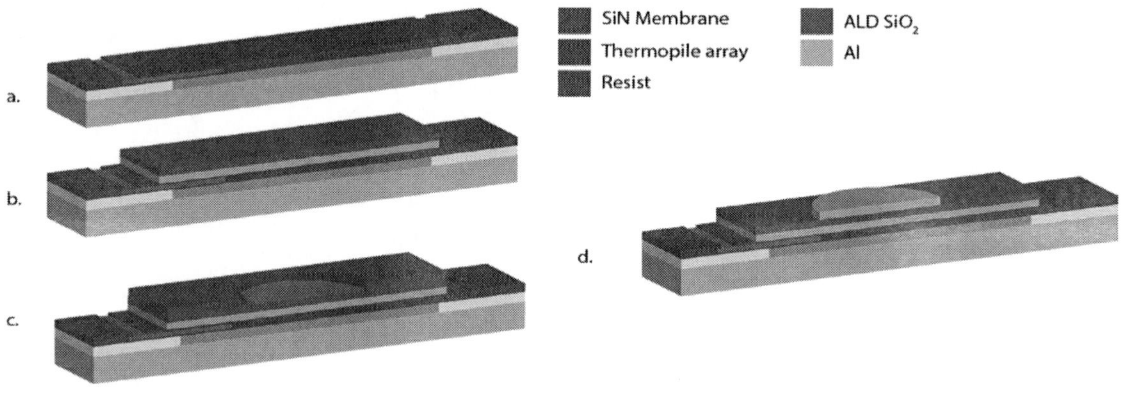

Figure 5. A cross-sectional view of the fabrication flow with lift-off: a. Substrate with detector array (thermopile) on a 0.9 μm Si_3N_4 membrane; b. Deposition and patterning of Al and SiO_2; c. Resist coating and e-beam lithography and development of patterns; d. Deposition of Al and lift-off.

4. Conclusion

The CMOS-compatible fabrication of mid-IR metamaterial-based absorbers was studied. This process involves deposition of Al-SiO_2-Al stack on top of a Si_3N_4 membrane using a lift-off process. Simulations indicate a 95±4 % absorption in the 3-4 μm mid-IR spectrum range using Al. It was also shown that the performance of the absorber structure is not significantly affected by variations in the the dielectric layer thickness. Although e-beam lithography is used in the first experiments, the relatively large disk dimensions (500-600 nm) and the limited susceptibility of the spectral absorption to shape distortion by limitations in the lithography would allow conventional lithography to be applied to the large-area absorbers. On-going experimentation should validate the robustness of the technique for variations in the circular shape. The masked lithography would be a significant advantage for high-volume wafer-level fabrication.

References

[1] Landy N I, Sajuyigbe S, Mock J J, Smith D R, and Padilla W J *2008, Phys. Rev. Lett. 100(20), 207402*

[2] Liu X, Tyler T, Starr T, Starr A F, Jokerst N M, and Padilla W J *2011, Phys. Rev. Lett. 107(4), 045901*

[3] Maier T and Brückl H *2009, Opt. Lett. 34(19), 3012–3014*

[4] Liu N, Mesch M, Weiss T, Hentschel M, and Giessen H *2010, Nano Lett. 10(7), 2342–2348*

[5] Karimi shahmarvandi E, Ghaderi M, Ayedren N P, De Graaf G and Wolffenbuttel R F *2016, Proc. SPIE 9883, Metamaterials X, 988309; doi:10.1117/12.2227800.*

[6] Zhong S, He S *2013, Scientific reports 3*

[7] Hokmabadi M, Wilbert D, Kung P, Seongsin K *2013, Terahertz Science and Technology 6.1 ; 40-58*

[8] Huang L, Chen H *2011, Progress In Electromagnetics Research 113: 103-110.*

[9] Ghaderi M, M, Ayedren N P, De Graaf G and Wolffenbuttel R F *2015, Journal of Micromechanics and Microengineering 25(7), 074010*

Integrated CoPtP Permanent Magnets for MEMS Electromagnetic Energy Harvesting Applications

Dhiman Mallick[1], **Saibal Roy**[2]

[1] Micro-Nano-Systems Center, Tyndall National Institute, Cork, Ireland

saibal.roy@tyndall.ie

Abstract. This work reports the development of integrated Co rich CoPtP hard magnetic material for MEMS applications such as Electromagnetic Vibration Energy Harvesting. We report a new method of electrodeposition compared to the conventional DC plating, involving a combination of forward and reverse pulses for optimized deposition of Co rich CoPtP hard magnetic material. This results in significant improvements in the microstructure of the developed films as the pulse reverse plated films are smooth, stress free and uniform. Such improvements in the structural properties are reflected in the hard magnetic properties of the material as well. The intrinsic coercivities of the pulse reverse deposited film are more than 6 times higher for both in-plane and out-of-plane measurement directions and the squareness of the hysteresis loops also improve due to the similar reasons.

1. Introduction

Microfabricated permanent magnets with thickness of the order of microns to hundreds of microns are the key building blocks for a number of magnetic MEMS (Micro-Electro-Mechanical Systems) applications. Development in this particular field is necessary for complete integration of electromagnetic transducers [1-3] in CMOS compatible processes for scavenging electrical energy from mechanical vibrations. Therefore, a number of fabrication techniques including powder-based fabrication methods [4, 5] (such as dry-packing, screen printing etc.) and conventional thin film deposition techniques [6, 7] (sputtering, electrochemical deposition etc.) have emerged over the years depending on the application requirements. Electrochemical deposition is an attractive choice among them due to its low cost and relatively high deposition rate at CMOS compatible temperature. Growth of transition metal alloys Co/Fe-Pt with equi-atomic ratios has received extensive attention [8] due to the high magneto-crystalline anisotropy which require high temperature, post-deposition annealing. On the other hand, Co rich $Co_{80}Pt_{20}P$ alloy exhibit a relatively high coercivity, squareness and remanent magnetization in its as deposited state making it a suitable choice for different Magnetic MEMS applications [7, 9].

In this work, we have reported the deposition, structural and magnetic characterization of Co rich CoPtP magnetic material for magnetic MEMS applications.

2. Magnetic Material Deposition

Co-rich CoPtP films are electrochemically deposited using a bath consisting of 0.1M Cobalt Sulfamate, 0.01M Diammine-Dinitro-Platinum, 0.1M Sodium Hypophosphite, 0.1M Dibasic Ammonium Citrate, 0.1M Glycine and 0.25M Saccharin. Among the used chemicals in the bath,

Cobalt Sulfamate, Diammine-Dinitro-Platinum and Sodium Hypophosphite act as the source of cobalt, platinum and phosphorous impurity ions. Dibasic Ammonium Citrate acts as the complexing agent to bring the reduction potentials of cobalt and platinum close together. Glycine reduces the surface tension between the electrolyte and the working electrode whereas saccharin is added in the electrochemical bath as stress-relieving additive. The pH of the solution was adjusted to 8. Cobalt pieces were used as anode and silicon pieces (1 cm^2 area) with sputtered (200/20 nm) Cu/Ti seed layer were used as cathode and deposition is carried at room temperature. First film is deposited using DC plating technique at a current density of 20 mA/cm^2 whereas a second film is electroplated using Pulse Reverse Plating (PRP) technique using a current waveform as shown in Fig. 1. This consists of forward and reverse current densities 20 mA/cm^2 (20 ms) and 10 mA/cm^2 (0.5 ms) respectively with intermediate off times of 0.5 ms. The cycle is repeated for the entire deposition duration. The magnetic measurements of the deposited films are carried out in Quantum Design SQUID magnetometer (MPMS-XL5) for applied field up to 5 Tesla.

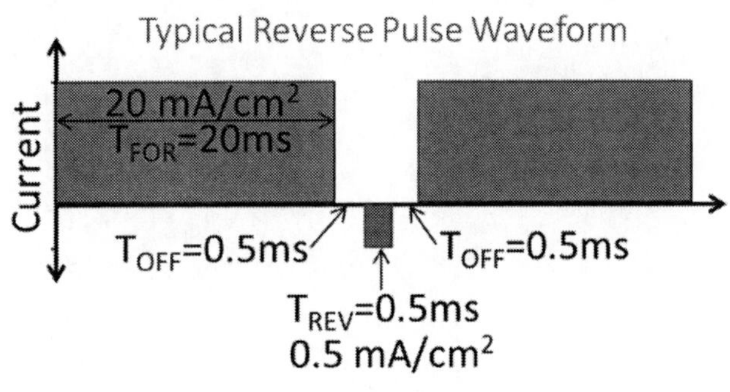

Figure 1. Current waveform for Pulse Reverse Plating (PRP) deposition of CoPtP.

3. Results & Discussions

The samples are electroplated for one hour each. The deposition rate for DC plated film is 1.7 μm/hr whereas that for the PR plated film is 2.9 μm/hr. The deposition current density is varied for both the electrodeposition techniques in order to observe the consequences on the developed films. In case of PRP technique, the reverse current density is always maintained to half of the forward current values with the previously mentioned durations. The variation of the plating rates with current density is shown in the Fig. 2.

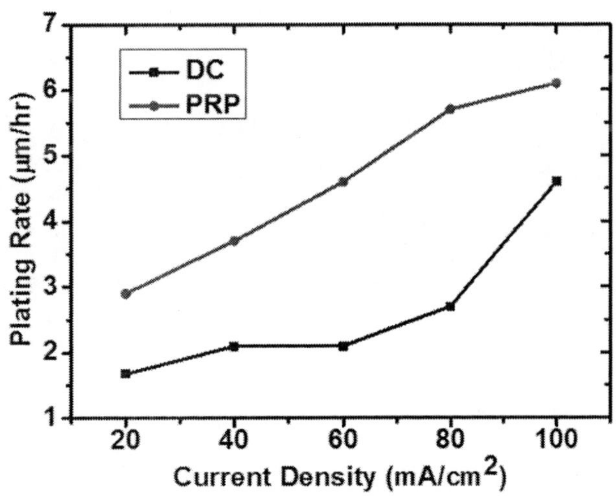

Figure 2. Variation of plating rate with the current density for DC and PRP plating.

Journal of Physics: Conference Series **757** (2016) 012034 doi:10.1088/1742-6596/757/1/012034

The surface morphology of the deposited films is characterized using Scanning Electron Microscopy (SEM) and shown in Fig. 3. The DC plated film is full of micro-cracks [Fig 3(a)] even without using the stress relieving agent. To overcome this and in order to develop smooth and stress free films of CoPtP, we adapted the PRP technique with the above mentioned parameters. The nano-scale roughness from the surface of the deposited film is etched away in each cycle due to the reverse current of the PRP process which results in smooth film with uniform composition [Fig 3(b)]. The cross-section images of the electrodeposited films are shown in Fig. 3 (c) and (d). The uniform columnar growth of the film in case of PRP technique can be easily observed even at a high current density whereas the DC plated film become more and more non-uniform and the grain size increases significantly with the deposition current.

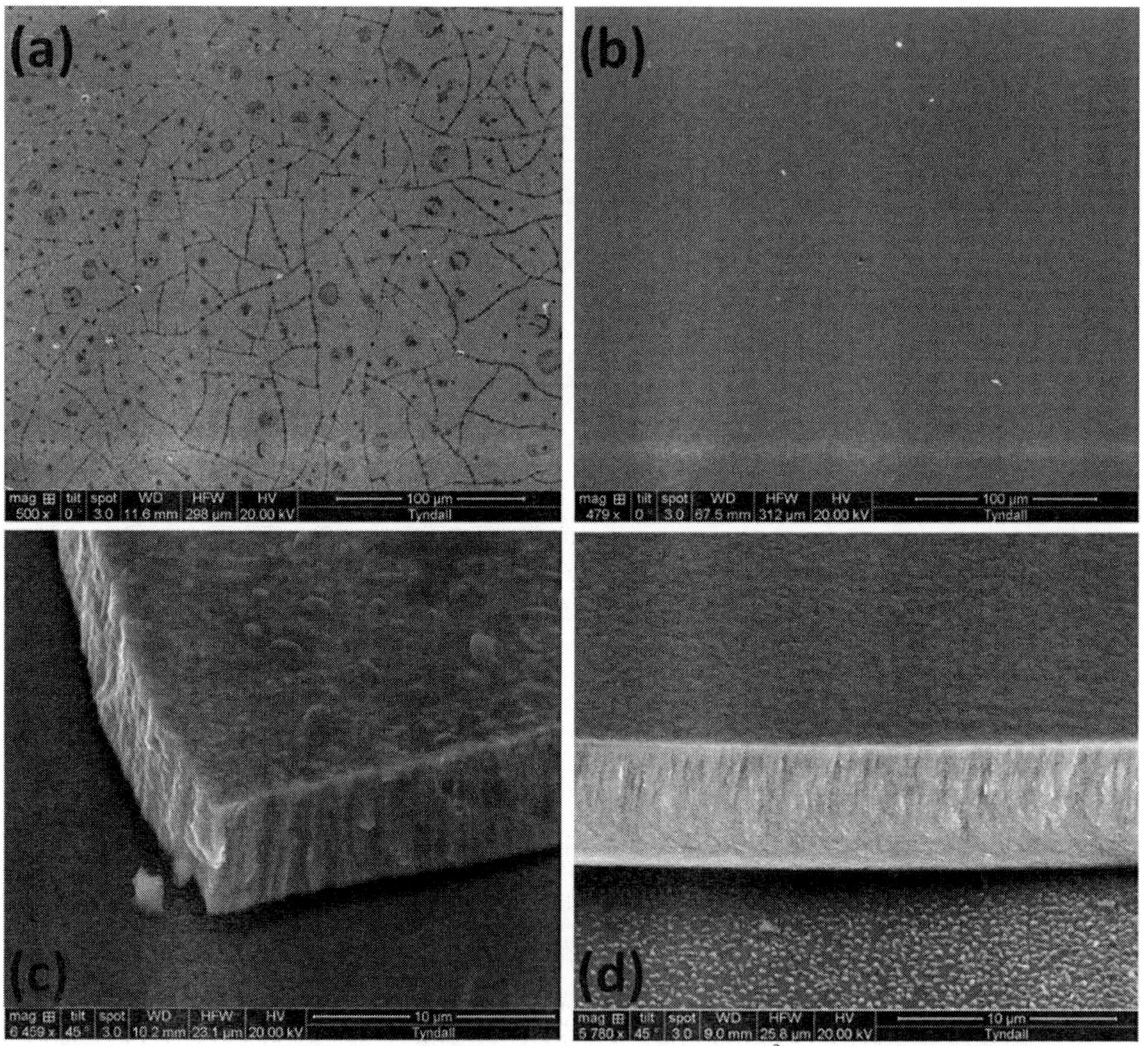

Figure 3. Surface morphology of the films deposited at 20 mA/cm^2 density (a) DC (b) PRP. The cross-section image of the films deposited at 100 mA/cm^2 density (c) DC (d) PRP.

As shown in Fig. 4, the Co percentage increases significantly (over 90%) along with further increase of current in both the deposition processes. As Co is soft magnetic in nature, the consequently developed films also lose their hard magnetic nature. Therefore, the above mentioned parameters are identified as the optimized deposition conditions.

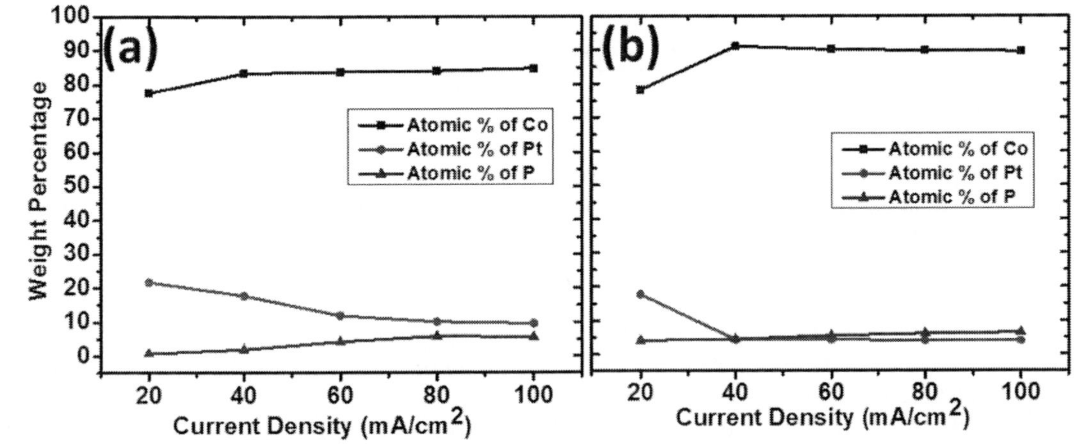

Figure 4. Effect of deposition current density on the alloy composition: (a) DC plated film (b) PRP plated film.

The PRP method improves the magnetic properties significantly too as compared in Fig 5. The perpendicular and parallel coercivities of the DC plated film are 515 Oe and 300 Oe respectively whereas the same values for the PRP deposited film are 3345 Oe and 1904 Oe. This huge improvement in the magnetic property can be attributed to the structural uniformity of the PRP deposited film. The squareness factor ($S=M_R/M_S$) of the PRP deposited film in perpendicular and parallel directions are 0.3 and 0.42 respectively. Corresponding values for the DC plated films are significantly low too. However, further improvements in the squareness and remanence properties of the developed films will be needed to increase the energy product which is essential for hard magnetic applications. Further study will also include development of thick films of CoPtP with preserved magnetic properties for MEMS electromagnetic energy harvesting applications.

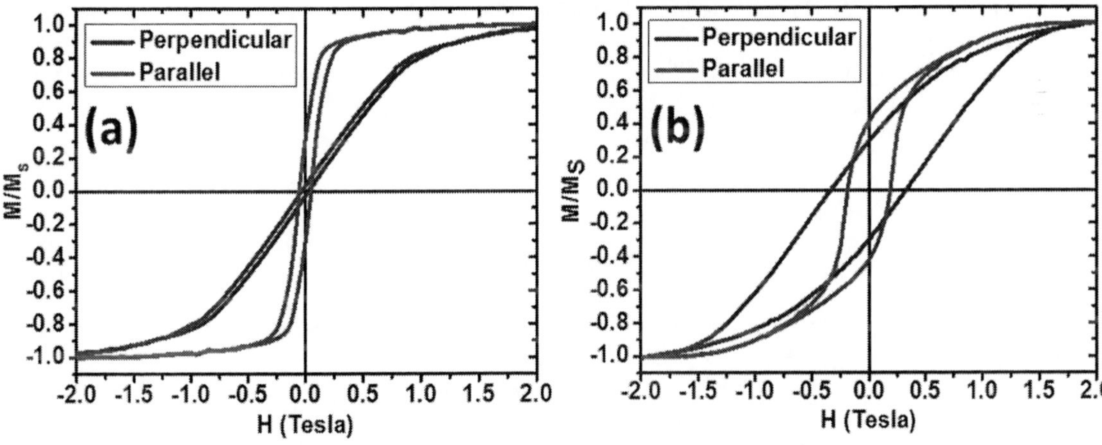

Figure 5. Magnetic hysteresis loop measurement of the deposited films at room temperature: (a) DC plated film (b) PRP plated film..

4. Conclusion

In this work, we report a new method of electrodeposition compared to the conventional DC plating, involving a combination of forward and reverse pulses for optimized deposition of Co rich CoPtP hard magnetic material. This results in significant improvements in the microstructure of the developed films as the PRP deposited films are smooth, stress free and uniform. Such improvements in the structural properties are reflected in the hard magnetic properties of the material as well. The intrinsic coercivities of the PRP deposited film are more than 6 times higher for both in-plane and out-of-plane measurement directions and the squareness of the hysteresis loops also improve due to the similar reasons. Further optimized and thick films of the developed CoPtP material will be used for complete integration of MEMS electromagnetic energy harvesting devices.

Acknowledgements
This work is financially supported by Science Foundation Ireland (SFI) Principal Investigator (PI) project on 'Vibration Energy Harvesting' -grant no SFI-11/PI/1201.

References

[1] S. P. Beeby, R. N. Torah, M. J. Tudor, P. Glynne-Jones, T. O'Donnell, C. R. Saha, S. Roy, A micro electromagnetic generator for vibration energy harvesting, J. Micromech. Microeng. 17:1257–1265, 2007.
[2] D. Mallick, A Amann, S Roy, Interplay between mechanical and electrical domains of a high performance nonlinear energy harvester, Smart Mat. Struc. (Fast Track Communication) 24 (12), 122001, 2015.
[3] D. Mallick, A. Amann, S. Roy, MEMS Based Nonlinear Monostable Electromagnetic Vibrational Energy Harvester for Wider Bandwidth, Journ. Phys.: Conf. Ser., 660 (012115), 2015.
[4] N. Wang, B. J. Bowers, D. P. Arnold, Wax-bonded NdFeB micromagnets for microelectromechanical systems applications, J. Appl. Phys. 103, 07E109, 2008.
[5] L. K. Lagorce, O. Brand, M. G. Allen, Magnetic microactuators based on polymer magnets, J. Microelectromech. Syst., vol. 8, no. 1, pp. 2–9, 1999.
[6] M. Ohtake, S. Ouchi, F. Kirino, M. Futamoto, Structure and Magnetic Properties of CoPt, CoPd, FePt, and FePd Alloy Thin Films Formed on MgO(111) Substrates, IEEE Trans. Mag., vol. 48, no. 11, pp. 3595-8, 2012.
[7] S. Kulkarni, S. Roy, Deposition of thick Co-rich CoPtP films with high energy product for magnetic microelectromechanical applications, J. Magn Magnetic Mat., 322, 1592–1596, 2010.
[8] O. D. Oniku, B. Qi, D. P. Arnold, Electroplated $L1_0$ CoPt thick-film permanent magnets, J. Appl. Phys., 115, 17E521, 2014.
[9] N. Wang, D. P. Arnold, Thick Electroplated Co-Rich Co-Pt Micromagnet Arrays for Magnetic MEMS, IEEE Trans. Mag., vol. 44, no. 11, pp. 3969-72, 2008.

Journal of Physics: Conference Series **757** (2016) 012035 doi:10.1088/1742-6596/757/1/012035

MEMS Based Micro Aerial Vehicles

Niranjan Joshi[+], **Elof Köhler and Peter Enoksson**

Department of Micro and Nanotechnology, Chalmers University of Technology, Sweden

E-mail: [+]`joshin@student.chalmers.se`

Abstract. Designing a flapping wing insect robot requires understanding of insect flight mechanisms, wing kinematics and aerodynamic forces. These subsystems are interconnected and their dependence on one another affects the overall performance. Additionally it requires an artificial muscle like actuator and transmission to power the wings. Several kinds of actuators and mechanisms are candidates for this application with their own strengths and weaknesses. This article provides an overview of the insect scaled flight mechanism along with discussion of various methods to achieve the Micro Aerial Vehicle (MAV) flight. Ongoing projects in Chalmers is aimed at developing a low cost and low manufacturing time MAV. The MAV design considerations and design specifications are mentioned. The wings are manufactured using 3D printed carbon fiber and are under experimental study.

1. Introduction

During the past decade the research on insect flight has substantially altered our understanding of aerodynamics, neuroscience and the flight mechanism of insects. This understanding promoted the research on Micro Aerial Vehicle(MAV) in the field of Micro Electromechanical Systems (MEMS). This article covers the basic principles of insect flight and various methods of achieving primitive MAV flight. It is intended for readers who are interested in gaining an overview of insect scaled MAVs as well as for readers interested in extending their individual research to developing MAVs. Discussion in the article is focused on the MAV flight by achieving higher lift than weight for hovering condition. Controllable untethered flight and turn behaviour of insects is not within scope of this article.

In the first section aerodynamics of the insect is briefly described, followed by the desired characteristics of the wings. Later, various methods for the actuator and the transmission are discussed with their strengths and weaknesses. In the last part some insights are provided on the ongoing research and design being carried out by the authors.

2. Aerodynamics

Since the flight aerodynamics is the driving force of the MAV system design, it is crucial to understand aerodynamic forces generated by the insect wings. This description is primarily focused on tethered hovering flight with objective to achieve higher lift than weight. In this section the kinematics of the wings are explained followed by the effects of the kinematic parameters on aerodynamic forces. For the parameter study, wings are assumed to be non flexible. The effect of finite wing stiffness is described in the wing section.

2.1. Kinematics

The insect wing kinematics can be described in 3-DOF system. The system consists of mean stroke plane in which the torsional axis of the wing rotates. The angle of the rotation measured with reference to the axis perpendicular to mean stroke plane, is the *stroke angle* θ. The difference between minimum and maximum stroke angle indicates the stroke amplitude. The angle between wing chord plane and the mean stroke plane is the angle of attack of the wing, also known as *rotational angle* ψ. However, the wing does not flap only in one plane, it rather has a characteristic deviation from the mean stroke plane. The *deviation angle* θ is angle between the instantaneous stroke plane and the mean stroke plane[1]. The angles are shown in Figure 1. This 3-DOF system can be represented by wing tip trajectory as shown in Figure 2. The angle of each of the lines to vertical axis is the angle of attack. The horizontal distance from the vertical center-line is proportional to stroke angle and the vertical displacement of the wing tip is result of the stroke plane deviation.

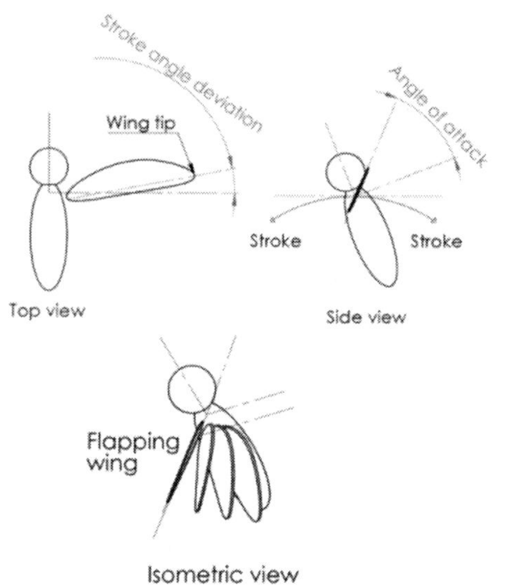

Figure 1. Kinematic degrees of freedom of wing motion.

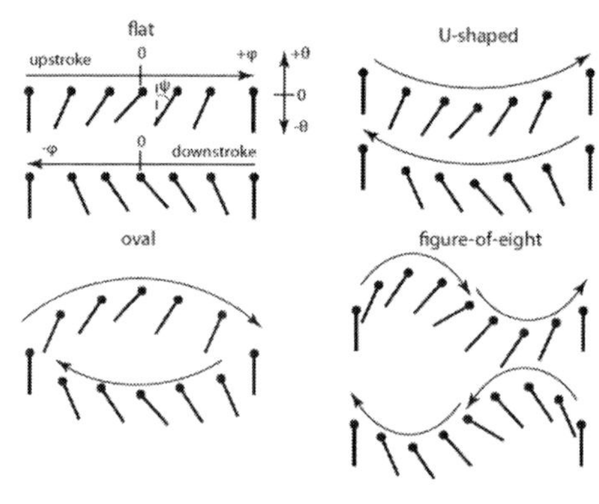

Figure 2. Illustration of wing tip trajectories. Courtesy of Wood [1]

2.2. Aerodynamic force generation

The forces generated on the wings are primarily governed by three aerodynamic phenomena; dynamic stall, rotational lift and wake capture [2, 3]. When an airfoil is translated through air with an angle of attack, a low pressure vortex is formed at the leading edge which is called as 'leading edge vortex (LEV)'. The LEV at high angle of attack keeps growing until it can no longer stay attached to the airfoil. This separation of airfoil is know as stall, which reduces the lift generation dramatically. Interestingly for the flapping wing motion, the flow stays attached to the wing producing high lift forces at high angle attack. This phenomenon is knows as "delayed stall". The "rotational lift" is effect of the wing rotation about its torsional axis at the end of each half-strokes (down-stroke or up-stroke). This maintains the the positive angle of attack and positive lift generation during both half-strokes. At the end of each half stroke when the wing changes its stroke direction, due to the fluid inertia the relative velocity between wing and the fluid is much larger than the absolute velocity of the wing. This higher relative velocity generates large amount of lift at each stroke reversal which is called "wake capture" [2, 3].

2.3. Flight parameter study

In the past two decades, to understand aerodynamic forces produced by the flapping wing, several simulation and experimental studies have been conduced by various scientists. Some of the insect flight parameters are mentioned here, along their effects on the force produced. The lift produced by the wings increases monotonically with the increasing stroke amplitude. Although the power required for the flapping is also increased, the lift-to-power ratio is observed to increase with the stroke amplitude [4] [5]. The angle of attack of the wing is not constant over the half-strokes. The mean angle angle of attack of $40^o - 50^o$ is observed to maximize the lift produced and the flight efficiency [4]. Increasing the mean angle of attack increases the drag, and reducing it produces low lift forces [4]. Several insects have been observed with the capability of actively controlling the angle of attack. However the wing rotation is largely aided by the inertial forces at the stroke reversal. This passive mechanism of the wing rotation is studied and utilized by MAV researchers [6] [7]. Sane [4] also shows that the time required by the wing rotation and the point at which the wing rotation starts, affect the aerodynamic forces largely. The short duration of wing rotation time and the rotation occurring before the stroke reversal increase the lift and the aerodynamic efficiency. These parameters needs to be considered while designing passive or active wing rotation. The stroke plane deviation alters the instantaneous angle of attack of the wing relative to the fluid flow. Experimental study of the scaled model [4] suggested that that the lift and lift-to-drag ratio was reduced with non zero stroke plane deviation. However, insect scale wing suggests the efficiency of the flight might be actually be higher with introduction of stroke plane deviation [1].

Strictly aerodynamically, the higher flapping frequency corresponds to higher wing velocity, increasing the lift and the power consumption. Hence as the frequency increases, the weight capacity of MAV increases, increasing the energy storage capacity. Since the power consumption is increased, the energy requirement for sustained flight also increases. Furthermore, in order to increase the aerodynamic performance, the flapping frequency also needs to satisfy the aerodynamic resonance of the wing motion, which depends on weight and geometry of the wing. Furthermore the actuator also has a high efficiency operating frequency region that needs to closer to the flapping frequency for an efficient system. Hence, the flapping wing frequency is a complex optimization problem which needs close attention in order to achieve successful insect scale flight [8][4][9][10][5].

3. Wings

Contrary to the previous assumption of the stiff wing, the real insect wings are complex and deformable structures. Some of the insects modulate the wing shape during the flight [3], however in most cases the insect wing deformation is a passive phenomenon. The recent research suggest that the flexibility is necessary and advantageous to flight aerodynamics. Studies have discovered that in some cases the mass of the wing is also not uniform over the wingspan, which improves the aerodynamic force balancing and flight stability [3]. Additionally, wings have been found to have camber in the wing profile, having convex upper surface and concave lower surface. This creates different stiffness values of the wing for upstroke and downstroke, producing different bending and twisting of the wing.

Numerical simulations of the flexible wings suggests that the flexibility improves the lift production, efficiency and other aspects of the wing aerodynamics. The flexible wing enhances the wake capture during the stroke reversal, reducing the drag and in some cases enhancing the lift produced [3]. Computational Fluid Dynamic (CFD) simulation studies for locust and butterflies have respectively revealed 12% and 4% reduction in lift to power ratio due to removal of dynamic wing camber. Whereas the removal of camber and twist cost 35% and 26% lift to power ratio respectively. These studies strongly indicates that the flexibility of the insect wing largely impacts the aerodynamic performance. Additionally the aspect ratio and the geometry of the wing also affect the lift production and the power consumption.

4. Actuator

There are two configurations of flight muscles found in insects. For direct flight muscles, each wing is connected to a pair of antagonistic muscles which provide power as well as independent control of wings. The indirect flight muscles are connected to an intermediate structure that powers both the wings with a single pair of antagonistic muscles and small muscles are utilized for control. The indirect muscle configuration is found in more evolved species and is also considered to be a superior configuration for reduction in weight and due to good wing synchronization. There are several properties of muscles that need to be satisfied by an artificial actuator such as high energy density, low volume, sufficiently high frequency of operation, high efficiency, high actuator stress-strain. In this section some relevant actuator technologies are mentioned along with their individual strengths and weaknesses.

Ionic polymer metal composite (IPMC): IPMC is an electrochemical actuator where polyeletrolyte is sandwiched between flexible metallic electrodes. Applying voltage across the electrodes produces a bending moment in the actuator. The prime advantage of IPMC is, it operates at low voltage typically less than 5V. Lower voltage has several advantages including simplicity and efficiency of electronics and energy storage. However being an electrochemical actuator the speed of operation is limited by the ion transportation, as well as practical issues such as leakage are present in IPMC.

Piezoelectric: The piezoelectric material is attached to an elastic substrate. When a voltage is applied across it the piezoelectric material changes its shape and dimensions, applying a bending moment on the actuators. The piezoelectric actuator can used as bending actuator in various configuration viz. unimorph, bimorph [11]. Piezoelectric actuator is a well established actuator with sufficient actuator properties in all the aspects such as energy density, stress-strain, operating voltage etc. However it has complicated manufacturing process and yet not up to the the mark of the natural muscle performance.

Dielectric elastomer (DE): An elastic polymer is sandwiched between the compliant electrodes, such that upon application of voltage across the electrodes the attraction between opposite charges generates Maxwell stress' in the material. This phenomenon can be utilized to generate linear or bending actuation. Although the large strain and energy density of the DE actuator are very attractive features, the high voltage requirement and the high weight of supporting material is a drawback. The DE actuator shows a large potential to achieve similar or better performance than the natural muscles, and hence chosen for further research and development.

There are several other types of actuators, but those are not considered due to the large difference between design specifications and their properties e.g. shape memory alloy with very low actuation efficiency. Interested readers can read about shape memory alloy, conducting polymer, molecular actuator, thermoelectric Carbon NanoTube (CNT) yarn actuator, CNT electrolytic actuator, electromagnetic actuators MEMS ICE.

5. Transmission

Transmission connects the actuator to the wings of the MAV, converting the actuator movement and force to the desired wing motion. A four-bar mechanism with a pivot point at the base of the wing is the simplest design to provide power to the flapping wings. Several designs have been studied that are suitable for various actuators such as bending type actuator, linear actuator etc.

Journal of Physics: Conference Series **757** (2016) 012035 doi:10.1088/1742-6596/757/1/012035

The dipteran insects are equipped with an additional feature in their transmission known as "click mechanism" [12][13]. This thorax like compliant mechanism stores elastic energy in the flexible hinges to create a bistable system that has stable position at the end of upstroke and downstroke. From start of the half-stroke the hinges store the energy until wing reaches equilibrium point. Beyond the equilibrium, actuator does not need to to expend energy, the wing snaps into it stable position using the energy stored in the hinges. This increases the instantaneous velocity of the wings and thereby increasing the lift.The experimental study in [12] found reduced power consumption in the click mechanism compared to the non clicking transmission.

6. Design and Experiments

The authors are currently investigating the field of MAVs in MEMS technology. In this section the design specification and the motivation for those specifications is mentioned. Then the manufacturing of wings and their specifications are also included. The general design target included low cost, low process/manufacturing time. The concept design of the MAV is shown in Figure 3

The Dielectric Elastomer (DE) actuator was chosen for the experimentation due to its high work density, high strain, high frequency operation, ease of manufacturing and low cost. The DE actuator was chosen despite of the requirement of the high voltage. Some studies have reported the possibility of very thin DE or stacking the DE's such that the operating voltage can be reduced to few 100 Volts. Also the DE actuator is a linear actuator and can be designed with the transmission to act as the insect thorax mechanism. Hence, considering the potential development in this technology the DE actuator was chosen for experimentation. The carbon black will be applied on thin silicone film, which will be subjected to high voltage of $\approx 5kV$. The actuator will be used in bow-tie configuration where supporting structure will be made from high stiffness carbon fiber laminate. Along with the actuator, the click mechanism is designed which by the design adds the lift production and efficiency.

The design constraints for the wings were chosen to be very low weight, minimum stiffness corresponding to the insect wing. The 3D printed carbon fiber was chosen for manufacturing of the wing skeleton along Mylar film as wing membrane. The $100\mu m$ thick 3D printed was epoxy glued with with $5\mu m$ thick Mylar film as shown in Figure 4. Further investigation of the stiffness properties of the wings is under experimentation and will be compared with the insect wing specifications.

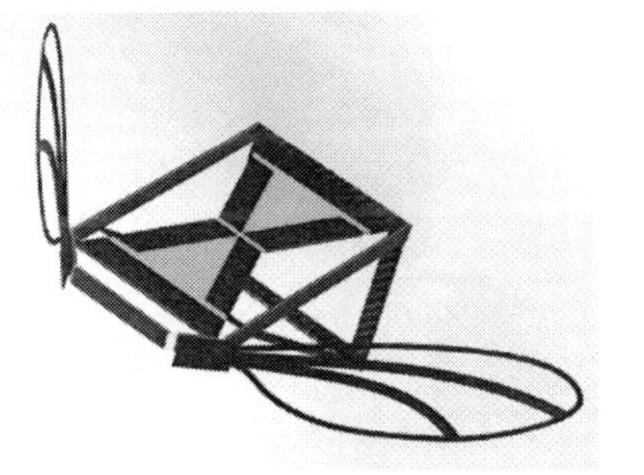

Figure 3. Concept design of the MAV

Figure 4. 3D printed wing skeleton

7. Conclusion

The aerodynamic study reveals that the at least 2-DOF wing motion is desired to achieve MAV flight. Stroke amplitude should be maximized and the mean angle of attack close to 45°. The stroke plane deviation is not mandatory to generate higher lift than weight, but it will improve the MAV's performance. The flapping frequency is a crucial parameter which depends upon and drives several aspects of the MAV design. The insect wing is a complex piece of engineering, but the a very stiff but light wing design is sufficient for the primitive/simple MAV flight.

3D printed carbon fiber shows promising properties and could be a good alternative for manufacturing complicated structure like insect wing. Piezoelectric actuator are proven for MAVs [11]. However, further research on dielectric elastomer could prove them a better choice for MAV actuator. Click mechanism improves lift and efficiency of the flight [12] and will be included in experimentation.

The insect scale flight is fascinating technological challenge and is an excellent project for MEMS research. Achieving the insect scale flight requires multidisciplinary study. Various aspects of the MAV are currently under investigation and the further research on the topic will aid to progress individual technologies such as unsteady aerodynamics, polymer actuators, energy storage etc.

Although Micro Aerial Vehicles is a thorough and complex topic of study, this project aims to prove it is possible to achieve the primitive flight without extensive research in the field. However natural insect capable flight is a very deep and exiting topic which is yet to be fully exploited. The attention from MEMS community will surely help the MAVs to get closer to natural insects.

References

[1] Finio B M, Whitney J P and Wood R J 2010 Stroke plane deviation for a microrobotic fly *Intelligent Robots and Systems (IROS), 2010 IEEE/RSJ International Conference on* pp 3378–3385 ISSN 2153-0858
[2] Deng X, Schenato L, Wu W C and Sastry S S 2006 *IEEE Transactions on Robotics* **22** 776–788 ISSN 1552-3098
[3] Hedrick T L, Combes S A and Miller L A 2015 *Canadian Journal of Zoology* **93** 925–943 (*Preprint* `http://dx.doi.org/10.1139/cjz-2013-0196`) URL `http://dx.doi.org/10.1139/cjz-2013-0196`
[4] Sane S P and Dickinson M H 2001 *Journal of Experimental Biology* **204** 2607–2626 ISSN 0022-0949 (*Preprint* `http://jeb.biologists.org/content/204/15/2607.full.pdf`) URL `http://jeb.biologists.org/content/204/15/2607`
[5] Gravish N, Chen Y, Combes S A and Wood R J 2014 High-throughput study of flapping wing aerodynamics for biological and robotic applications *Intelligent Robots and Systems (IROS 2014), 2014 IEEE/RSJ International Conference on* pp 3397–3403
[6] BERGOU A J, XU S and WANG Z J 2007 *Journal of Fluid Mechanics* **591** 321–337 ISSN 1469-7645 URL `http://journals.cambridge.org/article`$_S$0022112007008440
[7] Arabagi V and Sitti M 2008 Simulation and analysis of a passive pitch reversal flapping wing mechanism for an aerial robotic platform *2008 IEEE/RSJ International Conference on Intelligent Robots and Systems* pp 1260–1265 ISSN 2153-0858
[8] Whitney J P and Wood R J 2012 *Bioinspiration and Biomimetics* **7** 036001 URL `http://stacks.iop.org/1748-3190/7/i=3/a=036001`
[9] Ellington C 1999 *Journal of Experimental Biology* **202** 3439–3448 ISSN 0022-0949 (*Preprint* `http://jeb.biologists.org/content/202/23/3439.full.pdf`) URL `http://jeb.biologists.org/content/202/23/3439`
[10] Karpelson M, Whitney J P, Wei G Y and Wood R J 2010 Energetics of flapping-wing robotic insects: towards autonomous hovering flight *Intelligent Robots and Systems (IROS), 2010 IEEE/RSJ International Conference on* pp 1630–1637 ISSN 2153-0858
[11] Wood R, Steltz E and Fearing R 2005 *Sensors and Actuators A: Physical* **119** 476 – 488 ISSN 0924-4247 URL `http://www.sciencedirect.com/science/article/pii/S0924424704007757`
[12] Chin Y W and Lau G K 2012 x201c;clicking x201d; compliant mechanism for flapping-wing micro aerial vehicle *2012 IEEE/RSJ International Conference on Intelligent Robots and Systems* pp 126–131 ISSN 2153-0858
[13] Thomson A J and Thompson W A 1977 *Acta Biotheoretica* **26** 19–29 ISSN 1572-8358 URL `http://dx.doi.org/10.1007/BF00115924`

A refined model for piezoelectric composite beams

Luca Luschi[1], **Francesco Pieri**[1]

[1]Dipartimento di Ingegneria dell'Informazione, Università di Pisa, Pisa, Italy

luca.luschi@for.unipi.it

Abstract. This work presents and compares few simple one-dimensional models for the piezoelectric actuation and detection of beams. The 1D nature, which allows an easy embedding of the model in the classical Euler-Bernoulli beam equations, is obtained by adopting simplifying assumptions along directions of the cross-sectional plane. By changing such assumptions, different models can be built. Their validity is discussed and compared with results of FEM simulations for varying geometries. We show that commonly adopted models fail in a series of practical cases and propose a new model capable of accurately describing wide beams.

1. Introduction

Piezoelectricity is widely used in MEMS devices, such as resonators [1, 2], sensors [3] and energy harvesters [4], as well as in macroscopic harvesters [5]. In several applications, a bimorph beam with the structure shown in figure 1 is used, and for this reason compact descriptions of the electromechanical behavior of a piezoelectric beam have been developed in the past [6, 7]. Such models typically take the form of simple one-dimensional problems, taking advantage of the classic Euler-Bernoulli and Timoshenko beam equations. The reduction of the full 3D problem to a 1D model is obtained by adopting simplifying assumptions on the stress and strain distributions in the directions of the beam cross-sectional plane. These assumptions may be justified for specific geometries, or if one or more piezoelectric coefficients apart from e_{31} are assumed to be negligible, which is rarely the case in actual piezoelectric materials. In actual materials, other coefficients may play a significant role and cannot be neglected.

In this work we review some of the assumptions proposed in literature and investigate their range of validity by comparing them with FEM results. We show that, in the case of wide beams, established assumptions are not justified and a different model, taking more reasonable boundary conditions into account, accurately describe this case.

2. Model

We present here a complete model for the electromechanical behavior of piezoelectrically actuated beams. We consider the cantilever geometry of figure 1, with a mechanical actuation in the form of a vertical force T_0 applied at the cantilever tip. Problems of this form are frequency encountered in piezo actuators or energy harvesters.

The generic constitutive equations for the piezoelectric material (orange in figure 1) and beam structural material (grey in the same figure) can be written as:

$$T_I^B = C_{IJ}^B S_J \tag{1}$$

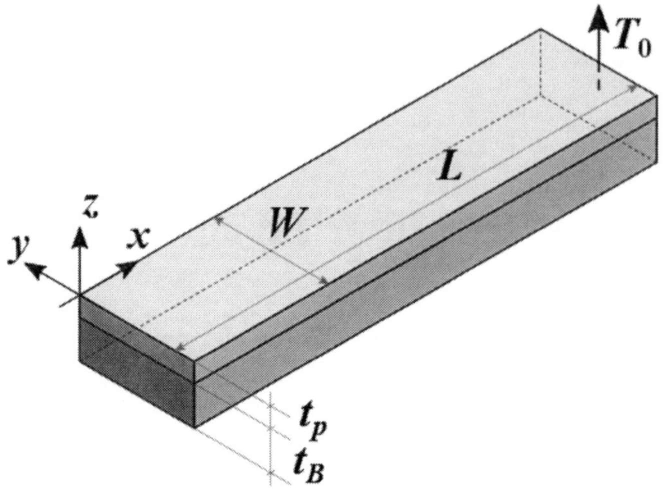

Figure 1. Schematic structure of the piezoelectric device. The area shaded in red is clamped.

$$T_I^p = C_{IJ}^p S_J - e_{IJ} E_j \tag{2}$$

where T_I and S_J are the stress and strain vectors in contracted (engineering) notation. C_{IJ} and e_{IJ} are the stiffness and piezoelectric stress matrices, respectively, and E_j is the electric field. Superscripts B and p stand for beam and piezo, respectively. The equation for the electric displacement D_i in the piezoelectric material is:

$$D_i = \varepsilon_{ij} E_j + e_{iJ} S_J \tag{3}$$

where ε_{ij} is the electrical permittivity matrix at zero strain.

These general three-dimensional equations can be (and have been) simplified to obtain a one-dimensional model for the system of figure 1. We will show that the choice of the simplifying assumptions, however, is crucial for the accuracy of the model. Different hypotheses have been proposed in the literature. Garcia and coworkers [6] assume no displacement in the structure either along y or z, i.e. $S_{yy} = S_{zz} = 0$. We call this approach Full Zero Displacement or FZD. On the other hand, Inman and coworkers [7] assume the dual hypothesis of Full Zero Force (FZF), i.e. the normal stresses along y and z are both zero ($T_{yy} = T_{zz} = 0$). These two choices correspond to the two limiting cases of a continuum, and as such neither is always justified.

Since beams are typically slender (i.e. $L \gg t_p + t_B$), the assumption of zero force along z ($T_{zz} = 0$) is reasonable. Along the y direction however, the situation is complicated by the fact that the two materials are clamped at their interface. This fact may play a significant role in blocking deformations along y if the in-plane dimensions L, w are comparable, as suggested by FEM simulations by the authors.

For these reasons, in this work we propose an alternative, yet still very simple, hybrid assumption, i.e. that the beam displacement is zero only along the y direction, and retain the assumption of zero vertical stress ($S_{yy} = 0,\ T_{zz} = 0$). We call this model In-Plane Zero Displacement or IPZD. The observations above suggest that FZF should be valid at small w/L, and IPZD at large w/L.

Reduction of equations (1)-(3) to the one-dimensional case, given one of the three cases, is straightforward and was already carried in the aforementioned papers. Equations (1)-(3) can be solved for T_{xx} and D_z, giving:

Table 1. Equivalent longitudinal stiffness and piezoelectric coupling coefficients for the three proposed models. Y is the Young's modulus, ν the Poisson's ratio and d the piezo strain coefficient.

equivalent parameter	FZD	FZF	IPZD
K_*	C_{11}^*	Y_x^*	$\frac{Y_x^*}{(1-\nu_{xy}^{*\,2})}$
P_c	e_{31}	$d_{31}Y_x = e_{31}(1-\nu_{xy}) - e_{33}\nu_{xz}$	$\frac{d_{31}Y_x}{1-\nu_{xy}} = \frac{[e_{31}(1-\nu_{xy}) - e_{33}\nu_{xz}]}{(1-\nu_{xy})}$

$$T_{xx}^B = K_B S_{xx} \tag{4}$$

$$T_{xx}^p = K_p S_{xx} - P_c E_z \tag{5}$$

$$D_z = \varepsilon_p E_z + P_c S_{xx} \approx \varepsilon_{zz} E_z + P_c S_{xx} \tag{6}$$

where K_B and K_p are equivalent axial stiffness constants for the structural and piezoelectric layers, respectively, P_c is a piezoelectric coupling coefficient, and ε_p is an equivalent dielectric permittivity constant. These constants depend on the assumptions along y and z directions, with only ε_p having a negligible variation which allows its substitution with ε_{zz}. The values of each equivalent constant for the three models are reported in table 1. For the derivation of these constants, hexagonal anisotropy, which is typical of piezoelectric materials such as PZT and AlN, has been assumed for the piezoelectric material. It is worth mentioning that, if e_{33} and e_{32} are negligible, the three models give the same P_c.

These models can be used in the classical Euler-Bernoulli beam equation, which, along with the electrical equation of the piezoelectric capacitor, leads to the following ODE system in the frequency domain:

$$\begin{cases} \frac{d}{dx}\begin{bmatrix} U(x) \\ \Psi(x) \\ M(x) \\ T(x) \end{bmatrix} = \begin{bmatrix} 0 & 1 & 0 & 0 \\ 0 & 0 & -1/(KI)_c & 0 \\ 0 & 0 & 0 & 1 \\ -\omega^2(\rho A)_c & 0 & 0 & 0 \end{bmatrix} \begin{bmatrix} U(x) \\ \Psi(x) \\ M(x) \\ T(x) \end{bmatrix} + \Gamma_p V \begin{bmatrix} 0 \\ 0 \\ \delta(x) - \delta(x-L) \\ 0 \end{bmatrix} \\ Q = C_0 V + \Gamma_p \left(\frac{dU(L)}{dx} - \frac{dU(0)}{dx} \right) \end{cases} \tag{7}$$

where $U(x)$, $\Psi(x)$, $M(x)$ and $T(x)$ are the displacement, rotation, bending moment and shear force along the beam, respectively. V, Q and C_0 are the voltage across the piezo, the charge over the upper electrode and the static capacitance of the piezo itself. $(KI)_c$ and $(\rho A)_c$ are the flexural stiffness and the mass per unit length of the composite structure. ω is the angular frequency, while Γ_p is the system piezoelectric coupling coefficient, with value:

$$\Gamma_p = P_c w \left(t_B + \frac{t_p}{2} - z_N \right). \tag{8}$$

In equation (8), z_N is the position of the neutral axis. Once that the equivalent axial stiffnesses K_B and K_p are known, $(KI)_c$ and z_N can be computed with composite beam theory [8].

3. Validation

To validate the model we consider both static and dynamic results and compare the outcomes of the three models with FEM simulations. Polysilicon is used as the structural material and PZT as the piezoelectric material.

Table 2. Properties of the materials.

Constant	Y_x, Y_y	Y_z	ν_{xz}, ν_{xz}	ν_{xy}	G_{xz}, G_{xz}	G_{xy}	ρ	ε_{zz}	e_{31}, e_{32}	e_{33}	e_{15}, e_{24}
(Unit)	GPa		(/)		(GPa)		(kg/m³)	(/)		(C/m²)	
PZT-5H	60	48.2	0.51	0.29	23	23.3	7500	1470	-6.5	23.3	17
Polysilicon	164		0.219		67.3		2330	/		/	

Properties used in simulations and models are summarized in Table 2. The beam and piezo thicknesses are 10μm and 1μm, respectively. The length L is fixed at 500 μm, while the width is swept from 25μm to 5000μm. FEM simulations were performed with ANSYS, using element SOLID226 for the piezoelectric layer, and SOLID186 for the structural layer. Both a static analysis and a modal analysis were performed for each geometry.

3.1. *Static behavior*

We solved the ODE system (7) in static conditions ($\omega = 0$) with clamped boundary conditions at one end of the beam and a constant vertical force at the other end (figure 1). Depending on the electrical boundary condition, different expressions for the tip deflection are obtained. For the two limiting cases of short-circuit (sc) and open-circuit (oc) conditions at the piezoelectric actuator, we found:

$$U_{sc}(L) = \frac{L^3}{3(KI)_c} T_0 , \qquad U_{oc}(L) = \frac{L^3}{3(KI)_c} \frac{\Gamma_p^2 L/4 + C_0 (KI)_c}{\Gamma_p^2 L + C_0 (KI)_c} T_0 \tag{9}$$

The short circuit solution coincides with the Euler-Bernoulli theory for a cantilever with a force load at its tip, while the open circuit solution includes the stiffening effect introduced by piezoelectric coupling. Static results are reported in figure 2 as a function of the ratio w/L for short circuit (dashed lines) and open circuit (solid lines) electrical boundary conditions. The simulated tip displacement decreases for increasing w/L, passing from the FZF approximation ($w/L \to 0$) to the IPZD approximation ($w/L \to \infty$). Limit values are very accurate in short circuit while presenting small errors in open circuit. Since $(KI)_c$ is the only parameter affecting the short circuit model (see equation (9)), we conclude that the error is due to inaccuracy on the Γ_p model. As expected from the discussion presented in the previous section, the FZD model is inaccurate for both electrical boundary conditions.

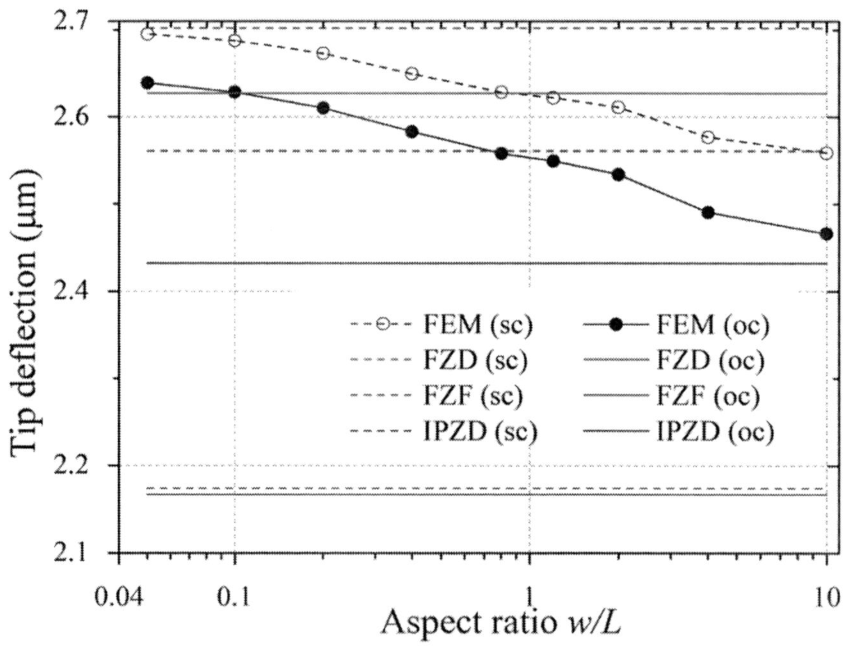

Figure 2. Tip displacement as a function of w/L ratio.

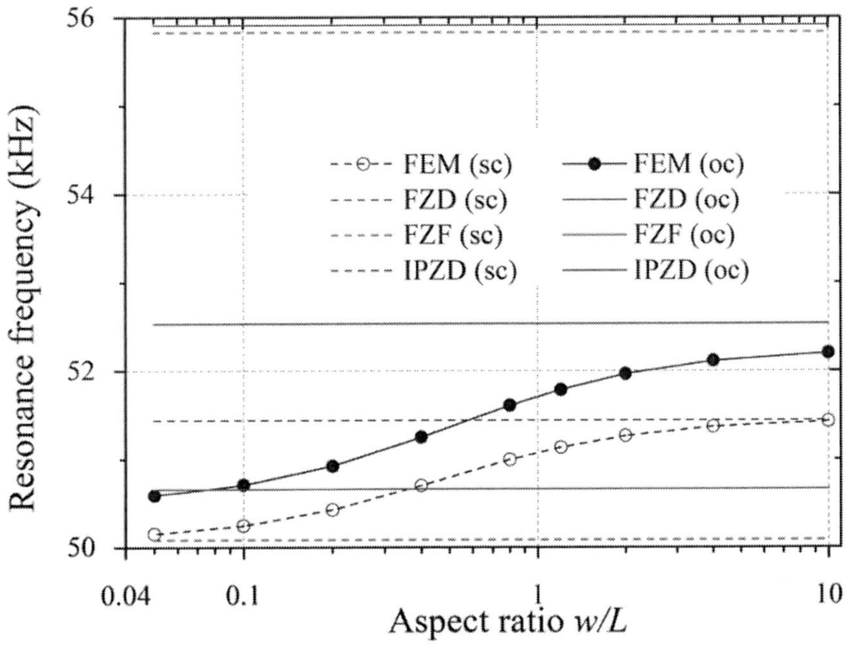

Figure 3. Resonance frequency as a function of w/L ratio.

3.2. Dynamic behavior

We compute the resonance frequency by writing the eigenproblem relative to the system (7) and then numerically solving for the roots of the characteristic equation, whose only variable is the angular frequency ω. A comparison between simulations and models is reported in figure 3. The behavior is very similar to the one observed in the static analysis, with the resonance frequency moving from the FZF limit to the IPZD limit as the w/L ratio increases. Small errors at the limit values for the open circuit case are observed as well.

4. Conclusions

In this work we discussed three one-dimensional models describing the electromechanical behavior of piezoelectrically actuated beams. The models are based on different hypotheses on the stress/strain state along the two directions of the beam cross-sectional plane. The hypothesis of zero displacement, present in the literature, is disproved by FEM simulations. The assumption of zero stress fails in the case of wide structures. We show that a new hypothesis developed by the authors, that of in-plane zero displacement, correctly models this latter case.

Acknowledgments

This work was supported in part by the ENIAC Joint Undertaking in the framework of project Lab4MEMS under grant n. 325622-2.

References

[1] Shahmohammadi M, Harrington B P and Abdolvand R 2013 Turnover temperature point in extensional-mode highly doped silicon microresonators *IEEE Trans. Electron Devices* **60** 1213–20

[2] Tabrizian R, Casinovi G and Ayazi F 2013 Temperature-stable silicon oxide (SilOx) micromechanical resonators *IEEE Trans. Electron Devices* **60** 2656–63

[3] Pettine J, Patrascu M, Karabacak D M, Vandecasteele M, Petrescu V, Brongersma S H, Crego-Calama M and Van Hoof C 2013 Volatile detection system using piezoelectric

micromechanical resonators interfaced by an oscillator readout *Sensors Actuators A Phys.* **189** 496–503
[4] Shi Q, Wang T and Lee C 2016 MEMS Based Broadband Piezoelectric Ultrasonic Energy Harvester (PUEH) for Enabling Self-Powered Implantable Biomedical Devices *Sci. Rep.* **6** 24946
[5] Reissman T, Wickenheiser A and Garcia E 2016 Generalized Solutions of Piezoelectric Vibration-Based Energy Harvesting Structures Using an Electromechanical Transfer Matrix Method *J. Vib. Acoust.* **138** 041001
[6] Dietl J M, Wickenheiser A M and Garcia E 2010 A Timoshenko beam model for cantilevered piezoelectric energy harvesters *Smart Mater. Struct.* **19** 055018
[7] Erturk A and Inman D J 2008 On Mechanical Modeling of Cantilevered Piezoelectric Vibration Energy Harvesters *J. Intell. Mater. Syst. Struct.* **19** 1311–25
[8] Lobontiu N and Garcia E 2005 *Mechanics of Microelectromechanical Systems* (Boston: Kluwer Academic Publishers)

Frequency adjustable MEMS vibration energy harvester

P Podder[1], P Constantinou[1], A Amann[2, 3] and S Roy[1, 4]

[1] Micro-Nano Systems Centre, Tyndall National Institute, Cork, Ireland
[2] Photonics Centre, Tyndall National Institute, Cork, Ireland
[3] School of Mathematical Sciences, University College Cork, Cork, Ireland
[4] A.S. Paintal Chair Professor in Engineering, Indian National Science Academy

Email: saibal.roy@tyndall.ie

Abstract. Ambient mechanical vibrations offer an attractive solution for powering the wireless sensor nodes of the emerging "Internet-of-Things". However, the wide-ranging variability of the ambient vibration frequencies pose a significant challenge to the efficient transduction of vibration into usable electrical energy. This work reports the development of a MEMS electromagnetic vibration energy harvester where the resonance frequency of the oscillator can be adjusted or tuned to adapt to the ambient vibrational frequency. Micro-fabricated silicon spring and double layer planar micro-coils along with sintered NdFeB micro-magnets are used to construct the electromagnetic transduction mechanism. Furthermore, another NdFeB magnet is adjustably assembled to induce variable magnetic interaction with the transducing magnet, leading to significant change in the spring stiffness and resonance frequency. Finite element analysis and numerical simulations exhibit substantial frequency tuning range (25% of natural resonance frequency) by appropriate adjustment of the repulsive magnetic interaction between the tuning and transducing magnet pair. This demonstrated method of frequency adjustment or tuning have potential applications in other MEMS vibration energy harvesters and micro-mechanical oscillators.

1. Introduction

Over the last couple of decades, the rapid advancements in the field of low-power integrated electronics have led to significant reduction in size and power consumption of electronic devices [1]. Subsequently, these advancements propelled the drive towards the development of the "Internet-of-Things (IoT)", a responsive network of billions of sustainable wireless sensor nodes (WSNs) that would facilitate seamless interaction between the physical and the digital world [2]. In order to ensure the sustainability and reliable operation of such enormous and expansive networks, the problem of long-term power supply to the WSNs must be addressed [3]. Conversion of the many forms of ambient energy (e.g. light, thermal gradient, wind flow, mechanical vibrations etc.) into usable electrical energy to power the WSNs offers a viable solution to this problem. The sheer abundance of ambient vibrations in the increasingly urbanized landscapes provides an untapped source of energy to power the WSNs [4].

Despite the wide availability, efficient conversion of ambient vibrations into electricity is a formidable challenge [5], primarily due to immense variability in the vibration spectrum (e.g. resonant, random, variable frequency, impulse etc.) and amplitude (e.g. very weak, perceptible, strong etc.). Many of the reported vibration energy harvesting devices (VEHs) rely on linear resonant oscillator systems that are designed to operate efficiently only within a very narrow frequency range near the resonance [6, 7]. In recent years, researchers attempted to widen the frequency range with nonlinear oscillator based energy harvesters and VEH frequency tuning [8, 9] to match that of the ambient vibrations. While the nonlinear oscillators have their advantages (e.g. broader bandwidth) and disadvantages (e.g. co-existing low and high energy branches) [10-13], the frequency adjustment or tuning by electrical (e.g. load variation, piezoelectric patch) [14, 15] and mechanical (e.g. magnetic force, mass variation etc.) [16, 17] methods offer a viable alternative. However, most of the reported frequency adjustment mechanisms are complicated and has been deployed only in macro or meso-scale devices.

In this work we develop a miniaturized frequency adjustable MEMS electromagnetic VEH where the spring stiffness can be controlled through manipulation of the repulsive magnetic interaction between a pair of tiny magnets. We have demonstrated through finite element analysis and numerical simulation that a frequency range of more than 25% of the initial resonance can be achieved. The silicon spring structure and copper planar micro-coils are micro-fabricated using standard MEMS processes, followed by assembly with an adjustable screw mechanism. The proposed mechanism of frequency adjustment can be exploited in other VEHs or micro-mechanical oscillators for post-fabrication frequency tuning.

2. Design of frequency adjustable MEMS electromagnetic vibration energy harvester

The architecture of the proposed frequency adjustable MEMS electromagnetic VEH comprising micro-fabricated silicon spring and double-layer planar coil and sintered NdFeB magnets is illustrated in figure 1. The spring structure consists of four folded-arm cantilever beams supporting a cubic NdFeB magnet attached to the central paddle. The four folded cantilevers are attached to the two diagonally opposite corners of the movable central paddle such that large amplitude out-of-plane oscillation of the suspended magnet is facilitated, while in-plane torsional movement around the vertical out-of-plane axis is eliminated (Figure 1(c)). The spiral planar double-layer copper coil is positioned underneath the suspended transducing magnet such that maximum flux linkage is established in the transducing magnet-coil assembly (Figure 1(b)). A second cubic NdFeB magnet is positioned above the transducing magnet such that they interact via repulsive magnetic force. The extent of the magnetic repulsion can be controlled by precise manipulation of the gap between the magnets through a screw-mechanism. By precise manipulation of the repulsive magnetic force, the potential energy profile and the spring stiffness can be controlled, leading to adjustability of the resonance frequency. The different design parameters for the micro-VEH are given in table 1.

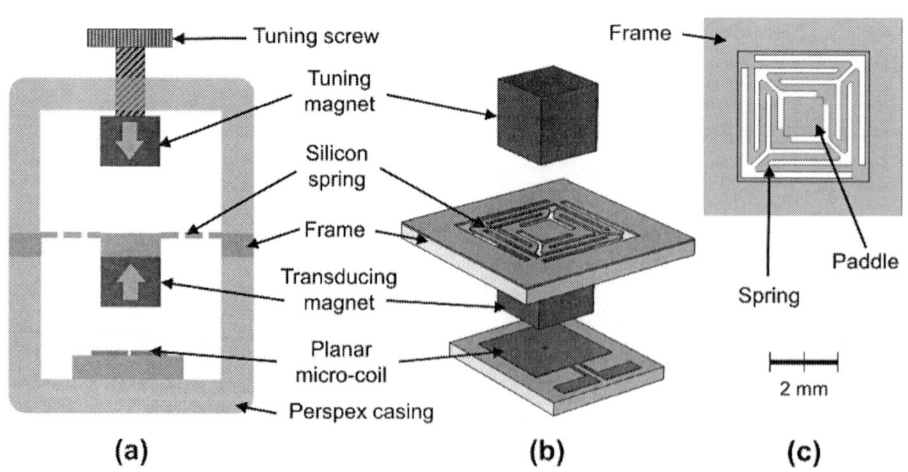

Figure 1. (a) Schematic diagram and (b) perspective view of the frequency adjustable micro-VEH. (c) The silicon spring structure with two diagonally clamped corners.

3. Finite element analysis on silicon spring and magnet-coil assembly

The vibrational modes and electromagnetic coupling factor are determined through finite element analysis using COMSOL Multiphysics and Ansoft Maxwell software tools. The first two modes (300.76 Hz and 318.9 Hz) are due to torsional motion of the combined paddle-magnet mass with axis along the diagonals in the spring plane (Figure 2(a)). The third mode (408.81 Hz) is the intended primary vibration mode, which represents a vertically oriented out-of-plane translational motion. The fourth mode (6291 Hz) is a complex combination of translational and torsional motions. Among all of these vibrational modes, the third mode produces the maximum relative displacement between the transducing magnet and coil. The distribution of magnetic flux densities in the device shows a substantial flux linkage between the vertically oriented transducing magnet and the planar coil (Figure 2(b)). Therefore, the vertically oriented motion (as in the third mode) of the transducing magnet will induce the optimal flux linkage gradient and generate maximum power.

Table 1. Micro-VEH device design parameters.

Parameter	Value
Silicon spring foot-print	6 mm × 6 mm
Spring thickness	50 μm
Spring arm width	200 μm
Planar micro-coil foot-print	4 mm × 5 mm
Number of coil turns	132 turns
Size of NdFeB magnets	2 mm × 2 mm × 2 mm

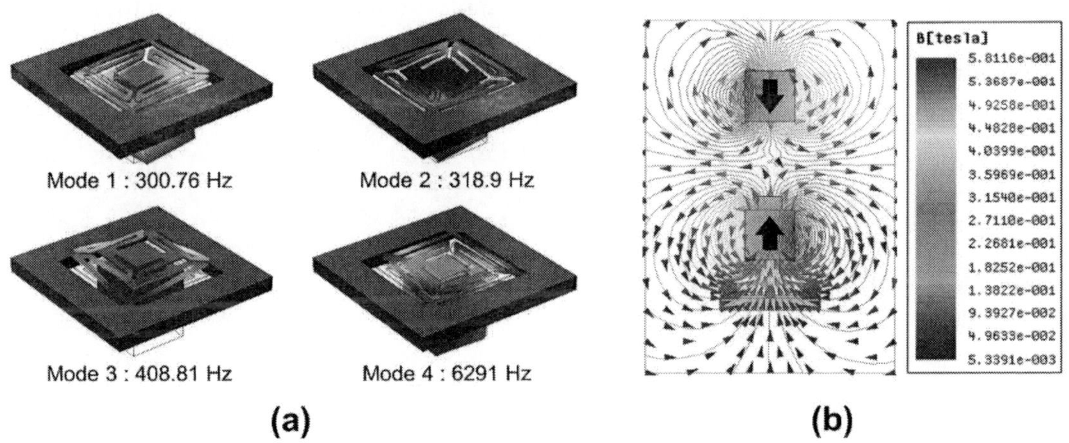

Figure 2. (a) The first four vibrational modes of the micro-VEH. Mode 3 shows the vertical translational motion, which is the intended vibration mode. (b) Magnetic flux distribution due to the tuning and transducing permanent magnets. Substantial flux linkage is established in the planar coil.

4. Numerical simulation results and discussion

A numerical model of the frequency adjustable micro-energy harvester is developed to study the effect of varying magnetic repulsion on the potential energy, spring stiffness and frequency response of the system. The total potential energy ($U(x)$) of the system, considering the spring stiffness (k) and the repulsive magnetic interaction between the transducing and tuning magnet can be expressed as [18],

$$U(x)=\frac{1}{2}kx^2+\frac{\mu_0 p_1 p_2}{2\pi(d-x)^3} \quad (1)$$

where x is the vertical displacement of the transducing magnet, μ_0 is the air permeability, p_1 and p_2 are the magnetic dipole moments of the repulsively positioned transducing and tuning magnets respectively, and d is the vertical distance between the spring frame and the tuning magnet. The associated restoring force can be obtained by differentiating (1) as given below,

$$F(x) = -\frac{\partial}{\partial x}U(x) = -kx - \frac{3\mu_0 p_1 p_2}{2\pi(d-x)^4} \tag{2}$$

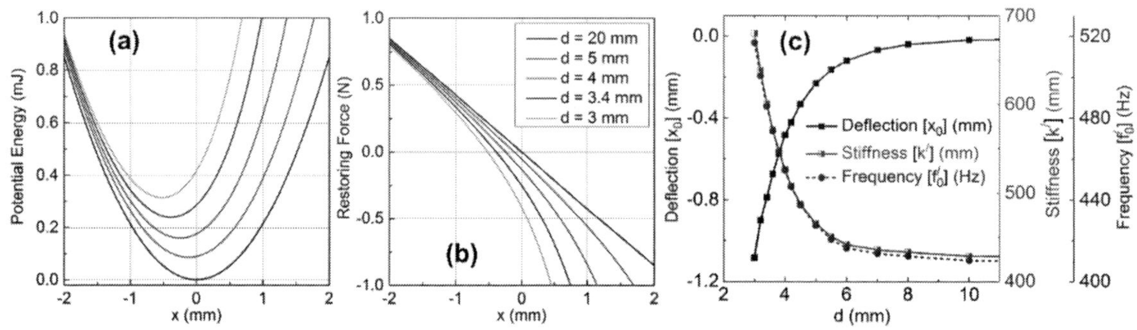

Figure 3. Variation of (a) potential energy profile and (b) restoring force with d. (c) With decreasing d, the spring deflects downward, and the stiffness and resonance frequency increases significantly.

The variation of the potential energy and restoring force profiles with d are illustrated in figure 3(a, b). With decreasing d, the potential energy profile become asymmetric, while the slope of restoring force become steeper, indicating increased linear spring stiffness which can be expressed as,

$$k'(d) = \left[-\frac{\partial}{\partial x}F(x)\right]_{x=x_0} \tag{3}$$

where x_0 is the downward deflection of the movable paddle due to the repulsive force between the tuning and transducing magnets. The increasing spring stiffness with decreasing d results in increasing resonance frequency (Figure 3(c)). The complete dynamical system equation is given by,

$$m\ddot{x} + kx + \frac{3\mu_0 p_1 p_2}{2\pi(d-x)^4} + D\dot{x} + \gamma I = -m\ddot{y} \tag{4}$$

where m is the oscillating equivalent mass, D is the mechanical damping factor, γ is the electromagnetic coupling factor, I is the current induced in the coil and $y = Y_0 \cos \omega t$ represents the vibratory displacement of the system. The simulation parameters are given in table 2. The numerical frequency sweep responses obtained by fourth-order Runge-Kutta method in Matlab shows increasing output power with increasing vibrational accelerations (figure 4(a)). At 1g, the device generates 18.7 µW power across 135 ohm resistive load. With decreasing d, the shift in the resonance frequency towards higher values are revealed (figure 4(b)), which shows more than 25% increase from the initial resonance frequency. Therefore, the frequency adjustment effect by manipulation of the repulsive magnetic interaction is validated through numerical simulations.

Table 2. Micro-VEH device simulation parameters.

Parameter	Symbol	Value
Equivalent mass	m	6.45×10^{-5} kg
Spring stiffness	k	425.5 N m^{-1}
Permeability of air	μ_0	1.256×10^{-6} H m^{-1}
Magnetic moments	p_1, p_2	7.56×10^{-3} A m^2
Mechanical damping factor	D	2.27×10^{-4} kg s^{-1}
Electromagnetic coupling factor	γ	3.74×10^{-2} Wb m^{-1}

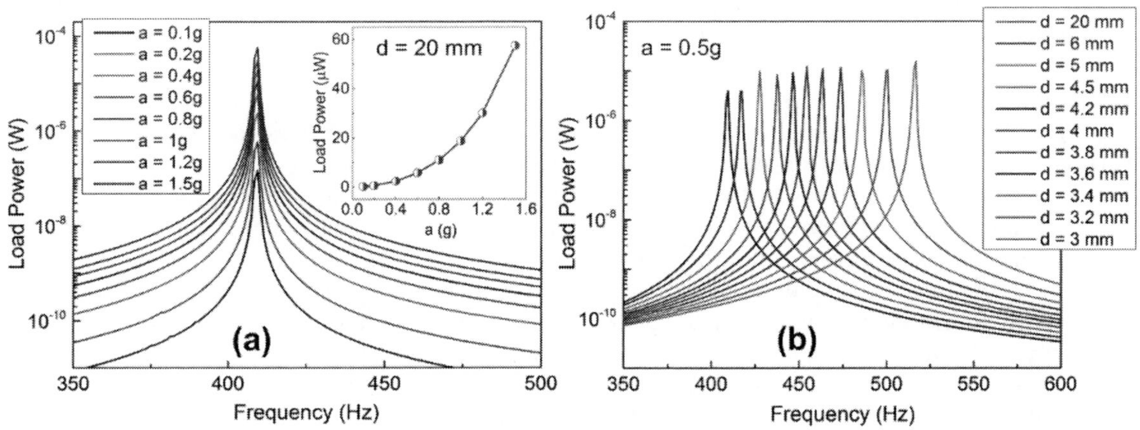

Figure 4. (a) Variation of load power with acceleration. At 0.4g and 1g, the micro-VEH produces 2.4 μW and 18.7 μW respectively. (b) Variation of the frequency response with decreasing d. The resonance frequency changes from 408.8 Hz to 516.2 Hz (> 25%) as d is varied from 20 mm to 3 mm

5. Fabrication of MEMS electromagnetic VEH

The principal components of the frequency adjustable MEMS electromagnetic vibration energy harvester are the silicon spring, the planar micro coil and the sintered NdFeB magnets. The silicon spring is fabricated on silicon-on-insulator (SOI) wafer with a 3 μm buried oxide layer in between the silicon device (50 μm) and handle (450 μm) layers. The spring patterns are first transferred onto the device layer by standard photolithography, followed by release of the springs via deep reactive ion etching (DRIE). The same process is applied on the backside (handle layer) to release the spring frame and the paddle and the individual springs (6mm × 6mm) are then diced off.

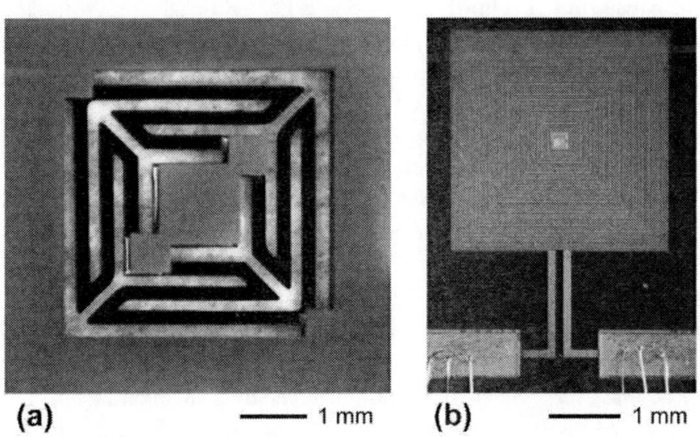

Figure 5. (a) The fabricated silicon spring structure, and (b) planar double-layer micro-coil

The planar double-layer micro-coils are fabricated using standard MEMS micro-fabrication processes, starting from silicon wafers with sputtered titanium-copper (20nm-200nm) seed layer. The first spiral coil layer pattern is then transferred photo-lithographically, and the copper coil layer is then electroplated (to a height of 15 μm) and seed layer is etched. Then, SU-8 photoresist layer is spin-coated on top of the first coil layer as insulation, and a central via is etched to connect to the second coil layer. A Ti-Cu seed layer is then sputtered on the insulating SU-8 layer, and the second coil layer is patterned and electroplated (12.5 μm). Finally, the seed layer is etched and the coil is spin-coated using SU-8 and connecting pads are opened, and the individual micro-coils are diced off.

The micro-fabricated silicon spring and double-layer planar coils (Figure 5(a, b)) are assembled together with the miniaturized NdFeB magnets and enclosed within a Perspex casing complete with the adjustable screw mechanism for stiffness and frequency adjustability.

6. Conclusion

We have shown the adjustability of the resonance frequency of an MEMS electromagnetic energy harvester through manipulation of magnetic repulsion force. This method of tuning the operational frequency of the micro-energy harvesters enable post-fabrication frequency adjustment for deployment in environments with different vibration frequencies. It has been demonstrated through finite element analysis and numerical simulations that precise manipulation of the repulsive magnetic interaction between the transducing and tuning magnet pair the spring stiffness and the resonance frequency can be adjusted to match any vibration frequency within a certain range (25% of initial resonance frequency). The frequency adjustable micro-generator prototype devices has been fabricated using standard MEMS micro-fabrication processes, which will be characterized to experimentally validate the results in the next phase. This frequency adjustment mechanism can potentially be deployed in other vibration energy harvesting devices and micro-mechanical oscillators.

Acknowledgements

This work is financially supported by Science Foundation Ireland (SFI) Principal Investigator (PI) project on 'Vibrational Energy Harvesting' grant no. SFI-11/PI/1201. The authors also would like to acknowledge the help provided by the Central Fabrication Facility (CFF) and their staff in Tyndall National Institute for silicon MEMS fabrication.

References

[1] T A C M Claasen, Proc. of the IEEE, Vol. 94, No. 6, June 2006.
[2] I F Akyildiz et al, Comput. Netw., vol. 51, pp. 921–960, 2007.
[3] H Sundmaeker et al., Cluster of European Research projects on the Internet of Things, 2010.
[4] S Roundy, J. Intell. Mater. Syst. Struct., vol. 16, pp. 809–823, Oct. 2005.
[5] S P Beeby et al, Meas. Sci. Technol. 17 (2006) R175–R195.
[6] P Glynne-Jones et al, Sens. Actuators A, vol. 110, pp. 344–349, 2004.
[7] S. P. Beeby et al, J. Micromech. Microeng., vol. 17, pp. 1257–1265, 2007.
[8] D Zhu et al, Meas. Sci. Technol. 21 (2010) 022001.
[9] L Tang et al, J. Int. Mat. Sys. Struct, Vol. 21, December 2010.
[10] A Erturk et al, Appl. Phys. Lett., vol. 94, art. no. 254102, (2009).
[11] B Marinkovic et al, Appl. Phys. Lett. 94, 103505 (2009).
[12] D Mallick, A Amann and S Roy, Smart Mater. Struct. 24 (2015) 015013.
[13] P Podder, D Mallick and S Roy, IoP Journal of Physics: Conference Series, 2014.
[14] V R Challa et al, Smart Mater. Struct. 17 (2008) 015035.
[15] D Mallick et al, Sens. Actuators A, 226, 154-162 (2015).
[16] D Guyomar et al, IEEE/ASME Trans. Mechatronics, Vol. 13, No. 5, October 2008.
[17] D Zhu et al, Sens. and Act. A: Physical Vol. 158, Issue 2, March 2010.
[18] P Podder et al, IEEE/ASME Transactions On Mechatronics, Vol. 21, NO. 2, April 2016.

Journal of Physics: Conference Series **757** (2016) 012038 doi:10.1088/1742-6596/757/1/012038

Characterization of piezoelectric device for implanted pacemaker energy harvesting

Sunny Jay, Manuel Caballero, William Quinn, John Barrett, Martin Hill

Cork Institute of Technology, Bishopstown, Cork, Ireland.

sunny.jay@mycit.ie

Abstract. Novel implanted cardiac pacemakers that are powered by energy harvesters driven by the cardiac motion and have a 40 year lifetime are currently under development. To satisfy space constraints and energy requirements of the device, silicon-based MEMS energy harvesters are being developed in the EU project (MANpower[1]). Such MEMS harvesters for vibration frequencies below 50 Hz have not been widely reported. In this paper, an analytical model and a 3D finite element model (FEM) to predict displacement and open circuit voltage, validated through experimental analysis using an off-the-shelf low frequency energy harvester, are presented. The harvester was excited through constant amplitude sinusoidal base displacement over a range of 20 to 70 Hz passing through its first mode natural frequency at 47 Hz. At resonance both models predict displacements with an error of less than 2% when compared to the experimental result. Comparing the two models, the application of the experimentally measured damping ratio differs for accurate displacement prediction and the differences in symmetry in the measured and modelled displacement and voltage data around the resonance frequency indicate the two piezoelectric voltage models use different fundamental equations.

1. Introduction

The human heart has a natural pacemaker called the sino-atrial (SA) node, which sends an electrical pulse through nerves around the muscle tissue causing them to contract [1]. However, if this begins to fail an artificial pacemaker is required to provide pacing. Conventional pacemakers require battery power to last the 5 to 10 year life expectancy of the device [2]. With the power densities available, these batteries are too large to suit an implanted intracardiac pacemaker. This necessitates pacing leads as shown in Figure 1-1 [3], these often cause discomfort and infection for the patient, along with electrical issues such as insulation deterioration and lead fractures [4]. This problem could be overcome if it were possible to harvest the mechanical energies from the heartbeat using an embedded energy harvester. Embedded energy harvesting has provoked more and more interest in recent years as the desire to have portable, wireless or embedded technology increases. Piezoelectric materials have received much attention due to their ability to convert mechanical strain directly into electrical energy. However, much of this work has focused on higher frequency ranges above 100 Hz as it is difficult to generate usable energy from lower frequency vibration sources as energy is an accumulation of power over a number of cycles and a low cycle rate means a low accumulation rate. Since the main energy in the vibration spectrum of the beating human heart is concentrated below 50 Hz, there has been little development of energy harvesters targeted at this application.

[1] The authors would like to thank the MANpower consortium for their contribution. The research leading to these results was funded from the European Union Seventh Framework Program (FP7-NMP-2013-SMALL-7) under the grant agreement No. 604360.

In addition to the low vibration frequencies, the confined space in the human heart, the shape of the ventricles and the requirement to implant the intracardiac pacemaker through a vein constrain the pacemaker shape to a cylinder less than 6mm diameter and 40mm in length. This cylinder has to also contain all the conventional pacing electronics, leaving only very limited space for an embedded energy harvester that needs to generate power in the 5-10μW range [5], [6]. The circular cross-section of the cylinder also limits the maximum displacement of a planar MEMS harvester. These combined constraints have demanded a novel harvester and new approaches to design, modelling, fabrication and testing [7].

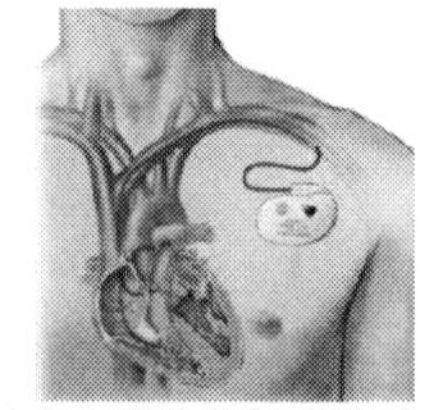

Figure 1-1 Conventional pacemaker location

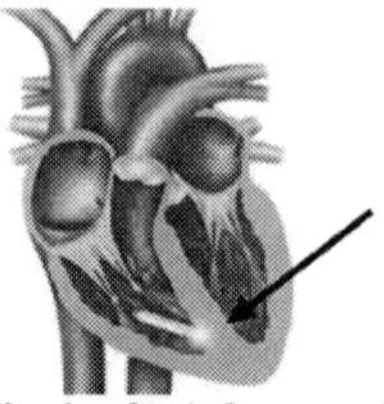

Figure 1-2 Intracardiac implanted pacemaker location [8]

To predict the power capabilities and reliability of such a device and to explore the full design space, accurate models are essential. This paper reports the successful development of two models, validated by experiment, capable of accurately predicting the displacement and open circuit voltage of a commercial off the shelf (COTS) clamped-free cantilever type piezoelectric energy harvester similar in size and operating frequency to the intracardiac harvester. These verified models can now be used to fully explore the design space of the required harvester.

2. Experimental Set Up

A thin film piezoelectric energy harvester was purchased for characterization. The device has a polyvinylidene fluoride (PVDF) core laminated in a sheet of polyester (Mylar) and is a simple cantilever beam type harvester.

Figure 2-1 Thin Film Energy Harvester

The device was subject to a harmonic vertical base excitation provided by an electromagnetic shaker (LDS V450 - Permanent Magnet Shaker). Apart from the physical dimensions of the harvester, the resonant frequency, damping, capacitance and impedance were measured before any response to base excitations was studied. The total thickness was measured using a digital micrometre, (MITUTOYO Series 293-344 Digimatic Micrometer (IP-65)). The damped resonant frequency was measured by clamping the base to a rigid support and subjecting the free end to an impulse. A laser displacement sensor (Micro Epsilon optoNCDT 1420 - Smart Laser Triangulation Displacement Sensor) was used in combination with a high speed camera (FASTCAM SA1.1 photron) to record the oscillations. From this the damped resonant frequency and damping ratio could be extracted. The capacitance and impedance of the energy harvester were measured using an impedance analyser (HP 4192A LF). With these values the thickness of the PVDF could be calculated using equation (2.1)

$$C = \frac{\mathcal{E}_0 \mathcal{E}_r A}{t_p} \quad \gg \quad t_p = \frac{\mathcal{E}_0 \mathcal{E}_r A}{C} \tag{2.1}$$

Where $\mathcal{E}_0$ is the permittivity of free space, $\mathcal{E}_r$ is the relative permittivity, C is the capacitance, A is the area of piezoelectric material and t_p is the PVDF thickness. As the harvester had two metal rivets at the base to attach the lead wires, a rubber sheath was fashioned in order to clamp it to the shaker without causing a short circuit across the terminals. The harvester was subjected to a base excitation of 1 mm peak to peak at frequencies sweeping through its first resonant mode. At incremental frequency steps the tip displacement and open circuit AC voltage produced were recorded, using laser displacement sensors and an oscilloscope (Aglient Technologies DS01022A).

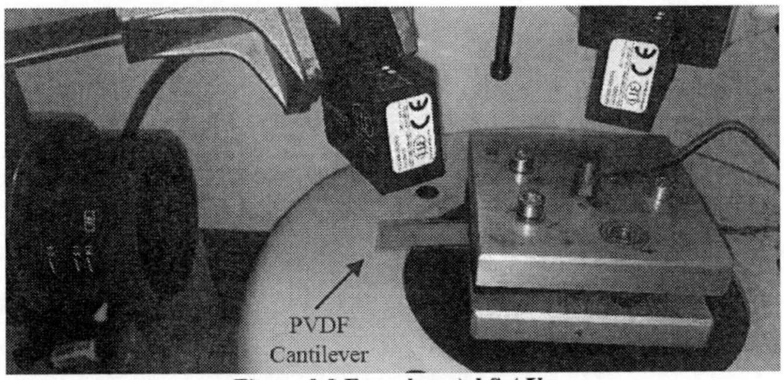

Figure 2-2 Experimental Set Up

3. Modelling PVDF Unimorph Sensor

Unimorph piezoelectric cantilever beams can be modelled using composite beam equations. Based on the mechanics of vibrating beams, piezoelectric energy conversions and fundamental circuit theory accurate models were previously reported [9]. The three most commonly used analytical models are Pin-Force, Enhanced Pin-Force and Euler-Bernoulli [10]. The Euler-Bernoulli method is reported as the most accurate of the three methods and for this reason is used for this work [11]. The PVDF and substrate are modelled to have a perfect bond and both bend about a common neutral axis. This neutral axis is calculated using a weighted Young's modulus approach. The strain is modelled to increase linearly, with thickness, through the composite beam [11]–[13].

Cantilever Beam Modelling

Table 3-1 Unimorph Cantilever beam Parameters

	Parameter	*Energy Harvester*
Beam	*Length (m) L*	*0.032*
	Width (m) b	*0.01219*
	Thickness (m) t_b	*0.000194*
	Material	*Polyester (Mylar)*
	Density (kg/m$^{3)}$	*1390*
	Young's Modulus (GPa) E_b	*4.9*
PVDF	*Length (m) l_p*	*0.03*
	Width (m) b	*0.01219*
	Thickness (m) t_p	*0.000045*
	Material	*PVDF*
	Density (kg/m$^{3)}$	*1780*
	Young's Modulus (GPa) E_p	*1.1*
	Voltage Constant g31 (Vm/N)	*0.281*

In order to use the voltage estimation model the harvester was first modelled as composite cantilever beam. A free clamped beam was modelled with a harmonic base excitation and tip mass. Although the harvester did not have a tip mass, this was satisfied by setting the mass value to 0. The properties of the harvester are listed in Table 3-1 and the cantilever beam model is shown in Figure 3-1. The governing equation of a clamped-free cantilever beam with a tip mass excited through base excitation with structural and air damping is

$$E_b I_b \frac{\partial^4 w_{rel}(x,t)}{\partial t^4} + c_s I_b \frac{\partial^5 w_{rel}(x,t)}{\partial x^4} + c_a \frac{\partial w_{rel}(x,t)}{\partial t} + m \frac{\partial^2 w_{rel}(x,t)}{\partial t^2} = -[m + M_t \delta(x - L_b)] \frac{\partial^2 w_b(x,t)}{\partial t^2} \quad (3.1)$$

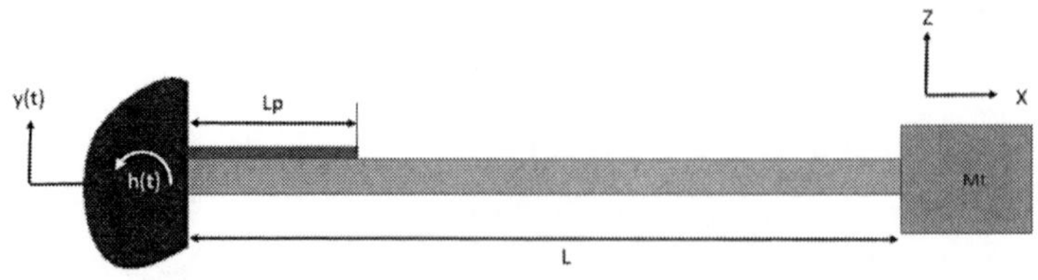

Figure 3-1 Cantilever Beam Model

Where $w_{rel}(x,t)$ is the relative displacement between the cantilever tip and the base, $w_b(x,t)$ is the displacement of the base, m is the mass per unit length and M_t is the tip mass. c_s is the Kelvin-Voigt or strain rate damping and c_a is the viscous air damping. The absolute displacement of the beam can be expressed as

$$w(x,t) = w_{rel}(x,t) + w_b(x,t) \tag{3.2}$$

The base displacement $w_b(x,t)$ is input as a sinusoidal function and the relative displacement with respect to time and position can be calculated using the modal shape and the modal coordinate equation.

$$w_{rel}(x,t) = \sum_{i=1}^{n} q_i(t)\, Z_i(x) \tag{3.3}$$

The *ith* modal shape is denoted as $Z_i(x)$ while the *ith* modal coordinate equation is $q_i(t)$. For a cantilever beam with a tip mass the modal shape can be calculated using the beam parameters and material properties [10], [11], [14]. The amplitude of the modal coordinate equation is heavily dependent on damping ratio ζ which was measured experimentally.

$$q_i(t) = \frac{1}{\omega_d} \int_0^t F_i(\tau) e^{-\zeta\omega(t-\tau)} \sin(\omega_d(t-\tau))\, d\tau \tag{3.4}$$

Since the base can be assumed to have very little rotation ($h(t) = 0$) [14]–[16], and only be excited by vertical vibrations the force equation can be simplified to

$$F(t) = -m\left(\frac{d^2y(t)}{dt^2}\int_0^L Z_i(x)dx\right) - M_t Z_i(L)\left(\frac{d^2y(t)}{dt^2}\right) \tag{3.5}$$

The relative displacement was then calculated using equation (3.3). The curvature of the beam was found by differentiating the displacement twice with respect to time. The average curvature was then evaluated as

$$k_{ave}(t) = \frac{1}{L_p}\int_0^{L_p} k(x,t)dx \tag{3.6}$$

Finally the average moment acting on the PVDF was calculated using

$$M(t) = E_b I_b k_{ave}(t) \tag{3.7}$$

Maximum Voltage Estimation

The maximum voltage will occur under open circuit conditions, this value was predicted using the Euler-Bernoulli voltage equation

$$Veb = -\frac{6g_{31}\left(\frac{E_b t_b}{E_p t_p}\right)\left(1+\frac{t_b}{t_p}\right)M}{bt_p\left[1+\left(\frac{E_b t_b}{E_p t_p}\right)^2\left(\frac{t_b}{t_p}\right)^2+2\frac{E_b t_b}{E_p t_p}\left(2+3\frac{t_b}{t_p}+2\left(\frac{t_b}{t_p}\right)^2\right)\right]} \tag{3.8}$$

Computer Simulation

The harvester was also modelled using the FEM software COMSOL Mutiphysics. The computer simulations gave a very clear visual representation of the cantilever beam response and gave a comparison between 2 dimensional and 3 dimensional modelling. The two models were developed with the same parameters and subjected to the same base excitation. The model was run in the frequency domain with a fixed support at the base and all other domains free.

4. Results

The damping was calculated both by fitting an exponential decay curve to the recorded displacement data, and using equation (4.1):

$$\frac{n_{th}+1}{n_{th}} = \exp\left[\frac{-2\zeta\pi}{\sqrt{1-\zeta^2}}\right] \tag{4.1}$$

Where n_{th} and $n_{th}+1$ are the amplitude of successive oscillations of the beam once the initial chaotic response from the impulse has settled and ζ is the damping ratio. Both methods gave the same damping ratio of 0.05.

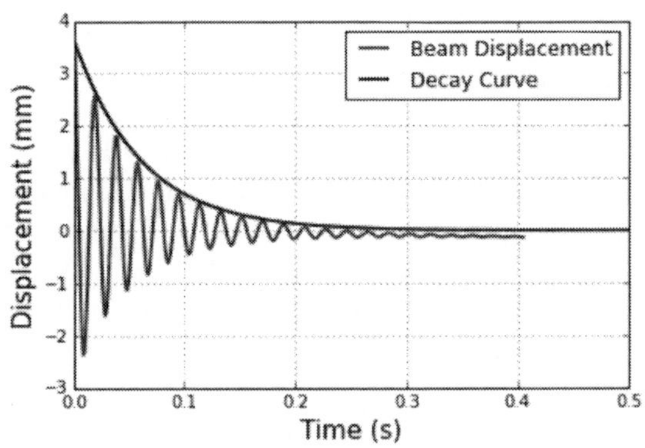

Figure 4-1 Response to impulse, Damping ratio = 0.05

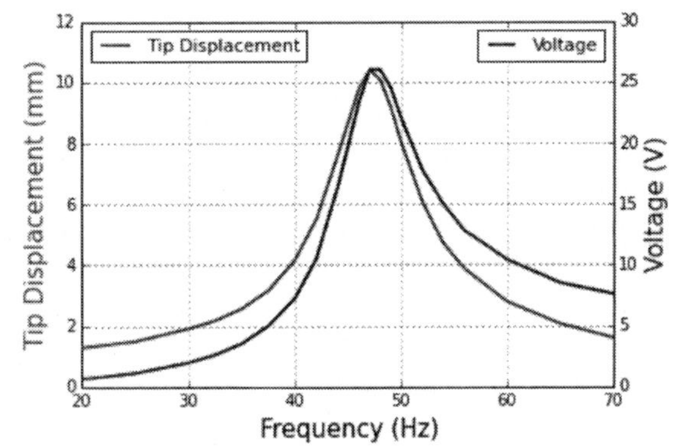

Figure 4-2 Response to 1 mm base excitation frequency sweep.

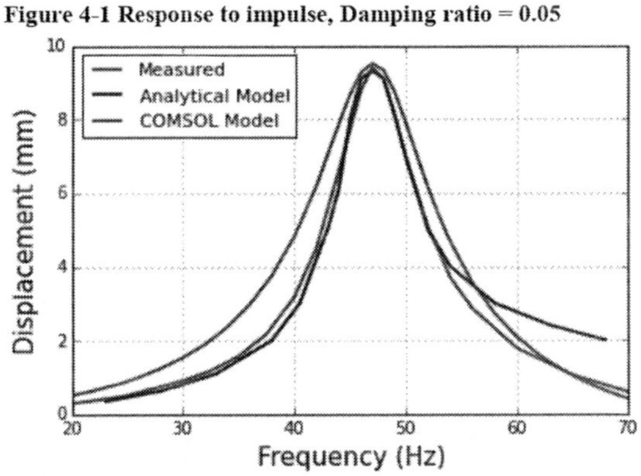

Figure 4-3 Measured and modelled displacements

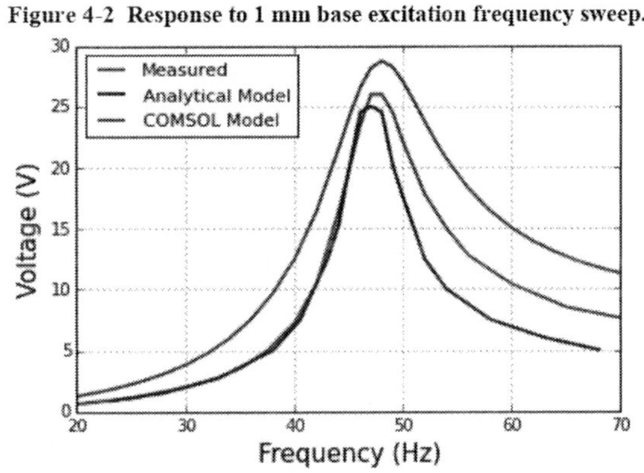

Figure 4-4 Measured and modelled Voltages

Figure 4-2 shows the raw recorded data of the energy harvesters' response to a 1 mm peak to peak base excitation frequency sweep. All results are peak to peak values. The measured displacements are total displacement, this was corrected to the relative tip displacement by subtracting the base displacement from the total tip displacement taking into account the phase shift. In order to match displacement amplitude, the COMSOL model required a damping ratio of 0.1, higher than the measured value.

5. Discussion

Figure 4-3 shows that both the analytical and COMSOL models are very accurate at predicting tip displacement at resonance. The COMSOL models accuracy decreases either side of the resonant frequency while the analytical model remains very accurate up to 53 Hz. A damping ratio cannot be input directly into COMSOL, Rayleigh damping coefficients had to be calculated in order to damp the model. The COMSOL model required higher damping than the measured values to match the displacement amplitudes and there is ongoing work trying to solve this issue. The measured data shows an increase in voltage for a given displacement at higher frequencies compared to lower frequencies. The COMSOL model also predicts this, but exceeds the recorded voltage. Conversely, the analytical model predicts voltages directly from displacements levels, indicating the two models work off different fundamental equations. The analytical model predicts the voltage accurately until resonance. After this it consistently predicts a lower voltage than the recorded value. This may be because the analytical model predicts that the beam will resonate only at its first mode shape, and in reality as it reaches higher frequencies there may be some torsional rotations causing higher strain levels on the device and in turn creating higher voltages. It was noted that overall both models are most accurate at resonance which suits the required application. For an implanted pacemaker the main source of available energy is an impulse from the heartbeat. This would result in a response similar to Figure 4-1.The device will most likely vibrate at its resonant frequency with a decaying amplitude. Future work will look at heartbeat acceleration profiles, and use modelling to predict how much power can be generated from each impulse.

6. Conclusion

In this paper analytical and 3D FEM electromechanical models of a microscale energy harvester were presented. The application for the modelled device is to operate on the low frequency range produced by cardiac motion on which little previous work was reported. The models make accurate displacement and voltage predictions at resonance compared to experimental COTS device measurements. However, for vibration inputs at non resonant frequencies both modelling approaches yield different output voltages suggesting they use different fundamental equations. In addition, the FEM model requires a damping ratio twice as large as the measured value in order to achieve the same displacement amplitude while the analytical model can have the measured value input unmodified. Both models will be developed further to predict the power capabilities of a MEMS energy harvester in an implanted cardiac pacemaker.

7. References

[1] R. W. Joyner and F. J. van Capelle, "Propagation through electrically coupled cells. How a small SA node drives a large atrium.," *Biophys. J.*, vol. 50, no. 6, pp. 1157–1164, Dec. 1986.

[2] V. S. Mallela, V. Ilankumaran, and N. S. Rao, "Trends in Cardiac Pacemaker Batteries," *Indian Pacing Electrophysiol. J.*, vol. 4, no. 4, pp. 201–212, Oct. 2004.

[3] V. L. Paone S, Trimaglio F, Migliore A, Maltoni S, "Transcatheter implantable miniaturised leadless pacemakers [Internet].," Agenas, Agnezia nazionale per i servizi sanitari regionali, No. 17, Dec. 2014.

[4] R. E. Kirkfeldt, J. B. Johansen, E. A. Nohr, M. Moller, P. Arnsbo, and J. C. Nielsen, "Risk factors for lead complications in cardiac pacing: A population-based cohort study of 28,860 Danish patients," *Heart Rhythm*, vol. 8, no. 10, pp. 1622–1628, Oct. 2011.

[5] M. Deterre, *Toward an energy harvester for leadless pacemakers*. Paris 11, 2013.

[6] M. Deterre, B. Boutaud, R. Dalmolin, S. Boisseau, J. J. Chaillout, E. Lefeuvre, and E. Dufour-Gergam, "Energy harvesting system for cardiac implant applications," in *2011 Symposium on Design, Test, Integration and Packaging of MEMS/MOEMS (DTIP)*, 2011, pp. 387–391.

[7] N. Jackson, R. O'Keeffe, F. Waldron, M. O'Neill, and A. Mathewson, "Evaluation of low-acceleration MEMS piezoelectric energy harvesting devices," *Microsyst. Technol.*, vol. 20, no. 4–5, pp. 671–680, Dec. 2013.

[8] V. Y. Reddy, R. E. Knops, J. Sperzel, M. A. Miller, J. Petru, J. Simon, L. Sediva, J. R. de Groot, F. V. Y. Tjong, P. Jacobson, A. Ostrosff, S. R. Dukkipati, J. S. Koruth, A. A. M. Wilde, J. Kautzner, and P. Neuzil, "Permanent Leadless Cardiac Pacing Results of the LEADLESS Trial," *Circulation*, vol. 129, no. 14, pp. 1466–1471, Apr. 2014.

[9] D. J. I. A. Erturk, "A Distributed Parameter Electromechanical Model for Cantilevered Piezoelectric Energy Harvesters," *J. Vib. Acoust.*, vol. 130, no. 4, 2008.

[10] R. R. Muppala, K. P. Raju, N.-M. Moon, and B.-H. Jung, "Analytical Models to Predict Power Harvesting with Piezoelectric Transducer," *J. Korean Inst. Electromagn. Eng. Sci.*, vol. 8, no. 1, pp. 6–11, Mar. 2008.

[11] T. Eggborn, "Analytical Models to Predict Power Harvesting with Piezoelectric Materials," thesis, Virginia Tech, 2003.

[12] K. L. Bing, T. Li, H. H. Hng, F. Boey, T. Zhang, and S. Li, *Waste Energy Harvesting: Mechanical and Thermal Energies*. Springer Science & Business Media, 2014.

[13] N. Jalili, *Piezoelectric-Based Vibration Control: From Macro to Micro/Nano Scale Systems*. Springer Science & Business Media, 2009.

[14] A. Erturk and D. J. Inman, *Piezoelectric Energy Harvesting*, 1 edition. Chichester: Wiley, 2011.

[15] A. Erturk and D. J. Inman, "An experimentally validated bimorph cantilever model for piezoelectric energy harvesting from base excitations," *Smart Mater. Struct.*, vol. 18, no. 2, p. 25009, 2009.

[16] F. Häggström, J. Gustafsson, and J. Delsing, "Piezoelectric energy harvesting modeled with SPICE," *Submitt. IEEE Sens. J. Vol.*

Journal of Physics: Conference Series **757** (2016) 012039 doi:10.1088/1742-6596/757/1/012039

A low frequency MEMS energy harvester scavenging energy from magnetic field surrounding an AC current-carrying wire

Oskar Z. Olszewski, Ruth Houlihan, Alan Mathewson, Nathan Jackson

Tyndall National Institute, University College Cork, Cork, Ireland

zbigniew.olszewski@tyndall.ie

Abstract. This paper reports on a low frequency piezoelectric energy harvester that scavenges energy from a wire carrying an AC current. The harvester is described, fabricated and characterized. The device consists of a silicon cantilever with integrated piezoelectric capacitor and proof-mass that incorporates a permanent magnet. When brought close to an AC current carrying wire, the magnet couples to the AC magnetic field from a wire, causing the cantilever to vibrate and generate power. The measured average power dissipated across an optimal resistive load was 1.5 μW. This was obtained by exciting the device into mechanical resonance using the electro-magnetic field from the 2 A source current. The measurements also reveal that the device has a nonlinear response that is due to a spring hardening mechanism.

1. Introduction

In recent years, much attention has been placed on the Internet of Things (IoT) [1]. The IoT refers to a large number of nodes, also called "things" that exchange the data wirelessly across the network. The individual nodes are typically powered by batteries with limited lifetime and currently the IoT urges for development of energy-autonomous nodes and energy harvesting is a key technology addressing such a demand [2, 3]. Multiple energy sources are available within the environment e.g. light, mechanical vibrations, electromagnetic fields, etc., and these can be converted into useable electrical power by means of various transduction methods, including photovoltaic, thermoelectric, electrostatic, and piezoelectric. Because of their numerous advantageous, vibrational piezoelectric harvesters have received much attention [3-8]. A key challenge of vibrational harvesters is that the devices typically resonate with a narrow frequency bandwidth. This is a major problem because the frequency of the vibration source may vary over time and, as a result, the power generation drops significantly. Many solutions to broaden the bandwidth of vibrational harvesters have been investigated. The bandwidth can be broadened by mechanical tuning of the device stiffness [9], connecting of multiple devices [10, 11], introducing a nonlinearity to the device behavior [12-17]. An alternative approach to bandwidth broadening is to select an application that provides a stable source of vibration [18-20]. One such source is the AC magnetic field surrounding an AC current-carrying wire such as the power mains of electrical equipment. In this paper, we report on a piezoelectric harvester that scavenges energy from an AC current-carrying wire. In previous work [18-20] a similar concept was shown; however the device was a macro-scale assembly to enable a low frequency operation (≈50 – 60 Hz). In this work, we demonstrate a smaller device fabricated in MEMS process that also operates at low frequency.

2. Device Concept, Description and Fabrication Process

2.1. Device Concept

A general concept used in this work is described in Figure 1 and has been proposed by Leland et al. [18]. The device consist a permanent magnet attached to the free-end of a piezoelectric cantilever with the magnetization direction aligned with the mechanical compliance direction of the cantilever. When the cantilever is brought in close proximity to an AC current-carrying wire the magnet couples to the AC magnetic field from the wire. The force acting on the magnet is proportional to the integral of the field gradient over the magnets volume [20-22]. This, in the most general form, can be described as:

$$F_M = B_r \int \frac{d(H_z)}{dz} dV, \tag{1}$$

where F_M is the vertical force acting on the magnet, B_r is the vertical remanence of the magnet, V is the magnet volume, and H_z is the vertical component of the magnetic field that, at a given distance from the wire, is proportional to the current magnitude. Since F_M is an alternating force, the cantilever vibrates at the frequency of the current and with a displacement that is proportional to its magnitude.

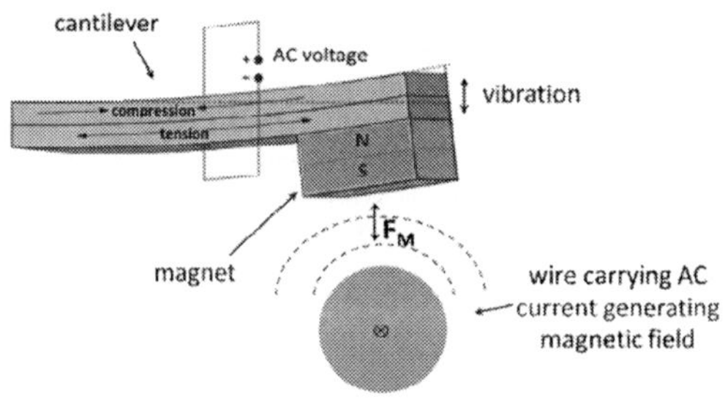

Figure 1. Schematic diagram of a device concept in this work.

From equation 1 it can be concluded that in order to maximize the coupling force F_M for a given remanence, magnet volume and current, the field gradient must be maximized. The studies [20, 22] showed that when coupling to a single-wire cord, the field gradient is maximized along a 45° line from the centre of the wire, whereas in case of a double-wire cord, the maximum gradient occurs above the cord centre. The vibrating piezoelectric cantilever acts as a transducer, whereby the induced stress and strain along the cantilever generates an electric potential across the material thickness, e.g. during displacement upwards the top part of the cantilever is under compression while the bottom part is under tension and vice versa for the downward displacement. This is a common operation mode (31) of piezoelectric cantilever-based harvesters. The voltage generated from the harvester depends on the exact geometry of the device and on the effective material properties. In general however, it increases with the mechanical strain developed in the piezoelectric film along the device length. This strain is proportional to the device displacement which depends on the magnitude of the excitation force (F_M) and which can be maximized by driving the device into mechanical resonance [7, 22-24].

2.2. Device Description and Fabrication Process

Figures 2(a), (b) and (c) show the top-view, bottom-view and cross-sectional view of the device respectively. The harvester is comprised of a thin device silicon cantilever that supports an extended silicon mass. On the top surface of the cantilever, the piezoelectric capacitor is formed of a titanium (Ti) bottom electrode, an aluminium nitride (AlN) piezoelectric layer, and an aluminium (Al) top electrode. As described above, the capacitor converts the mechanical vibration into a voltage that can be measured across the capacitor. The total length of the device is 10 mm and that includes a 1.5 mm cantilever and 8.5 mm mass. The width of the device is 7 mm and the cavity etched in the mass is 6.5

mm along the device length by 5 mm along the width. The harvesters were fabricated from an SOI wafer with a 17 μm thick device silicon layer, a 1 μm thick buried oxide, and a 535 μm thick bulk silicon. Details of the fabrication process were described elsewhere [25].

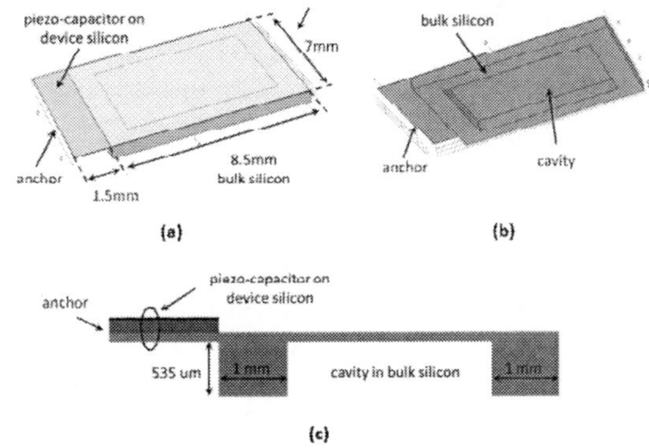

Figure 2. Schematic representation of the harvester (b) top-view, (b) bottom-view, and (c) cross-sectional view.

Figure 3(a) illustrates the top-view of the device attached to a PCB. The bottom-view with the assembled magnet is shown in Figure 3(b). Future devices will use an integrated micro-magnet which will be deposited in the cavity of the mass [26]. For the purposes of concept demonstration, a commercial permanent neodymium magnet (*Br* = 1.3) was used here.

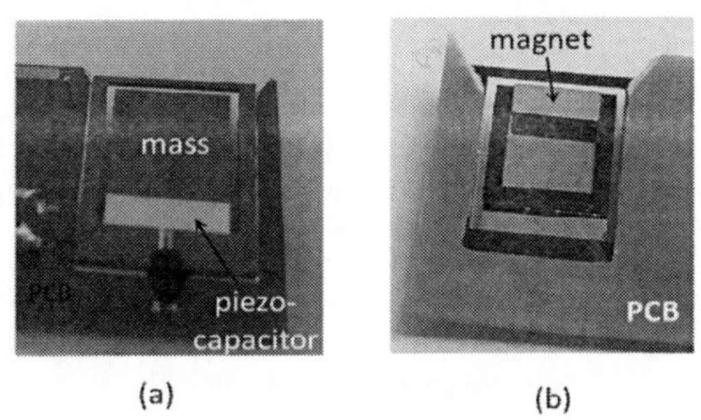

Figure 3. Photograph of the device, (a) top-view, (b) bottom-view.

3. Experimental Results and Discussion

3.1. Initial Characterisation

The first resonance of the device, before and after the magnet assembly was measured using Laser Doppler Vibrometry (LDV). The device was excited into vibration by applying a mechanical shock pulse to the clamp holding the device PCB under the vibrometer head. The test results along with the simulated values are summarized in Figure 4(a). The simulated values of resonance frequency were obtained using a modal analysis in Comsol (model shown in Figure 2). The default material properties in Comsol for silicon and AlN were used. The model assumed a 535 μm thick bulk silicon, a 5 x 2.0 x 0.5 mm^3 magnet volume and a 7500 kg/m^3 magnet density. The thicknesses of the device silicon and AlN layer used in the model were 15 μm and 0.4 μm (values measured using SEM). The metal layers and silicon nitride were not considered during simulation. Figure 4(a) shows the measured resonance of the device before and after the magnet was attached. The results show that the resonance drops from around 75 Hz to 39 Hz due to added mass of the magnet. This agrees well with the results obtained from the model.

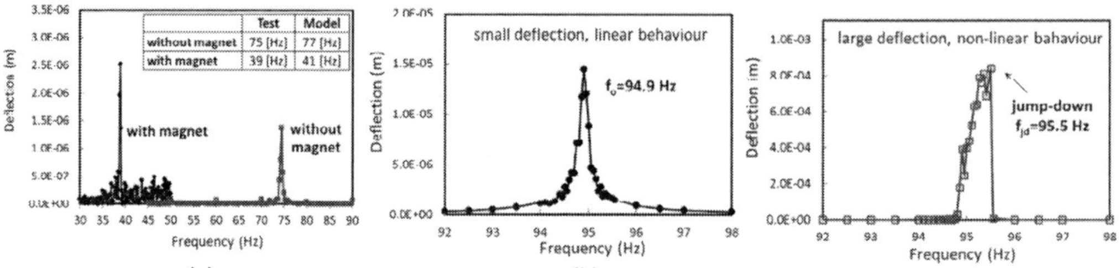

(a) (b) (c)

Figure 4. (a) The resonance curve with and without a magnet, (b) resonance curve at small displacements and (c) resonance curve at large displacements.

Figures 4(b) and 4(c) show the measured displacement of a device similar to that described above but without the piezoelectric capacitor and with no permanent magnet. This all-silicon device can therefore be considered a purely mechanical test structure. The thickness of the device silicon was 20 μm and the total length was 8.2 mm with the beam and mass lengths equaling 1.2 mm and 7 mm, respectively. The device was excited by a mechanical shaker into two distinct vibration levels. When the device vibrates with low deflections relative to the thickness of the cantilever, the response is linear and symmetrical around a single frequency of 94.9 Hz. In contrast, as can be seen in Figure 4(c), when the device vibrates with large deflections, the response of the device is nonlinear and asymmetrical about the center frequency: the deflection increases steadily from linear resonance of 94.9 Hz before it jumps down at 95.5 Hz. This is a typical characteristic of a resonator exhibiting spring hardening and can be modeled by a spring element featuring a cubic term in its force-displacement relationship as [16, 17, 27-30]

$$F = F_b + F_s \approx k_b x + k_s x^3, \tag{3}$$

where F_b is the linear spring component due to the bending strain in the beam that dominates the device stiffness at low displacements and F_s is the nonlinear spring component due to the stretching strain that comes into play at large displacements. The k_b and k_s are the linear and nonlinear spring constants, respectively and x is the displacement. In such a system, the stretching strain causes the spring hardening whereby the stiffness of the cantilever increases and shifts the resonance frequency to the right, as in Figure 4(c). The full solution of Duffing equation [29] reveals that resonators with nonlinear compliance feature a hysteretic frequency response that depends on whether the frequency is being increased or decreased. The harvesters in this work operate with displacements in the range of millimeters and the mechanical nonlinearity is also evidenced on the device electrical characteristics.

3.2. Electrical Test Setup and Experimental Results

In the experiments a custom-built AC current source was used capable of supplying the AC current of various frequencies and amplitudes. Figure 5 shows the image of the device under test. As explained in [20, 22], to maximize the device displacement and charge generation, the device in its rest position is located such that the magnet is at the angle of approximately 45° with respect to the wire center.

Figure 5. The device image taken with a high speed camera during operation when it is displaced upwards.

To evaluate the average power from the harvester, the RMS current, in a resistive optimal load of 2 MΩ was measured versus the frequency of the source current of 2 A (RMS). This level of source current actuated the device to the peak amplitude of 3 mm. The frequency sweep-up and sweep-down were performed and results are shown in Figure 6(a). The average power shown in Figure 6(b) was calculated according to the formula $P_{RMS} = I_L^2 \times R_L$. It can be seen that the electrical characteristics of the harvester have strong nonlinear behavior due to nonlinear mechanical spring as observed on structure in Figure 4(d). From the sweep-up data, the power increases steadily with the source frequency up to the peak value of approximately 1.5 µW at frequency of 42.3 Hz, after which the power suddenly drops. This is the jump-down frequency at which the feedback from the stretching strain that causes the beam stiffening can no longer compensate for the increase in the source frequency. By considering the sweep-down response of the device, it is evident that the full characteristic has the hysteresis typical for the Duffing resonators [16, 17, 27-30].

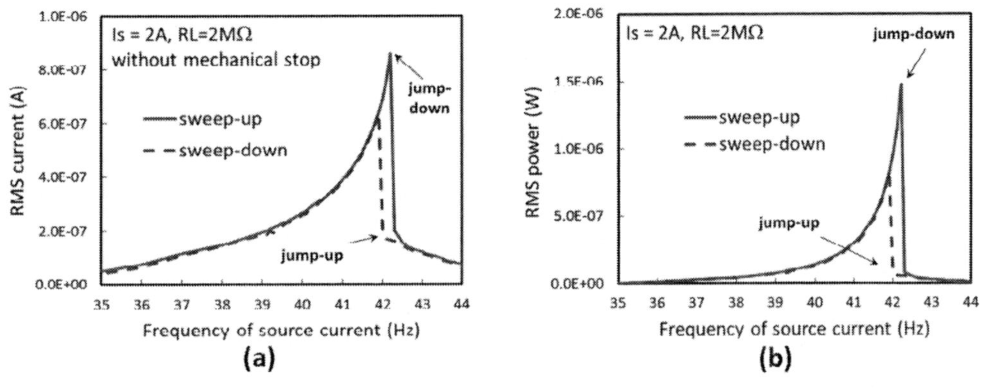

Figure 6. (a) RMS current measured in the load, (b) RMS power calculated from the measured current.

4. Conclusions and future work

This paper has described a MEMS piezoelectric energy harvester that targets scavenging the energy from the magnetic field surrounding a power line. The device concept, fabrication and test results were presented. To maximize the generated power the device was tested such that its mechanical resonance matches the frequency of the source current. The results revealed that the harvester has a nonlinear mechanical compliance and the average power output of 1.5 µW at 2 A of source current. Further characterisation of the devices should be performed in order to understand the device performance in terms of frequency bandwidth, stability around the jump-down frequency, and reliability. It is anticipated that incorporation of mechanical stops as proposed in earlier studies could broaden the frequency response of the device leading to better device stability, and this will be investigated in subsequent research.

Acknowledgements

The authors would like to thank Enterprise Ireland and Analog Devices for supporting this research.

References

[1] M. H. Miraz, et al., "Review on Internet of Things (loT), Internet of Everything (IoE) and Internet ofNano Things (IoNT)", ITA 2015, Wrexham, 8-11 Sept. 2015, 219 - 224.

[2] R. J. M. Vullers, et al., "Energy Harvesting for Autonomous Wireless Sensor Networks, IEEE Slid-State

Circuits Magazine, Spring 2010, 29-38.

[3] S. Kim, et al., "Ambient RF Energy-Harvesting Technologies for Self-Sustainable Standalone Wireless Sensor platforms ", Proc. of IEEE, Vol. 102, No. 11, Nov. 2014, 1649 – 1666.

[4] S-G. Kim, et al., S. Priya, I. Kanno, "Piezoelectric MEMS for energy harvesting", Material Research Society Bulletin, Vol. 37, November 2012, 1039 – 1050.

[5] H. Li, et al., "Energy harvesting from low frequency applications using piezoelectric materials", Applied Physics Reviews 1, 041301 (2014).

[6] P. D. Mitcheson, et al., "Energy Harvesting From Human and Machine Motion for Wireless Electronic Devices" Proc. of IEEE, 96, No. 9, 2008, 1457 – 1486.

[7] S. Roundy, et al., "A piezoelectric vibration based generator for wireless electronics, Smart Mater. Struct., 13 (2004) 1131–1142.

[8] C-B. Eom, et al., "Thin-film piezoelectric MEMS", Material Research Society Bulletin, Vol. 37, November 2012, 1007 – 1017.

[9] Y. Hu, et al., "A piezoelectric power harvester with adjustable frequency through axial preloads," Smart Mater. Struct., Vol. 16, No. 5, Oct. 2007, 1961 – 1966.

[10] Z. Yang, "Connected Vibrating Piezoelectric Bimorph Beams as a Wide-band Piezoelectric Power Harvester", Journal of Intell. Mater. Syst. Struct., Vol. 20, No. 5, November 2008, 569 – 574.

[11] S. M. Shahruz, "Design of mechanical band-pass filters for energy scavenging", Journal of Sound and Vibration, Vol. 292, No. 3–5, May 2006, 987 – 998.

[12] L. Gammaitoni, et al., "Nonlinear oscillators for vibration energy harvesting", Appl. Phys. Lett. 94, 164102 (2009).

[13] F. Cottone, et al., "Nonlinear energy harvesting", Physics Review Letters, 102, 080601, (2009).

[14] V. R. Challa, et al., "A vibration energy harvesting device with bidirectional resonance frequency tenability, Smart Mater. Struct., 17 (2008) 015035 (10pp).

[15] M. S. M. Soliman, et al., "A wideband vibration-based energy harvester" J. Micromech. Microeng. 18 (2008) 115021 (11pp).

[16] A. Hajati, et al., "Ultra-wide bandwidth piezoelectric energy harvesting", Appl. Phys. Lett. 99, 083105 (2011).

[17] A. Hajati, et al., "Wide-bandwidth MEMS-scale piezoelectric energy harvester", PowerMEMS'09, Washington DC, USA, December 1-4, 2009.

[18] E. S. Leland, et al., "Energy scavenging power sources for household electrical monitoring", Micro and Nanotechnology for Power Generation and Energy Conversion Applications, USA, Nov. 29, 2006.

[19] W. He, et al., "Energy harvesting from electric power lines employing the Halbach arrays", Rev. Scient. Instr. 84, 105004 (2013).

[20] I. Paprotny, et al., "Electromechanical energy scavenging from current-carrying conductor", Sensors, Vol. 13, No. 1, January 2013.

[21] B. Wagner, et al., "Microfabricated actuator with moving permanent magnet", Micro Electro Mechanical Systems, MEMS'91, 30 January - 2 February, Nara, 1991.

[22] E. S. Leland, at al., "A MEMS AC current sensor for residential and commercial electricity end-use monitoring", J. Micromech. Microeng. 19 (2009) 094018 (6pp).

[23] H. Liu, at al., "Piezoelectric MEMS Energy Harvester for Low-Frequency Vibrations With Wideband Operation Range and Steadily Increased Output Power", JMEMS, Vol. 20, No. 5, October 2011.

[24] T. M. Kamel, at al., "Modeling and characterization of MEMS-based piezoelectric harvesting Devices", J. Micromech. Microeng. 20 (2010) 105023 (14pp).

[25] N. Jackson, at al., "Influence of aluminium nitride crystal orientation on MEMS energy harvesting device performance", J. Micromech. Microeng. 23 (2013) 075014 (9pp).

[26] N. Jackson, at al., "Integration of Thick-Film Permanent Magnets for MEMS Applications", JMEMS, 14 June 2016.

[27] R. Ramlan, at al., "On the performance of a dual-mode non-linear vibration energy harvesting device", Journal of Intelligent Material Systems and Structures, 0(0) 1–10, 2012.

[28] G. Gafforelli, et al., "Experimental verification of a bridge-shaped, non-linear vibration energy harvesters", Appl. Phys. Lett. 105, 203901 (2014).

[29] S. G. Buttow, et al., "A resonant generator with non-linear compliance for energy harvesting in high vibrational environment", International Electric Machines & Drives Conference, Antalya, 3-5 May 2007.

[30] D. Mallick, et al., "Interplay between electrical and mechanical domains in a high performance nonlinear energy harvester", Smart Mater. Struct., 24 (2015) 122001 (9pp).

Design, modelling and preliminary characterisation of microneedle-based electrodes for tissue electroporation *in vivo*

Conor O'Mahony[1], Ruth Houlihan[1], Konstantin Grygoryev[1], Zhenfei Ning[2], John Williams[2] and Tom Moore[2]

[1]Tyndall National Institute, University College Cork, Cork, Ireland.

[2] School of Biochemistry and Cell Biology, University College Cork, Ireland.

conor.omahony@tyndall.ie

Abstract. We analysed the use of microneedle-based electrodes to enhance electroporation of mouse testis with DNA vectors for production of transgenic mice. Different microneedle formats were developed and tested, and we ultimately used electrodes based on arrays of 500 μm tall microneedles. In a series of experiments involving injection of a DNA vector expressing Green Fluorescent Protein (GFP) and electroporation using microneedle electrodes and a commercially available voltage supply, we compared the performance of flat and microneedle electrodes by measuring GFP expression at various timepoints after electroporation. Our main finding, supported by both experimental and simulated data, is that needles significantly enhanced electroporation of testis.

1. Introduction

Transgenic rodents, whose genomes have been altered through the use of genetic engineering techniques, are used to study normal development and for producing models of human disease to understand pathogenesis and to underpin development of therapies [1, 2]. Transgenic rodents are generally produced by delivery of DNA constructs to fertilised oocytes using microinjection, followed by embryo transfer to pseudopregnant mice. These are laborious, expensive and inefficient techniques that have not changed significantly in thirty years.

Recently, efforts have been made to use electroporation techniques as an alternative to this procedure. Electroporation involves the application of voltage pulses to create an electric field in a tissue sample. This leads to a temporary increase in permeability of cell membranes, facilitating the passage of DNA to cells. DNA injection into testis, followed by electroporation, has been reported as a novel method of producing transgenic rodent offspring [3, 4]. The ability to produce transgenic rodents by testis electroporation would greatly reduce the costs and skill sets needed to produce this essential resource required for the study of developmental biology and disease pathogenesis.

Electroporation pulses are applied using a pair of flat electrodes, applied to either side of the tissue region of interest. The outermost layer of the skin surrounding the testis, the stratum corneum, is only 10 - 20μm thick and is largely composed of dead cells. It is therefore highly resistive and a large percentage of the applied voltage is dropped across this barrier, reducing the electric field inside the tissue and diminishing the efficacy of the electroporation procedure [5]. Skin and the stratum corneum also exhibit considerable variability in thickness and electrical characteristics, leading to uncertainly regarding optimum procedures.

Microneedle devices were originally intended for use in transdermal drug and vaccine delivery [6, 7], but have also generated significant interest for use in electroporation [5, 8-10]. Electrodes formed from arrays of electrically conductive microneedles can pierce the outermost skin layer and make direct contact with the moist, highly conductive epidermis immediately underneath, thereby bypassing the high-impedance stratum corneum and enhancing the electric field within the tissue. The microneedles also produce an increased electric field at the needle tips, which may contribute to increased efficacy. Furthermore, microneedle lengths can be such that they do not stimulate nerve endings within the skin, and therefore their use is painless [11].

This paper addresses the development of microneedle-based electrodes for the production of transgenic rodents using testicular electroporation. Finite-element modelling software was used to simulate the strength and distribution of electric fields established inside the testis. A double-sided wet etching process was then used to create microneedle arrays, which were affixed to tweezers to create prototype electrodes. Preliminary *in vivo* experiments were carried out and showed a strong increase in efficacy over conventional flat electrodes.

These findings may aid in developing an effective and reproducible protocol for the production of transgenic rodents by testis electroporation.

2. Finite element modelling and electrode design

A finite element model was constructed to compare the electric field across the testes achieved using (a) flat electrodes placed either side of the testes with (b) microneedle electrodes placed either side of the testes. A three-dimensional half-model was generated in COMSOL with the testes modelled as an ellipsoid with a major axis of length 19 mm and minor axes of length 11 mm, Figure 1. These dimensions correspond to measurements taken on rat testes. The stratum corneum was modelled as a thin, low permittivity layer surrounding the testes. The relative permittivity of the testes tissue was assumed in this case to be two orders of magnitude greater than that of the outer membrane. The actual permittivity is difficult to measure and in our case an exact value was not necessary since the desired output of the modelling exercise was to establish the relative improvement in the magnitude of the field by incorporating microneedles onto the electrodes. In the case of the electrodes with microneedles, a 2 x 2 array of 300 μm long needles was assumed.

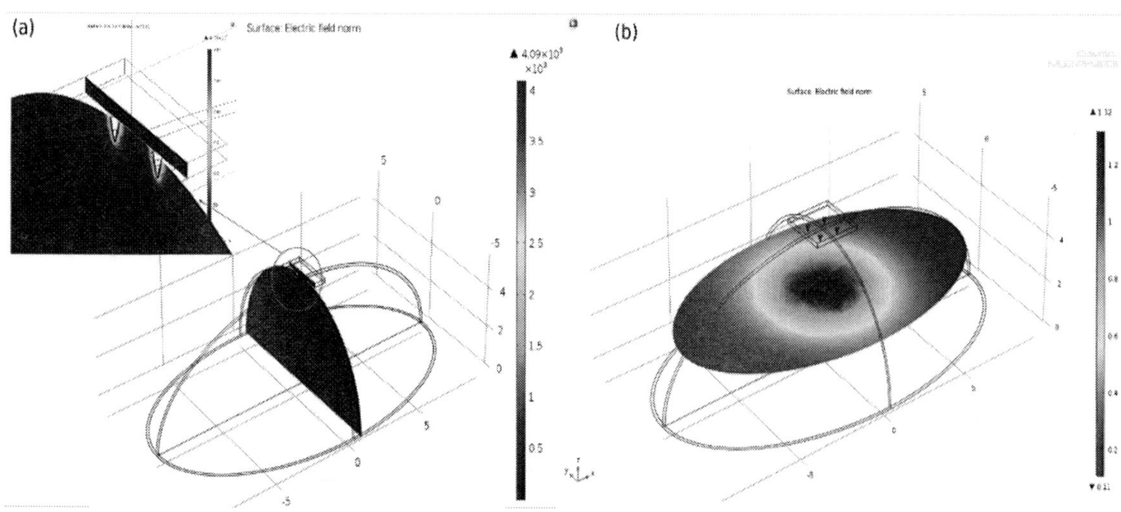

Figure 1. (a) Electric field in a plane perpendicular to the electrode and through the centre of two needles and (b) electric field in the plane mid-way between the electrodes and the centre of the testis ellipsoid.

The applied bias to the electrode was 40 V and the symmetry plane was held at ground. The results of the model with the microneedles show a maximum field three orders of magnitude higher (1.3 V/mm) than that of the flat electrodes (1 x 10^{-3} V/mm, data not shown), in which case the electric field is largely dropped across the layer of skin surrounding the testis.

3. Fabrication and assembly

3.1. Microneedle fabrication
Microneedle fabrication is based on potassium hydroxide (KOH) wet etching of single crystal silicon, where double-sided wafer patterning and an anisotropic etch process is used to simultaneously create a microneedle on the front side of the wafer, and a through-silicon via (TSV) from the back side. Metal deposition on both the front and back of the wafer then establishes electrical contact through this via. This rapid, single-step process eliminates the need for additional processing in order to establish electrical contact between the front and rear sides of the array, and full details are available in [12]. The wafer was then coated (front and back) with a layer of 20nm Ti/200nm Ag before dicing, and electrical contact was made between front and rear surfaces through the KOH-etched TSV openings.

3.2. Assembly
Based on testes measurement as outlined above, arrays of 5 x 4 needles were used. Needle height was 500 mm, pitch was 1725 μm, and the array base measured 8.62 mm x 6.90 mm. To generate prototype electrodes, arrays were affixed to the flat tips of plastic tweezers (Lerloy 6, Farnell, Ireland) using double-sided tape, and interfaced to a voltage source with ribbon cable and wire-to-board connectors. A stopper screw was added to the assembly to limit electrode travel and provide a consistent application force, Figure 2.

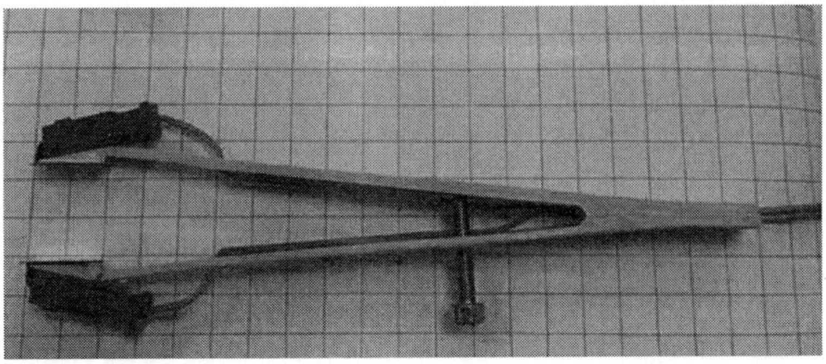

Figure 2. Completed device showing electrical connections and stopper screw.

4. Electroporation of adult mouse testis *in vivo*

4.1. Animal model.
Procedures were performed on eight C57Bl/6 male mice aged 35 days with body weight of approximate 20 g. Mice were anaesthetised, scrotal region hair was removed and each testis was injected with 20 μl (0.5 μg/μl) of circular plasmid pCX-EGFP DNA vector. 50 μl of electrogel was coated onto each electrode prior to electroporation to enhance electrical conductivity. Four mild square wave electric pulses (130 V, 5 ms, 50 ms interval) followed by 8 mild square wave electric pulses (30 V, 50 ms, 50 ms interval) were generated by an electroporator (NEPA21, Sonidel, Ireland) and delivered to each testis using flat or needle electrodes, with inter-electrode distance of 4 mm. A new electrode was first applied to the left testis, and reused once on the right testis of the same mouse. After electroporation, the testis area was cleaned with 70% ethanol.

Seventy-two hours post-procedure, the mice were euthanised and testes were harvested and imaged using a Leica DMI3000 fluorescence microscope. The testes were then homogenised and lysed using the M-PER mammalian protein extraction reagent (Thermo Fisher Scientific, Ireland). Protein sample concentration was determined by Bradford reagent (Sigma, UK), as per manufacturer's instructions. Anti-EGFP antibody (Takara Bio, USA) was used at 1:1000 dilution in western blotting. Anti-Gapdh antibody (Sigma, UK) was used at 1:2000 dilution in western blotting as an internal control. Western blotting method was performed as described in [13].

4.2. Results

Inspection of whole testis under microscopy indicated extensive fluorescence in both flat and needle electrode-treated testes as shown in Figure 3.

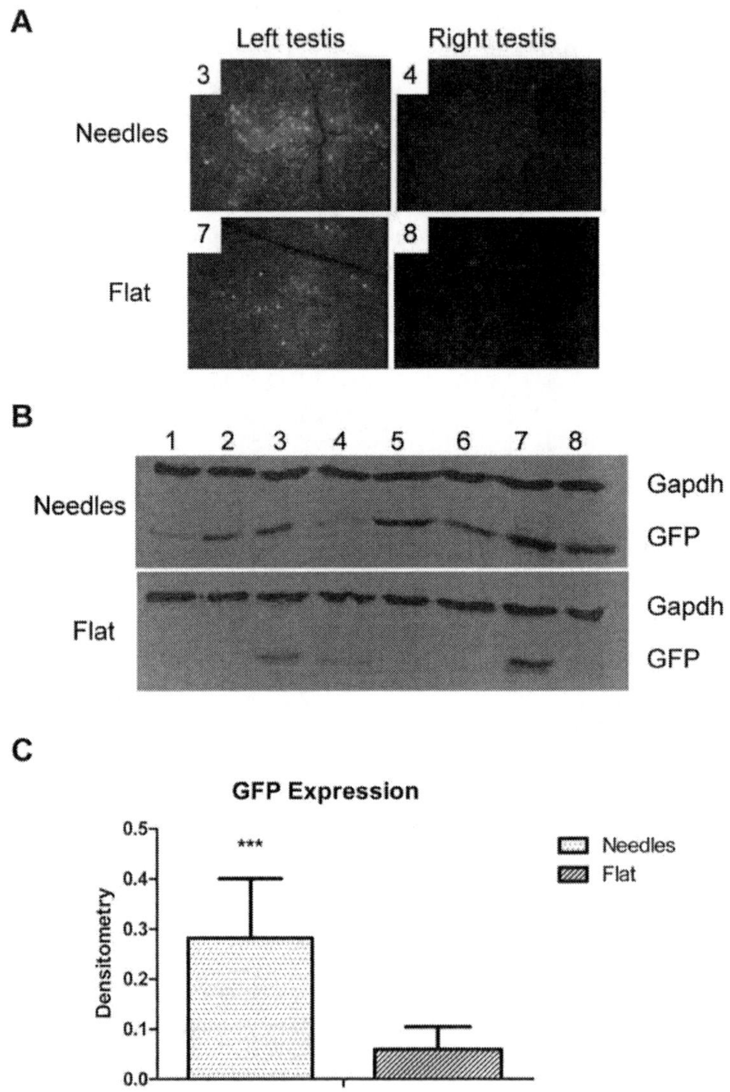

Figure 3. A) Selected examples of fluorescent images of whole testis transfected with GFP from two mice. B) Western blot of all eight testes treated with needle (upper panel) and flat (lower panel) electrodes. C) GFP expression from samples measured by densitometry.

However, there was a striking reduction in fluorescence intensity on second use of electrode on right testis (Figure 3A), which may be due to heat damage after first use on left testis. This difference between first and second use of electrodes was also evident after western blot analysis and densitometry. To provide quantitative data, densitometry was performed following western blotting and probing with antisera raised against GFP and Gapdh (loading control). Results are combined data for eight testes each for flat and needle electrodes and indicated that needle electrodes were significantly superior to flat electrodes as measured using GFP expression normalised to within-sample Gapdh expression (Figure 3C).

5. Discussion and Conclusion

This work has developed microneedle-based electrodes for the production of transgenic rodents using DNA microinjection and electroporation. Preliminary experiments *in vivo* in mouse models suggest that needle electrodes are significantly superior to flat electrodes for electroporation of testis cells. Histological analysis of some testes indicated that both somatic and germ cells express GFP following testis electroporation (data not shown). However, we did not observe GFP transgenics among one hundred offspring from mating testis-electroporated mice. One possible explanation is that we used the C57Bl/6 strain, which may not be optimal for this procedure, and so future work will involve testing the needle electrodes on mice from other genetic backgrounds [4]. Future work will also solve an issue of apparent heat damage to electrodes following a single use, by using different metallic coatings on electrodes.

Acknowledgements

Animal work was funded by the Higher Education Authority PRTLI5 Irish Transgenic Network infrastructure grant to TM. We thank Ms. Melanie Ball for technical assistance. This work has also been supported by Science Foundation Ireland under the National Access Programme (NAP 374).

References

[1] M. J. Justice, L. D. Siracusa, and A. F. Stewart, "Technical approaches for mouse models of human disease," *Disease Models and Mechanisms,* vol. 4, pp. 305-310, 2011.

[2] C. Leung and Z. Jia, "Mouse Genetic Models of Human Brain Disorders," *Frontiers in genetics,* vol. 7, 2016.

[3] S. Dhup and S. S. Majumdar, "Transgenesis via permanent integration of genes in repopulating spermatogonial cells in vivo," *Nature methods,* vol. 5, pp. 601-603, 2008.

[4] A. Usmani, N. Ganguli, H. Sarkar, S. Dhup, S. R. Batta, M. Vimal, *et al.*, "A non-surgical approach for male germ cell mediated gene transmission through transgenesis," *Scientific reports,* vol. 3, 2013.

[5] N. Wilke, J. O'BRIEN, G. Casey, T. Doody, D. Soden, and A. Morrissey, "Influence of electrode design on electric field distribution during electroporation," in *The 3rd European Medical and Biological Engineering Conference, IFMBE Proc Prague*, 2005, pp. 20-25.

[6] R. J. Pettis and A. J. Harvey, "Microneedle delivery: clinical studies and emerging medical applications," *Ther. Deliv.*, vol. 3, pp. 357-371, 2012.

[7] K. van der Maaden, W. Jiskoot, and J. Bouwstra, "Microneedle technologies for (trans) dermal drug and vaccine delivery," *J. Control. Rel.*, vol. 161, pp. 645-655, 2012.

[8] S.-O. Choi, Y. C. Kim, J.-H. Park, J. Hutcheson, H. S. Gill, Y.-K. Yoon, *et al.*, "An electrically active microneedle array for electroporation," *Biomedical Microdevices,* vol. 12, pp. 263-273, 2010.

[9] Z. Wei, S. Zheng, R. Wang, X. Bu, H. Ma, Y. Wu, *et al.*, "A flexible microneedle array as low-voltage electroporation electrodes for in vivo DNA and siRNA delivery," *Lab on a Chip,* vol. 14, pp. 4093-4102, 2014.

[10] N. Wilke, C. Hibert, J. O'Brien, and A. Morrissey, "Silicon microneedle electrode array with temperature monitoring for electroporation," *Sensors and Actuators A: Physical,* vol. 123, pp. 319-325, 2005.
[11] C. O'Mahony, "Structural characterization and in-vivo reliability evaluation of silicon microneedles," *Biomedical Microdevices,* pp. 1-11, 2014/02/02 2014.
[12] C. O'Mahony, F. Pini, A. Blake, C. Webster, J. O'Brien, and K. G. McCarthy, "Microneedle-based electrodes with integrated through-silicon via for biopotential recording," *Sens. Actuators A Phys.,* vol. 186, pp. 130-136, 2012.
[13] D. K. Shanley, P. A. Kiely, K. Golla, S. Allen, K. Martin, R. T. O'Riordan, *et al.*, "Pregnancy-specific glycoproteins bind integrin αIIbβ3 and inhibit the platelet—fibrinogen interaction," *PloS one,* vol. 8, p. e57491, 2013.

Journal of Physics: Conference Series **757** (2016) 012041 doi:10.1088/1742-6596/757/1/012041

Location Dependence of a MEMS Electromagnetic Transducer with respect to an AC Power Source

Ruth Houlihan, Oskar Olszewski, Finbarr Waldron, Mike O'Neill, Alan Mathewson, Nathan Jackson

Tyndall National Institute, Lee Maltings, Cork, Ireland

E-mail: ruth.houlihan@tyndall.ie

Abstract. A MEMS, silicon based device with a piezoelectric layer and an integrated magnet is presented for magnetic to electrical transduction. The cantilever structure can be configured either as an energy harvester to harvest power from an AC power line or as an AC current sensor. The positioning of the transducer with respect to the AC conductor is critical in both scenarios. For the energy scavenger, correct positioning is required to optimize the harvested power. For the current sensor, it is necessary to optimise the sensitivity of the sensor. This paper considers the effect of the relative position of the transducer with respect to the wire on the resulting electromagnetic forces and torques driving the device. It is shown here that the magnetic torque acting on a cantilever beam with an integrated magnet and in the vicinity of an alternating electromagnetic field is a very significant driver of the cantilever oscillations.

1. Introduction

A major challenge to vibration based energy harvesting systems is achieving alignment in terms of resonance frequency between the source and the transducer. On the one hand, the inherently narrow bandwidth of the mechanical transducer makes matching to the source problematic. On the other hand, the stability of the source excitation can be unreliable and the amplitude very low. In the harvester application targeted here, the energy source is the AC electromagnetic field surrounding a current carrying mains conductor. The source signal frequency is therefore defined by the mains supply (50 Hz in Europe) and cannot, by regulation, deviate by more than $\pm 1\%$ [1]. In reality, the deviation is typically much less than that (<0.1 Hz). The stability of the vibration source means that energy harvested from an AC power cord is a truly viable source of power for wireless sensor nodes. In an extended application, the resonating harvester structure can act both as a energy source and as a proximity current sensor. For wireless sensor nodes with limited energy budget, the advantage of a passive AC current sensor is self-evident.

The positioning of the MEMS electromagnetic transducer is critical in both the harvester and sensor applications, as it defines the forces and torques acting on the cantilever magnet and thereby indirectly determines the power available for harvesting and/or sensitivity of the sensor. This paper presents the theory describing the driving forces and torques that act on a magnet in an alternating electromagnetic field. This effect has been studied for energy harvesting [2, 3] and for current sensing applications [4, 5, 6]. In prior work however, only the vertical force exerted by the field on the harvester was taken into account. In reality, the harvester is driven by a combination of torque components and the omission, in particular, of the magnetic torque in the these studies has rendered them inaccurate. In this work, all of the torque components driving the harvester are considered and the effect of their omission considered.

2. Theoretical Background

Consider a piezoelectric cantilever-type harvester, oriented such that the width of the beam runs parallel to a conductor carrying an alternating current. The axis of the wire is designated the z-axis. The beam length is aligned with the x-axis and its thickness with the y-axis as shown in Figure 1. In its first resonance mode, the cantilever will experience a maximum displacement in the y-direction at its tip. A permanent magnet, magnetized along the y-axis, and positioned near the cantilever tip will experience alternating electromagnetic forces that will cause the cantilever to oscillate with the same frequency as that of the AC signal in the wire. If the cantilever is designed such that this frequency matches its natural resonance, then mechanical amplification of the electromagnetic response will occur. A piezoelectric film disposed on the beam can then be used to convert the mechanical oscillations into an electrical output signal either for harvesting and/or sensing.

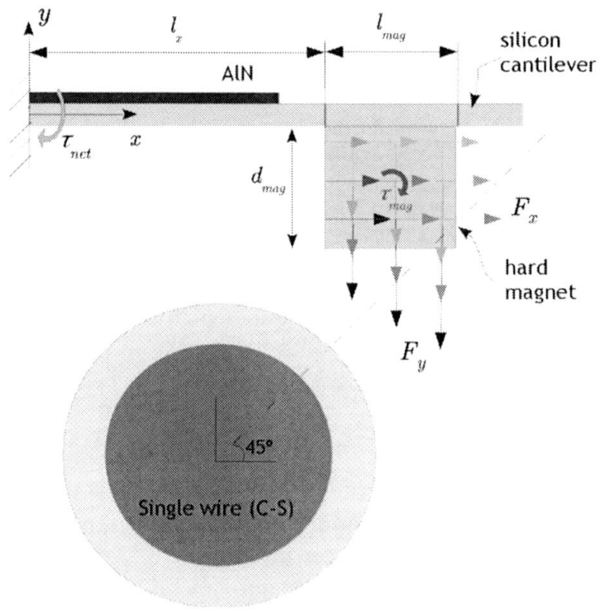

Figure 1. Schematic showing the lateral and vertical forces acting on a cantilever magnet that is in close proximity to a current carrying conductor.

Figure 1 shows the x-axis and y-axis forces that act on a magnet positioned on a cantilever beam and in close proximity to a magnetic field source. If it is considered that it is torque rather than force that drives the electromagnetic transducer then it is the net torque acting at the beam anchor, τ_{net}, that is of interest when determining the cantilever response. τ_{net} is given by Equation 1

$$\tau_{net} = \tau(F_x) + \tau(F_y) + \tau_{mag} \tag{1}$$

where $\tau(F_x)$ is the torque at the beam anchor resulting from the lateral magnetic force F_x; $\tau(F_y)$ is the torque resulting from the vertical magnetic force F_y; and τ_{mag} is the magnetic torque on a magnet in an electromagnetic field. Note that while the magnetic torque, τ_{mag}, acts through the center of the magnet, in a stiff beam it results in a moment about the anchor equal in magnitude to the moment about the magnet centre.

Prior work in this area analysed the response of the beam to the vertical force component, F_y, only. The contribution of the lateral force, F_x, was considered negligible and for thin magnets, this is indeed the case. The contribution of the magnetic torque, τ_{mag}, has not been considered at all previously and this omission is non-negligible. In the following, each of these torque components in Equation 5 is individually addressed and their impact on the mechanical response of the cantilever determined.

The net force on a magnetic dipole moment, $\boldsymbol{F_p}$, is given by Equation 2 [7].

$$\boldsymbol{F_p} = (\boldsymbol{p}\cdot\nabla)\boldsymbol{B} = \nabla(\boldsymbol{p}\cdot\boldsymbol{B}) \tag{2}$$

where $\boldsymbol{p}$ is the dipole moment and $\boldsymbol{B}$ is the magnetic field vector. A permanent magnet with magnetisation, $\boldsymbol{M}$, can be considered to have an associated dipole moment, $\boldsymbol{p_M}$, where $\boldsymbol{p}_M = \int \boldsymbol{M} dV$ and so the the total force on a permanent magnet in an electromagnetic field can be expressed by Equation 2 with $\boldsymbol{p} \to \boldsymbol{p}_M$.

$$\boldsymbol{F_{mag}} = \int_V \nabla(\boldsymbol{M}\cdot\boldsymbol{B})dV = \int_V \left[\boldsymbol{M}\cdot\frac{\partial\boldsymbol{B}}{\partial x}\right]\hat{\imath} + \left[\boldsymbol{M}\cdot\frac{\partial\boldsymbol{B}}{\partial y}\right]\hat{\jmath} + \left[\boldsymbol{M}\cdot\frac{\partial\boldsymbol{B}}{\partial z}\right]\hat{\boldsymbol{k}}\, dV \tag{3}$$

If the source of the electromagnetic field is a very long, current carrying conductor aligned with the z-axis, then the magnitude of the field at any point has a spatial dependence only on r, the radial distance between the wire and the point, where $\boldsymbol{r} = x\boldsymbol{i} + y\boldsymbol{j}$. If we further assume that our magnet is magnetized in the y-direction only, Equation 3 reduces to

$$\boldsymbol{F_{mag}} = F_x\boldsymbol{i} + F_y\boldsymbol{j} = \int_V \left[M_y\frac{\partial B_y}{\partial x}\right]\boldsymbol{i} + \left[M_y\frac{\partial B_y}{\partial y}\right]\boldsymbol{j}\, dV \tag{4}$$

It is clear from Equation 4 that the magnet experiences force contributions in both the lateral and the vertical directions. These forces, F_x and F_y are proportional to the gradient of the y-component of the magnetic field, B_y, with respect to their respective axes. The absolute value of $\partial B_y/\partial y$ is shown in a contour log plot in Figure 2. The circular area in the centre, shown by a block yellow colour, represents the wire in which the field is constant. A plot of $\partial B_y/\partial x$ is the same as the plot of $\partial B_y/\partial y$ but rotated by 45°.

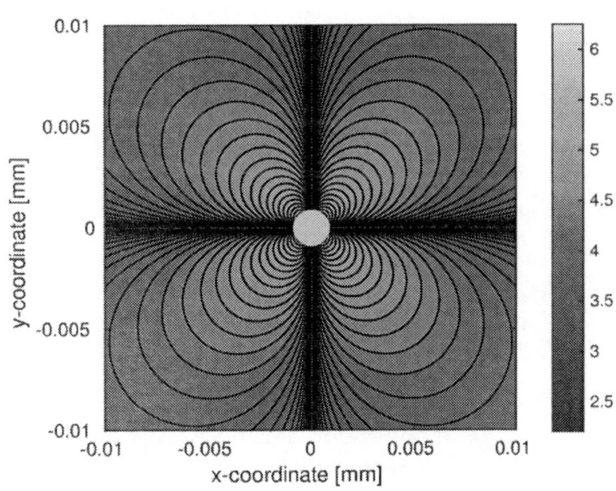

Figure 2. Contour plot showing the absolute value of the gradient with respect to y of the y-component of the $\boldsymbol{B}$ field.

From Figure 2, it appears that in order to optimize the vertical component of the magnetic force, the magnet should be positioned at 45° to the wire and indeed, this is the conclusion drawn by several authors who have considered this problem [3]-[6]. The x-component of the force acting on the magnet, F_x, is in most instances justifiably neglected since, although both F_y and F_x contribute to the total torque τ_{net} at the beam anchor, the contribution resulting from the axial force component, $\tau(F_x)$, will typically be much smaller than that from vertical force component, $\tau(F_y)$, except where a very thick magnet is used and/or the magnet is positioned very close to the beam anchor.

The force described by Equation 4 is the force on a magnet in a non-uniform electromagnetic field. There are, however, an additional set of forces acting on the magnet that are not accounted for by Equation 4 and that act to rotate the magnet into the field generated by the wire. The torque resulting from these forces acts about the center of the magnet and is given by Equation 5.

$$\tau_{mag} = \int_v \boldsymbol{M} \times \boldsymbol{B} \, dV \tag{5}$$

As with the lateral forces generated by the gradient of the field, the torque τ_{mag} acts over the thickness of the magnet and its effect is therefore more prominent in thicker magnets. The magnitude of τ_{mag} is much greater than $\tau(F_x)$ however, and so even for relatively thin magnets, it cannot be omitted from calculations.

3. Results

Figure 3 plots the three torque components versus magnet thickness for a 1 mm square magnet positioned mid-way along the length of an 8 mm long cantilever beam. The cantilever thickness is 25 μm. It can be seen from the figure that $\tau(F_x)$ is much less than $\tau(F_y)$ and that, for a magnet less than 1 mm thick, can be justifiably neglected. The field-aligning magnetic torque component, τ_{mag}, however, is comparable in magnitude to $\tau(F_y)$ and in fact, for a magnet thickness of approximately 2.5 mm, begins to dominate it.

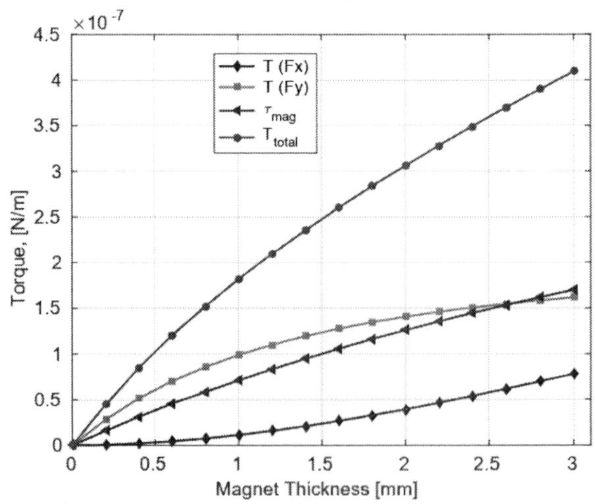

Figure 3. Plot of the three torque components acting at the beam anchor versus the magnet thickness. The cantilever was assumed to be 25 μm thick and 8 mm long. The magnet is a 1 mm x 1 mm square magnet. The net torque acting on the cantilever is shown by T_{total}.

Each of the three individual torque components that contribute to the displacement of the cantilever have a spatial dependence and their respective magnitudes varies as the magnet is

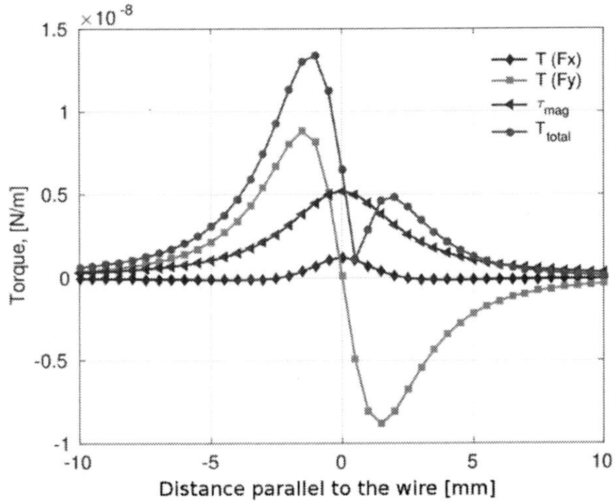

Figure 4. Torque generated at the anchor of a 8 mm long and 25 μm thick cantilever based harvester with 1 mm cube magnet located mid-way along the beam. A current of 1 A was used for the calculation and a gap of 2 mm between the magnet and the wire was assumed.

moved horizontally above the wire. Figure 4 plots $\boldsymbol{\tau}(F_x)$, $\boldsymbol{\tau}(F_y)$ and $\boldsymbol{\tau}_{mag}$ as the cantilever with its integrated magnet is passed over the current carrying conductor. The vertical component of the magnetic force is zero directly above the wire and increases to its maximum value as the magnet approaches the 45° axes, as indicated in Figure 2. The direction of the vertical force switches as the magnet passes the 90° line. By comparison, the horizontal component of the force and the magnetic torque are both at a maximum when the magnet is directly above the wire and the direction of these stays the same as the magnet passes over the wire. The result shows that, contrary to the conclusion drawn by Leland et al. [8, 4], positioning the magnet at 45° to the wire is not optimal in terms of maximizing displacement, nor indeed does locating it directly above the wire minimize displacement. The best location for the transducer will depend on the magnet dimensions and on its position on the cantilever.

4. Conclusions

The location dependence of an electromagnetic transducer with respect to its source, an AC power line, is analysed. The transducer is a silicon based MEMS cantilever with an integrated piezoelectric layer and with a permanent magnet disposed along its length. Placed in proximity to an AC current source, the cantilever oscillates with a frequency equal to that of the source and an electrical signal is generated across the piezoelectric capacitor. The position of the transducer with respect to the AC conductor is optimised here in order to maximise the tip displacement of the cantilever and consequently, the electrical output signal. Compared with previous work, where only the vertical component of the electromagnetic force is considered, here the horizontal component of the force is also considered as is the magnetic torque. It is found in particular, that the latter is a very significant driver of the cantilever oscillations and cannot be neglected for thick magnets.

References

[1] UK N G `http://www2.nationalgrid.com/uk/services/balancing-services/frequency-response/`
[2] Paprotny I, Xu Q, Chan W W, White R M and Wright P K 2013 *IEEE Sensors Journal* **13** 500 – 501
[3] He W, Li P, Wen Y, Zhang J, Yang A, Lu C, Yang J, Wen J, Qiu J, Zhu Y and Yu M 2013 *Sensors and Actuators, A* **193** 59 – 68

[4] Leland E S, Sherman C T, Minor P, Wright P K and White R M 2010 *IEEE Sensors Journal* 500 – 501
[5] Xu Q, Paprotny I, Seidel M, White R M and Wright P K 2013 *IEEE Sensors Journal* **13** 500 – 501
[6] Lao S B, Chauhan S S, Pollock T E, Schröder T, Cho I S and Salehian A 2014 *Actuators* **3** 162 – 181
[7] Knoepfel H E *Magnetic Fields: A Comprehensive Theoretical Treatise for Practical Use* (John Wiley & Sons)
[8] Leland E S, Wright P K and White R M 2009 *Journal of Micromechanics and Microengineering* **19** 094018

Journal of Physics: Conference Series **757** (2016) 012042 doi:10.1088/1742-6596/757/1/012042

A novel CE microchip with micro pillars column & double-L injection design for Capacitance Coupled Contactless Conductivity detection technology

Yineng Wang[1], Walter Messina[1], Xi Cao[1], Anna Hogan[1], Ed van Zalen[2], Eric Moore[1*]

[1]Tyndall National Institute, University College Cork
[2]CBRN Program Director, Netherlands Forensic Institute

*eric.moore@tyndall.ie

Abstract. This novel capillary electrophoresis microchip, or also known as μTAS (micro total analysis system) was designed to separate complex aqueous based compounds, similar to commercial CE & microchip (capillary electrophoresis) systems, but more compact. This system can be potentially used for mobile/portable chemical analysis equipment. Un-doped silicon wafer & ultra-thin borofloat glass (Pyrex) wafers have been used to fabricate the device. Double-L injection feature, micro pillars column, bypass separation channel & hybrid-referenced C4D electrodes were designed to achieve a high SNR (signal to noise ratio), easy-separation, for a durable and reusable μTAS for CE use.

1. Project background

The use of Chemical, Biological, Radiological and Nuclear (CBRN) reagents is an ever growing threat to today's society, and this drives the need for innovative tools capable of fast and efficient analysis techniques at incident crime scenes [1]. CBRN warfare has been developed since the early 1900's [2] and incidents related to their use can also be classified as accidental or intentional. Accidental incidents are those caused by human error, natural occurrences or technological reasons examples being spills, leakages or accidental release of agents [3]. These are referred to as HAZMAT (hazardous materials) or DG (dangerous goods) accidents [4]. Naturally occurring biological incidents are those involving outbreaks of diseases such as SARS, Ebola or influenza pandemics. Intentional CBRN incidents are those that involve the:

- Illegal release/ dumping of hazardous materials to avoid regulatory requirements.
- The ill-intentioned poisoning of one or more individuals [3-4].
- Acts of terrorism

Other consequences caused by CBRN incidents and: mass casualty/loss of life; potential long term consequences; formation of a dangerous environment [5]; narrow time frames for life saving treatment/interventions; potential to use a variety of specialised detection equipment; need of specifically trained and equipped service personnel. This research has been developed within the EU FP-7 project GIFT-CBRN which aims to fabricate a forensic toolbox for investigating CBRN incidents. This toolbox provide:

- Procedures, sampling methods and detection of CBRN agents at the crime scene.
- Traditional forensic laboratory methods for contaminated evidence.
- Laboratory methods for profiling the CBRN agents released at the incident.

The procedures and methods will be set up and validated according to ISO17025 and the system validation will be performed by a final exercise. Procedures for chain of custody and quality control (QC) to ensure the integrity of the investigations done on the evidences collected at the crime scene will be developed. An education and training curriculum will be designed and progressed to implementation.

In response to terrorist activities there is an urgent demand for rapid and reliable methods for the determination of chemical warfare agents (CWA) and HAZMAT (hazardous materials) and their degradation products. Tyndall National Institute is participating to the GIFT project consortium and is responsible for the fabrication of devices for chemical agents (Chemical substance of CBRN) detection. The objective is to develop an integrated portable μTAS device capable of performing analysis of suspicious substances at the crime scene. CE (capillary electrophoresis), especially in its miniaturized design (lab-on-a-chip/microfluidics), offers great possibilities to create portable, field deployable, rapidly responding and reliable devices, allowing the authorities to take important decisions with respect to the public safety.

Figure 1. EU FP-7 project GIFT.

2. Target reagents & detection technique

A micro-capillary electrophoresis (MCE) microfluidic system for project GIFT was designed at Tyndall National Institute for sensing chemical weapons (mainly nerve agents) at the crime scene. Nerve agents affect the nervous system; nerve agents relevant to the group of organophosphorus compounds will be tested during this project. They are stable and they easily diffuse in air or water, highly toxic and have rapid influence the human body through the skin or respiration [6]. The development of on-site techniques and devices capable of detecting organophosphate nerve agents and their degradation products is a challenge that needs to be undertaken sooner rather than later. Three typical CWA (chemical weapons agents), Sarin, Soman, and VX, contain the alkyl methyl-phosphonate moiety [7-8]. Rapid identification of their degradation products will allow first responders to make important decisions concerning the evacuation or the decontamination of a terrorist attack site and prevent becoming victims themselves [9]. UV fluorescence and Ion chromatography have been traditionally used for monitoring these CWA products [7-8]. More recently, conventional capillary electrophoresis (CE) systems based on the C4D (Capacitively-Coupled Contactless Conductivity Detection) [10-12] technique have been introduced for monitoring the breakdown products of organophosphate nerve agents.

Unlike the traditional UV detection method which needs UV lamps, reflection mirror sets, sophisticated semi-conductor sensor for UV sensing and its power supply system, C4D technology make all the components compact and more efficient. The C4D technology has proved to be able to achieve similar performances to those of the conventional UV detection [13].

Joseph Wang et al. have described the use of a low-cost PDMS MCE chip using the contactless conductivity method to separate organophosphate materials and their bread-down substance in early stage. His experiments described the combination of a CE microchip with a contactless conductivity detection method for monitoring organophosphate nerve agent degradation products. He also demonstrated a well-defined concentration dependency for CWA degradation products. The ability to sense the nerve agents and their breakdown-products is beneficial in terms of speed, efficiency, portability, sensitivity, cost and sample volume compared to conventional capillary electrophoresis or ion chromatography systems [9].

A series of organophosphate compounds like methyl paraoxon (MPX), ethyl paraoxon (EPX), methyl parathion (MPT), fenitronthion (FT), and ethyl parathion (EPT) will be used as reagents to test the MCE devices fabricated at Tyndall.

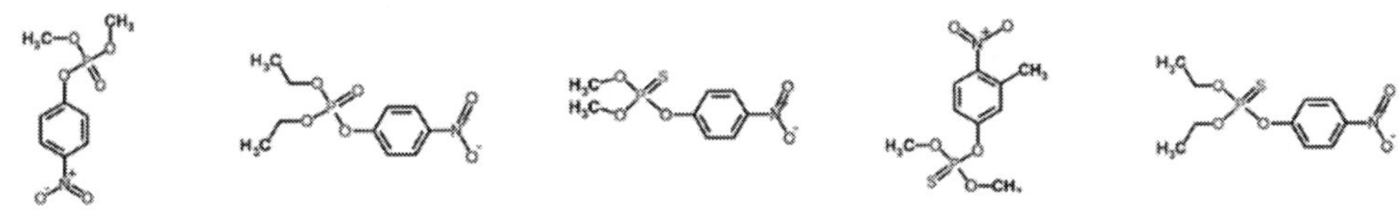

methyl paraoxon ethyl paraoxon methyl parathion fenitronthion fenitronethyl parathionthion

Figure 2. Target reagents for Tyndall's MCE system.

3. Device features

It was decided to fabricate the MCE chip using silicon wafers, because the surface of the oxidised silicon microchip (silicon dioxide) can guarantee high stability from the chemical perspective. Moreover, the silica capillary channel is able to generate the surface double-electric layer, which is similar to the conventional CE-UV detection technique [14].

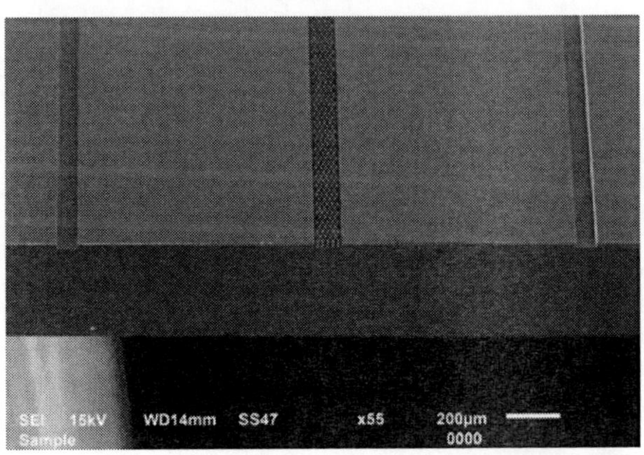

Figure 3. Micro-Pillars column via SEM observation.

Dry etching technique was applied to fabricate a clean, sharp, low-defects and high-quality "Micro-Pillars" column structure with an ultra-thin Pyrex glass wafer bonding for seal all the channels. This offers a 3D design which increases the total surface area inside the microfluidic channel (as the channel with "Micro-Pillars" columns has increased about 2.75 times its effective surface area).

Around 4500 pillars have been built within 4cm length channel on an un-doped silicon wafer. Only 3 pillars were damaged during the fabrication process (0.000666% damage rate), after an observation via a CCD microscope. Based on the novel micro-pillars column, a larger surface area-to-volume ratio is beneficial for Joule heat transfer which can be compared with the conventional capillary under same dimension [14].

According to the capillary zone electrophoresis theory, a simple 2D-model has been built to estimate the differential electrophoresis performance between this novel MCE system and conventional capillary system. From Debeye-Huckel-Henry's Capillary zone electrophoresis (CZE) theory, the electrophoretic mobility in a homogeneous buffer solution and constant field strength throughout the length of the capillary can be approximated.

$$\mu_{ep} = \frac{v_{ep}}{E} = \frac{q}{6\pi\eta R} \tag{1}$$

The electrophoretic mobility (μ_{ep}) , the electrophoretic velocity (v_{ep}) , and the electric field strength (E) , the net charge (q) , the capillary radius (**R**) , and (η) is the viscosity. These relationships between these factors are shown in Equation 1.

If we consider c as the perimeter of this 2D-model's section, we have Equation 2 and 3:

$$\mu_{ep} = \frac{q}{2\pi R \cdot \eta} = \frac{q}{c \cdot \eta} \tag{2}$$

$$v_{ep} = \frac{qE}{c\eta} \tag{3}$$

Comparison of an MCE (section area 90um*40um) and an ordinary capillary (50um diameter):

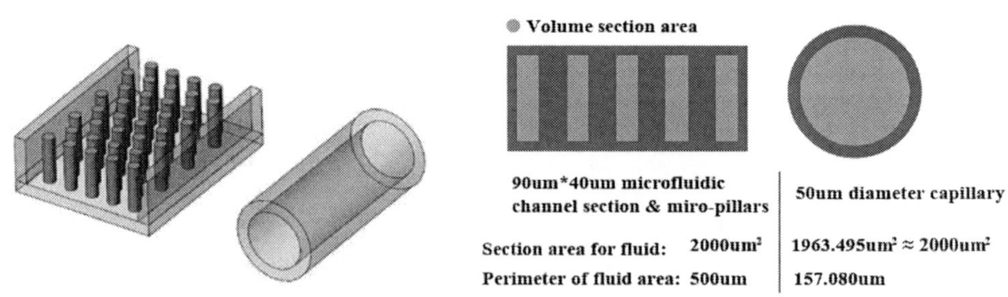

Figure 4. 2D-model of MCE (with micro-structure) and CE comparison.

$$\frac{v_{ep\ MCE}}{v_{ep\ Capillary}} = \frac{157.1}{500} = \frac{1}{3.15} \tag{4}$$

This 2D-model shows that under the same net charge (q), electric field strength (E), and viscosity (η) condition; higher total surface area (with the 3d-micro structure) will bring lower electrophoretic velocity or longer separation time. A MCE device with "Micro-Pillars" column is equivalent to a longer capillary under same volume condition.

This MCE chip has two different designs for the separation cross: one is designed as a conventional simple cross, while the other one includes a double L injection technique which reduces sample leakage in comparison to the traditional one [15]. In future experiments, the authors will compare the performances of the two of types:

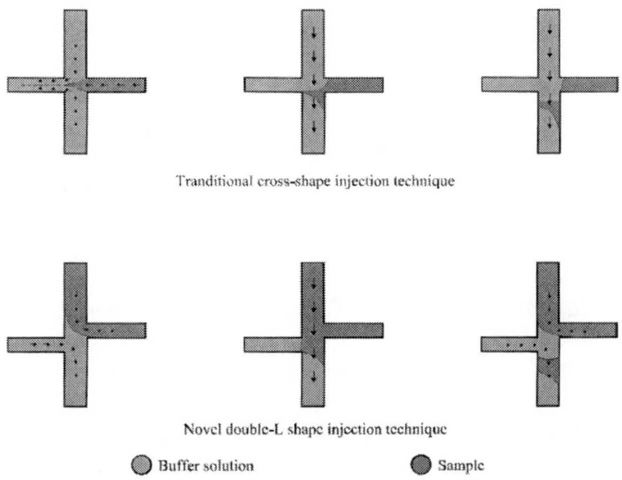

Figure 5. Traditional & novel injection technique.

In this novel MCE system the electrodes won't be in direct contact with the liquid, which will avoid any contamination to the samples or buffer solutions. "Lab-on-chip" technology offers tremendous potential for obtaining the desired forensic information in a faster, simpler, and less costly manner compared to traditional laboratory-based instruments. Particularly attractive for on-site counter-terrorism and compliance applications is the high-performance, small-footprint, high throughput, versatility of microchip devices.

4. Preliminary test

"NanoPort" assembly connector (Model: N-333, IDEX Health & Science LLC, USA) were bonded to the MCE chip to connect the chip to a 1/16" tube. A 5ml syringe (TERUMO, Leuven, Belgium) was used for manual aspiration (Figure 6.) to purge the DI water throughout the channel. A CCD microscope (BX51 microscope, DP71 CCD digital camera, Olympus, Japan) was used for observation. Figure 7. shows that, after keeping continuous aspiration for 10 minutes, most of the water has gone through the channel with the micro-structure; however there are still some air bubbles between the "Micro-Pillar" columns. Despite this bubble defects, all the liquid (DI water) has flown through without blockages.

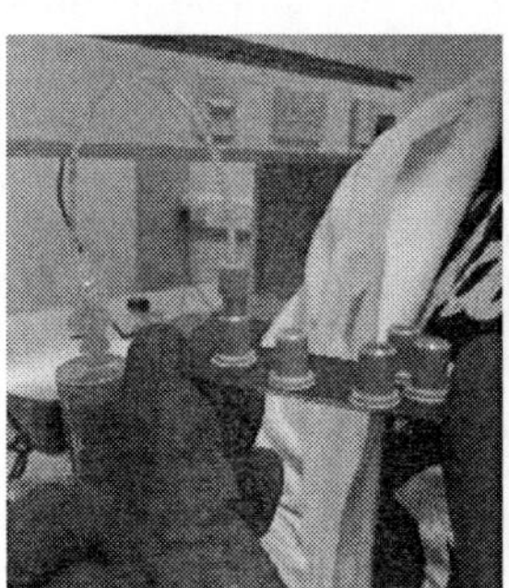

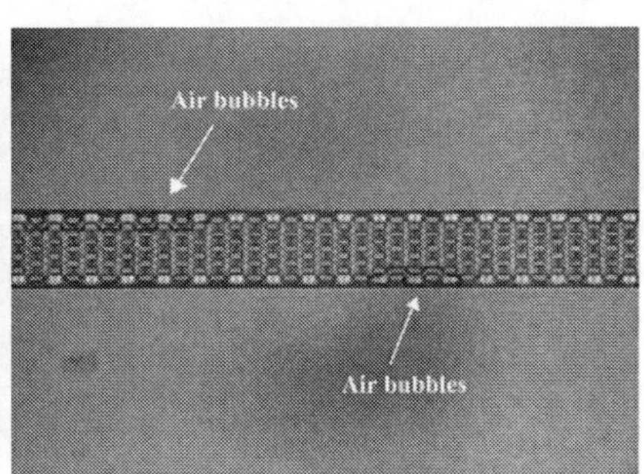

Figure 6 (left) and **7** (right). DI water purges test and observation under persistent aspiration.

5. Discussion

The microfluidic platforms developed here uses a novel "Micro-Pillar" column to enhance the electrophoresis separation performance and gives longer separation time based on the Debeye-Huckel-Henry's Capillary zone electrophoresis (CZE) theory calculation. During the preliminary DI water purge test, bubbles appeared in the "Micro-Pillars" column. Author will attempt to use sonication to solve these negative effects from the "Micro-Pillars" column in the future experiments.

Acknowledgements

The 7th Framework Programme, European Commission are acknowledged for the research funding. The Tyndall National Institute and the Netherlands Forensic Institute are acknowledged for hosting this research in their laboratories and for the consultations. The help and kindness of Dr. Anna Hogan at the Department of Chemistry, University College Cork was highly appreciated.

References

[1] B Sandstram, H, Eriksson, L Norlander, M Thorsensson, G Cassel, "Training of public health personnel in handling CBRN emergencies: A table-top exercise card concept." in *Environment International*, 72:164-169, 2014.
[2] R Harris, J Paxman, "New York: Random House Trade Paperbacks" in *A Higher Form of Killing: The Secret History of Chemical and Biological Warfare*, 2002.
[3] A Ramesh, S Kumar, "Triage, monitoring, and treatment of mass casualty events involving chemical, biological, radiological, or nuclear agents." in *Journal of Pharmacy Bioallied Science*, 2(3):239, 2010.
[4] M Christian, D Kollek, B Schwartz, "What is CBRN." in *The Centre for Excellence in Emergency Preparedness*, cited from online, http://www.ceep.ca/education/CBRNintro sheet.pdf [Accessed 14 Sep. 2016].
[5] A Silke. "Psychological perspectives on terrorism and its consequences." in *Terrorists, victims and society*, 2003.
[6] U Ivarsson, H Nilsson, J Santesson, eds. "A FOA briefing book on chemical weapons: threat, effects, and protection." in *Umeå, National Defence Research Establishment,* 1992.
[7] P Bossle, D Reutter, E Sarver, *Journal of Chromatography A*, 407, 397, 1987.
[8] A Kingery, H Allen, *Analytical Chemistry*, 66, 155, 1994.
[9] J Wang, M Pumera, G Collins, A Mulchandani. "Measurements of Chemical Warfare Agent Degradation Products Using an Electrophoresis Microchip with Contactless Conductivity Detector" in *Analytical Chemistry*, 74, 23, 2002.
[10] A Nassar, S Lucas, W Jones, L Hoffland. *Analytical Chemistry*, 70, 1085, 1998.
[11] A Nassar, S Lucas, C Myler, W Jones, M Campisano, L Hoffland. *Analytical Chemistry*, 70, 3598, 1998.
[12] T Rosso, P Bossle. *Journal of Chromatography A*, 824, 125, 1998.
[13] "Contactless Conductivity Introduction" in *edaq*, http://www.edaq.com, [Accessed 20 Aug 2016].
[14] H Lauerand, Gerard Rozing, *High Performance Capillary Electrophoresis, Second completely revised edition.* page 3, 2014.
[15] C Lin, R Yang, C Tai, C Lee, L Fu. "Double-L injection technique for high performance capillary electrophoresis detection in microfluidic chips." in *Journal of Micromechanics and Microengineering*. volume 14, number 4, 2004.

Institute of Physics
Dirac House, Temple Back
Bristol BS1 6BE UK

ISSN: 1742-6588
ISBN 978-1-5108-3225-1

9 781510 832251

6th ITB International Geothermal Workshop (IIGW 2017)

IOP Conference Series: Earth and Environmental Science
Volume 103

Bandung, Indonesia
22 – 23 March 2017

IOP
Institute of Physics